More Than Just a Textbook

Internet Resources

Step 1 Connect to Math Online glencoe.com

Step 2 Connect to online resources by using
You can connect directly to the chapter you want.

IM7048c1

Enter this code with the appropriate chapter number.

For Students

Connect to ***StudentWorks Plus Online*** which contains all of the following online assets. You don't need to take your textbook home every night.

- Personal Tutor
- Chapter Readiness Quizzes
- Multilingual eGlossary
- Concepts in Motion
- Chapter Test Practice
- Test Practice

For Teachers

Connect to professional development content at glencoe.com and ***eBook Advance Tracker*** at AdvanceTracker.com

For Parents

Connect to glencoe.com for access to ***StudentWorks Plus Online*** and all the resources for students and teachers that are listed above.

Glencoe McGraw-Hill

NSF

IMPACT Mathematics

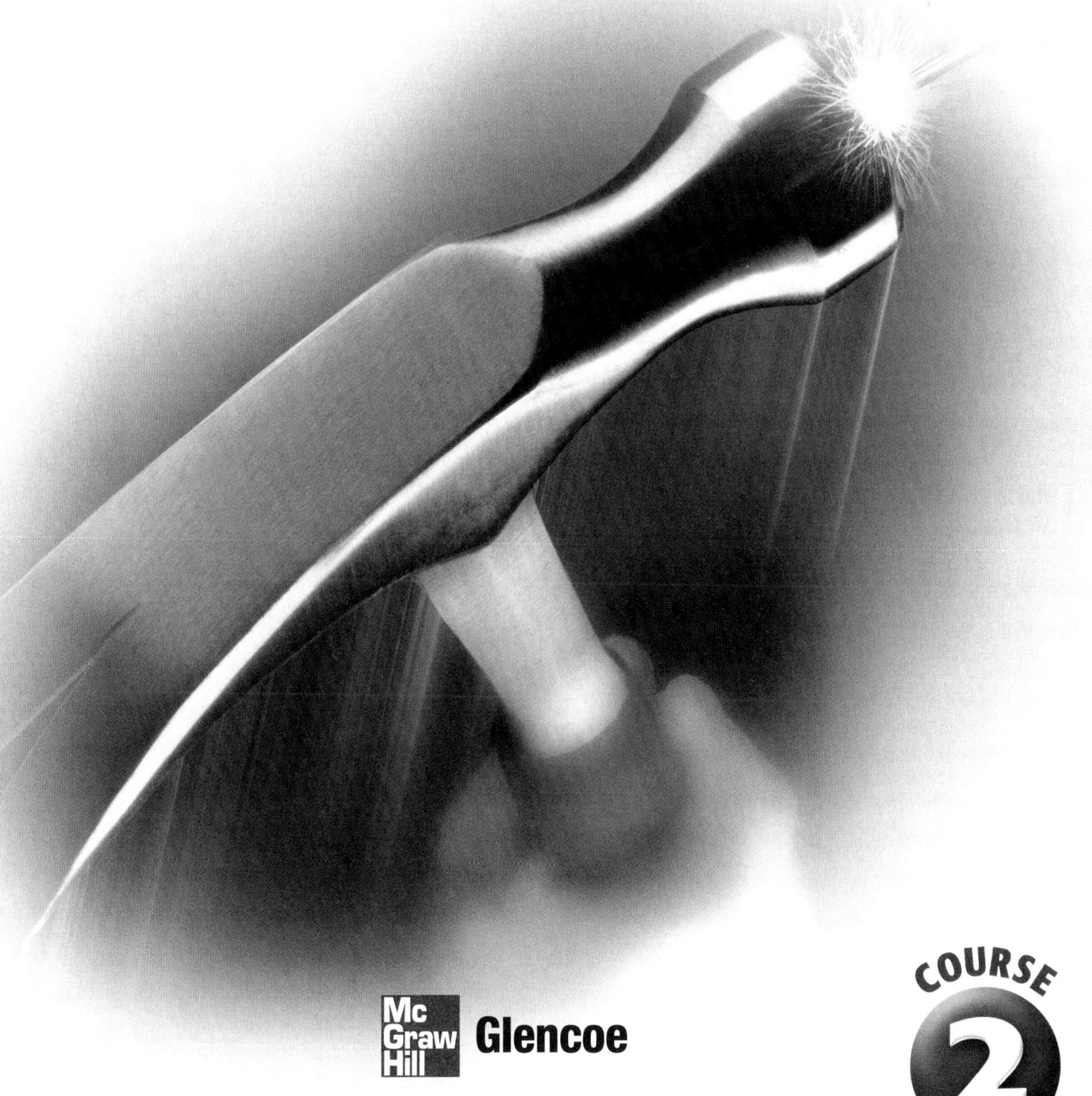

New York, New York   Columbus, Ohio   Chicago, Illinois   Woodland Hills, California

About the Cover

Hammers can be used to drive nails. They can also be used to shape metal that is struck against an anvil.

In ancient times, the renowned Greek mathematician Pythagoras noticed the exceptional sounds of hammers striking anvils. He determined that the weights of different-sized hammers were proportional, with weights of 12, 9, 8, and 6 pounds. Upon further examination, he determined that the harmonious sounds were a result of the numerical relationships among the hammers' weights and discovered a connection among numbers, ratios, and music.

These materials include work supported in part by the National Science Foundation under Grant No. ESI-9726403 to MARS (Mathematics Assessment Resource Service). Any opinions, findings, and conclusions or recommendations expressed in this material are those of the authors and do not necessarily reflect the views of the funding agencies. For more information on MARS, visit http://www.nottingham.ac.uk/education/MARS.

The McGraw·Hill Companies

Macmillan/McGraw-Hill
Glencoe

The algebra content for *IMPACT Mathematics* was adapted from the series *Access to Algebra,* by Neville Grace, Jayne Johnston, Barry Kissane, Ian Lowe, and Sue Willis. Permission to adapt this material was obtained from the publisher, Curriculum Corporation of Level 5, 2 Lonsdale Street, Melbourne, Australia.

Send all inquiries to:
Glencoe/McGraw-Hill
8787 Orion Place
Columbus, OH 43240-4027

ISBN: 978-0-07-888704-8
MHID: 0-07-888704-6

Printed in the United States of America.

1 2 3 4 5 6 7 8 9 10 055/079 17 16 15 14 13 12 11 10 09 08

COURSE 2

Contents in Brief

Focal Points and Connections
See pages vi and vii for key.

Chapter		Page	Focal Points
1	Expressions	2	G7-FP3
2	Exponents	72	G7-FP3
3	Signed Numbers	124	G7-FP3
4	Magnitude of Numbers	172	G7-FP3
5	Geometry in Three Dimensions	210	G7-FP2, G7-FP4C
6	Data and Probability	260	G7-FP6C, G7-FP7C
7	Real Numbers	318	G7-FP3, G7-FP5C
8	Linear Relationships	366	G7-FP1, G7-FP3, G7-FP5C
9	Equations	434	G7-FP3, G7-FP5C
10	Proportional Reasoning and Percents	492	G7-FP1, G7-FP4C

Principal Investigator

Faye Nisonoff Ruopp
Brandeis University
Waltham, Massachusetts

Consultants and Developers

Consultants

Frances Basich Whitney
Project Director, Mathematics K–12
Santa Cruz County Office of Education
Santa Cruz, California

Robyn Silbey
Mathematics Content Coach
Montgomery County Public Schools
Gaithersburg, Maryland

Dr. Selina Vásquez Mireles
Associate Professor of Mathematics
Texas State University—San Marcos
San Marcos, Texas

Teri Willard
Assistant Professor
Central Washington University
Ellensburg, Washington

Special thanks to:

Peter Braunfeld
Professor of Mathematics Emeritus
University of Illinois

Sherry L. Meier
Assistant Professor of Mathematics
Illinois State University

Judith Roitman
Professor of Mathematics
University of Kansas

Developers

Senior Project Director
Cynthia J. Orrell

Senior Curriculum Developers
Michele Manes, Sydney Foster, Daniel Lynn Watt, Ricky Carter, Joan Lukas, Kristen Herbert

Curriculum Developers
Haim Eshach, Phil Lewis, Melanie Palma, Peter Braunfeld, Amy Gluckman, Paula Pace

Special Contributors
Elizabeth D. Bjork, E. Paul Goldenberg

Project Reviewers

Glencoe and Education Development Center would like to thank the curriculum specialists, teachers, and schools who participated in the review and testing of the first edition of *IMPACT Mathematics*. The results of their efforts were the foundation for this second edition. In addition, we appreciate all of the feedback from the curriculum specialists and teachers who participated in review and testing of this edition.

Debra Allred
Math Teacher
Wiley Middle School
Leander, Texas

Tricia S. Biesmann
Retired Teacher
Sisters Middle School
Sisters, Oregon

Kathryn Blizzard Ballin
Secondary Math Supervisor
Newark Public Schools
Newark, New Jersey

Linda A. Bohny
District Supervisor of Mathematics
Mahwah Township School District
Mahwah, New Jersey

Julia A. Butler
Teacher of Mathematics
Richfield Public School Academy
Flint, Michigan

April Chauvette
Secondary Mathematics Facilitator
Leander ISD
Leander, Texas

Amy L. Chazaretta
Math Teacher/Math Department Chair
Wayside Middle School, EM-S ISD
Fort Worth, Texas

Franco A. DiPasqua
Director of K–12 Mathematics
West Seneca Central
West Seneca, New York

Mark J. Forzley
Junior High School Math Teacher
Westmont Junior High School
Westmont, Illinois

Virginia G. Harrell
Education Consultant
Brandon, Florida

Lynn Hurt
Director
Wayne County Schools
Wayne, West Virginia

Andrea D. Kent
7th Grade Math & Pre-Algebra
Dodge Middle School, TUSD
Tucson, Arizona

Russ Lush
6th Grade Teacher & Math Dept. Chair
New Augusta–North
Indianapolis, Indiana

Katherine V. Martinez De Marchena
Director of Education 7–12
Bloomfield Public Schools
Bloomfield, New Jersey

Marcy Myers
Math Facilitator
Southwest Middle School
Charlotte, North Carolina

Joyce B. McClain
Middle School Mathematics Consultant
Hillsborough County Schools
Tampa, Florida

Suzanne D. Obuchowski
Math Teacher
Proctor School
Topsfield, Massachusetts

Michele K. Older
Mathematics Instructor
Edward A. Fulton Jr. High
O'Fallon, Illinois

Jill Plattner
Math Program Developer (Retired)
Bend La Pine School District
Bend, Oregon

E. Elaine Rafferty
Retired Math Coordinator
Summerville, South Carolina

Karen L. Reed
Math Teacher–Pre-AP
Chisholm Trail Intermediate
Fort Worth, Texas

Robyn L. Rice
Math Department Chair
Maricopa Wells Middle School
Maricopa, Arizona

Brian Stiles
Math Teacher
Glen Crest Middle School
Glen Ellyn, Illinois

Nimisha Tejani, M.Ed.
Mathematics Teacher
Kino Jr. High
Mesa, Arizona

Stefanie Turnage
Middle School Mathematics
Grand Blanc Academy
Grand Blanc, Michigan

Kimberly Walters
Math Teacher
Collinsville Middle School
Collinsville, Illinois

Susan Wesson
Math Teacher/Consultant
Pilot Butte Middle School
Bend, Oregon

Tonya Lynnae Williams
Teacher
Edison Preparatory School
Tulsa, Oklahoma

Kim C. Wrightenberry
Math Teacher
Cane Creek Middle School
Asheville, North Carolina

The Curriculum Focal Points identify key mathematical ideas for this grade. They are not discrete topics or a checklist to be mastered; rather, they provide a framework for the majority of instruction at a particular grade level and the foundation for future mathematics study. The complete document may be viewed at www.nctm.org/focalpoints.

KEY

G7-FP1 Grade 7 Focal Point 1

G7-FP2 Grade 7 Focal Point 2

G7-FP3 Grade 7 Focal Point 3

G7-FP4C Grade 7 Focal Point 4 Connection

G7-FP5C Grade 7 Focal Point 5 Connection

G7-FP6C Grade 7 Focal Point 6 Connection

G7-FP7C Grade 7 Focal Point 7 Connection

G7-FP1 **Number and Operations and Algebra and Geometry: Developing an understanding of and applying proportionality, including similarity**

Students extend their work with ratios to develop an understanding of proportionality that they apply to solve single and multistep problems in numerous contexts. They use ratio and proportionality to solve a wide variety of percent problems, including problems involving discounts, interest, taxes, tips, and percent increase or decrease. They also solve problems about similar objects (including figures) by using scale factors that relate corresponding lengths of the objects or by using the fact that relationships of lengths within an object are preserved in similar objects. Students graph proportional relationships and identify the unit rate as the slope of the related line. They distinguish proportional relationships ($\frac{y}{x} = k$, or $y = kx$) from other relationships, including inverse proportionality ($xy = k$, or $y = \frac{k}{x}$).

G7-FP2 **Measurement and Geometry and Algebra: Developing an understanding of and using formulas to determine surface areas and volumes of three-dimensional shapes**

By decomposing two- and three-dimensional shapes into smaller, component shapes, students find surface areas and develop and justify formulas for the surface areas and volumes of prisms and cylinders. As students decompose prisms and cylinders by slicing them, they develop and understand formulas for their volumes (*Volume* = *Area of base* × *Height*). They apply these formulas in problem solving to determine volumes of prisms and cylinders. Students see that the formula for the area of a circle is plausible by decomposing a circle into a number of wedges and rearranging them into a shape that approximates a parallelogram. They select appropriate two- and three dimensional shapes to model real-world situations and solve a variety of problems (including multistep problems) involving surface areas, areas and circumferences of circles, and volumes of prisms and cylinders.

G7-FP3 **Number and Operations and Algebra: Developing an understanding of operations on all rational numbers and solving linear equations**

Students extend understandings of addition, subtraction, multiplication, and division, together with their properties, to all rational numbers, including negative integers. By applying properties of arithmetic and considering negative numbers in everyday contexts (e.g., situations of owing money or measuring elevations above and below sea level), students explain why the rules for adding, subtracting, multiplying, and dividing with negative numbers make sense. They use the arithmetic of rational numbers as they formulate and solve linear equations in one variable and use these equations to solve problems. Students make strategic choices of procedures to solve linear equations in one variable and implement them efficiently, understanding that when they use the properties of equality to express an equation in a new way, solutions that they obtain for the new equation also solve the original equation.

Connections to the Focal Points

G7-FP4C Measurement and Geometry: Students connect their work on proportionality with their work on area and volume by investigating similar objects. They understand that if a scale factor describes how corresponding lengths in two similar objects are related, then the square of the scale factor describes how corresponding areas are related, and the cube of the scale factor describes how corresponding volumes are related. Students apply their work on proportionality to measurement in different contexts, including converting among different units of measurement to solve problems involving rates such as motion at a constant speed. They also apply proportionality when they work with the circumference, radius, and diameter of a circle; when they find the area of a sector of a circle; and when they make scale drawings.

G7-FP5C Number and Operations: In grade 4, students used equivalent fractions to determine the decimal representations of fractions that they could represent with terminating decimals. Students now use division to express any fraction as a decimal, including fractions that they must represent with infinite decimals. They find this method useful when working with proportions, especially those involving percents. Students connect their work with dividing fractions to solving equations of the form $ax = b$, where a and b are fractions. Students continue to develop their understanding of multiplication and division and the structure of numbers by determining if a counting number greater than 1 is a prime, and if it is not, by factoring it into a product of primes.

G7-FP6C Data Analysis: Students use proportions to make estimates relating to a population on the basis of a sample. They apply percentages to make and interpret histograms and circle graphs.

G7-FP7C Probability: Students understand that when all outcomes of an experiment are equally likely, the theoretical probability of an event is the fraction of outcomes in which the event occurs. Students use theoretical probability and proportions to make approximate predictions.

COURSE 2

Table of Contents

1 Expressions 2

Lesson 1.1 Variables and Expressions **4**

Investigation 1: Sequences and Rules 5

Investigation 2: Write Expressions 9

Investigation 3: Evaluate Expressions 12

Investigation 4: Flowcharts 17

On Your Own Exercises 21

Lesson 1.2 Expressions and Formulas **30**

Investigation 1: What's the Variable? 32

Investigation 2: Formulas 35

▶ **Inquiry Investigation 3: Formulas and Spreadsheets** 40

On Your Own Exercises 43

Lesson 1.3 The Distributive Property **49**

Investigation 1: Bags and Blocks 51

Investigation 2: The Same and Different 53

Investigation 3: Use Parentheses 56

Investigation 4: Combine Like Terms 60

On Your Own Exercises 63

Review & Self-Assessment **69**

Focal Points and Connections
See pages vi and vii for key.

G7-FP3

2 Exponents 72

Lesson 2.1 Factors and Multiples 74

Investigation 1: Factors 75

Investigation 2: Prime Numbers 77

Investigation 3: Common Factors 80

Investigation 4: Multiples 83

On Your Own Exercises 86

Lesson 2.2 Exponent Machines 92

Investigation 1: Model Exponents 93

Investigation 2: Multiply Expressions with the Same Base 96

Investigation 3: Multiply Expressions with the Same Exponent 100

On Your Own Exercises 104

Lesson 2.3 More Exponent Machines 107

Investigation 1: Division Machines with Exponents 107

Investigation 2: Divide Expressions with Exponents 109

Investigation 3: Power Law of Exponents 111

▶**Inquiry Investigation 4: The Tower of Hanoi** 114

On Your Own Exercises 117

Review & Self-Assessment **120**

Focal Points
and Connections
See pages vi and vii for key.

G7-FP3

3 Signed Numbers 124

Lesson 3.1 Add and Subtract with Negative Numbers **126**

Investigation 1: The Two-Color Chip Model 128

▶ **Inquiry Investigation 2: A Number Line Model** 131

Investigation 3: Add and Subtract on the Number Line 134

Investigation 4: Equivalent Operations 140

Investigation 5: Inequalities and Negative Numbers 143

Investigation 6: Predict Signs of Sums and Differences 146

On Your Own Exercises 148

Lesson 3.2 Multiply and Divide with Negative Numbers **154**

Investigation 1: Multiply a Positive and a Negative 155

Investigation 2: Multiply Two Negatives 157

Investigation 3: Divide with Negative Numbers 160

Investigation 4: Signed Numbers and Data 162

On Your Own Exercises 165

Review & Self-Assessment **169**

Focal Points and Connections
See pages vi and vii for key.

G7-FP3

4 Magnitude of Numbers 172

Lesson 4.1 Scientific Notation **174**

Investigation 1: Powers of 10 175

Investigation 2: Work with Scientific Notation 178

Investigation 3: Scientific Notation on a Calculator 182

▶ **Inquiry Investigation 4: Relative Error** 185

On Your Own Exercises 188

Lesson 4.2 Negative Exponents **194**

Investigation 1: Model Negative Exponents 194

Investigation 2: Evaluate Expressions with Negative Exponents 198

Investigation 3: Laws of Exponents and Scientific Notation 200

On Your Own Exercises 202

Review & Self-Assessment **207**

Focal Points
and Connections
See pages vi and vii for key.

G7-FP3

5 Geometry in Three Dimensions 210

Lesson 5.1 Surface Area and Volume **212**

Investigation 1: Measure Prisms 213

Investigation 2: Volume of Prisms and Cylinders 216

▶**Inquiry Investigation 3: Package Design** 220

On Your Own Exercises 223

Lesson 5.2 Nets and Solids **228**

Investigation 1: Use a Net 229

Investigation 2: Use Nets to Investigate Solids 230

Investigation 3: Is Today's Beverage Can the Best Shape? ... 233

On Your Own Exercises 235

Lesson 5.3 Mass and Weight **240**

Investigation 1: Measure Mass 241

Investigation 2: Estimate Mass 244

Investigation 3: Measure Weight 247

On Your Own Exercises 251

Review & Self-Assessment **256**

Focal Points
and Connections
See pages vi and vii for key.

G7-FP2

G7-FP4C

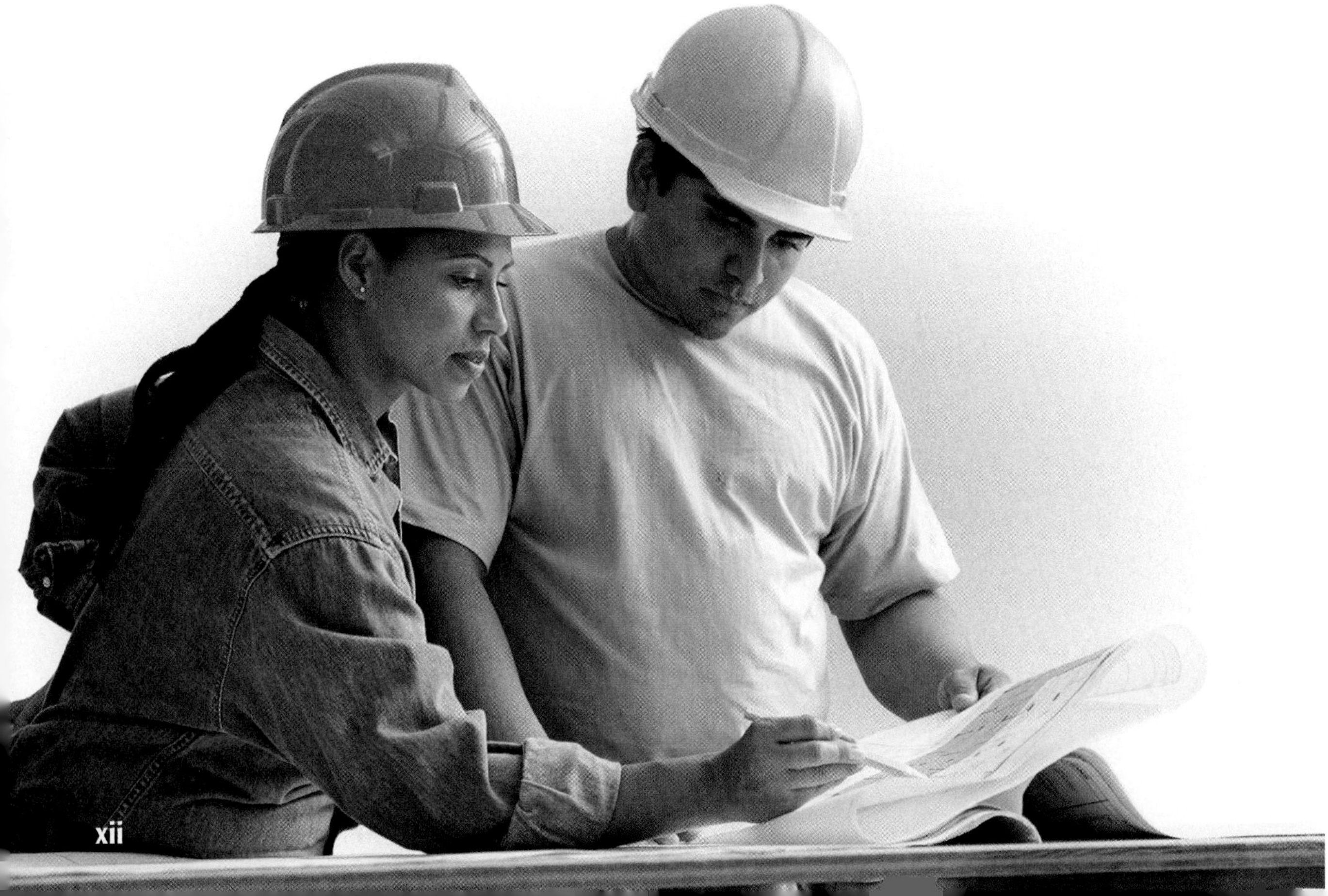

Dear Family,

The class is about to begin an exciting year of *IMPACT Mathematics.* Some of the topics that will be studied include exponents, three-dimensional geometry, ratios, probability, and data analysis. Throughout the year, your student will also develop and refine skills in algebra.

The class will begin by looking at algebraic expressions and variables. Once the students are familiar with variables and expressions, they will create flowcharts to represent equations and to solve equations. The class will also explore the use of formulas in real-world situations, including computer spreadsheets.

Key Concept–Algebraic Expressions and Variables

Algebraic expressions are combinations of numbers, letters, and mathematical symbols.

$$b + 2 \qquad 5m^2 - 4 \qquad 10(9y - 2) \qquad \frac{3x + 2}{4}$$

Variables are letters or symbols that can change or that represent unknown quantities. In the expression $b + 2$, the variable is b. The value of an expression depends on the number substituted for its variable(s). The following are examples.

When $b = 0$, $b + 2$ equals $0 + 2$, or 2.
When $b = 1$, $b + 2$ equals $1 + 2$, or 3.
When $b = 100$, $b + 2$ equals $100 + 2$, or 102.

Chapter Vocabulary

backtracking	**flowchart**
distributive property	**formula**
equivalent expressions	**like terms**
exponent	**polynomial**

Home Activities

- Identify the variable in each Key Concept algebraic expression.
- Use the thermometer illustration in Investigation 1 to convert daily temperature readings. Use the formula $F = \frac{9}{5}C + 32$ to check your conversions.
- Look for examples of expressions and formulas in everyday situations.
- Create a spreadsheet of expenses for a family outing, such as a trip to the movies.

Variables and Expressions

Inv 1 Sequences and Rules 5
Inv 2 Write Expressions 9
Inv 3 Evaluate Expressions 12
Inv 4 Flowcharts 17

Have you ever been given directions, only to find yourself lost? Most people would agree that being clear and precise is important in giving directions, as well as in other situations in and outside of mathematics. Consider this situation. Jing drew a rectangle. She then wrote a rule for generating a sequence of shapes starting with this rectangle.

Rule: Draw a rectangle twice the size of the preceding rectangle.

By following Jing's rule, Caroline drew this sequence.

Jahmal followed Jing's rule and drew this sequence.

Think & Discuss

Could both sequences above be correct? Explain your answer.

Rosita also followed Jing's rule. The sequence she drew was different from both Caroline's and Jahmal's. What might Rosita's sequence look like?

Rewrite the rule so that Caroline's sequence is correct but Jahmal's is not. Try to make your rule clear enough that anyone following it would get the sequence Caroline did.

Investigation 1 Sequences and Rules

You have just seen how three students could follow the same rule yet draw three different sequences. This is because Jing's rule is *ambiguous.* It can be interpreted in more than one way. In both mathematics and everyday life, it is often important to state rules in such a way that everyone will get the same result.

Develop & Understand: A

1. Create a sequence of shapes in which each shape can be made by applying a rule to the preceding shape. On a blank sheet of paper, draw the first shape of your sequence and write the rule. Try to make your rule clear enough that anyone following it will get the sequence that you have in mind.

2. Exchange starting shapes and rules with your partner. Follow your partner's rule to draw at least the next three shapes in the sequence.

3. Compare the sequence you drew in Exercise 2 with your partner's original sequence. Are they the same? If not, describe how they are different and why. If either your rule or your partner's rule is ambiguous, work together to rewrite it to make it clear.

Rules are often used to describe how two quantities are related. For example, a rule might tell you how to calculate one quantity from another.

Example

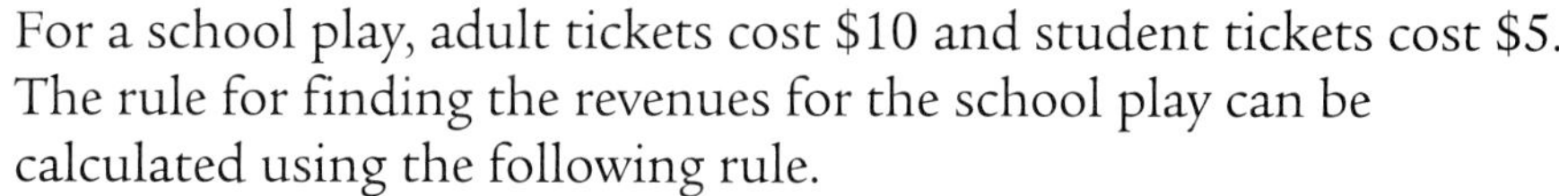

For a school play, adult tickets cost \$10 and student tickets cost \$5. The rule for finding the revenues for the school play can be calculated using the following rule.

Multiply the number of adults in attendance by 10. Multiply the number of students in attendance by 5. Add the two products.

Suppose 210 adults and 103 students attend the play. You can apply the rule to calculate the play's revenues.

$$\begin{aligned}\text{Revenues} &= 210 \cdot 10 + 103 \cdot 5 \\ &= 2{,}100 + 515 \\ &= 2{,}615\end{aligned}$$

So, the play revenues would be \$2,615.

In Exercises 4–6, you will look at some common rules for finding one quantity from another.

Develop & Understand: B

4. You can use the following rule to estimate how many miles away a bolt of lightning struck.

   Count the seconds between seeing the lightning flash and hearing the thunder. Then divide by 5.

   Use the rule to estimate how far away a bolt of lightning struck if you counted 15 seconds between the flash and the thunder.

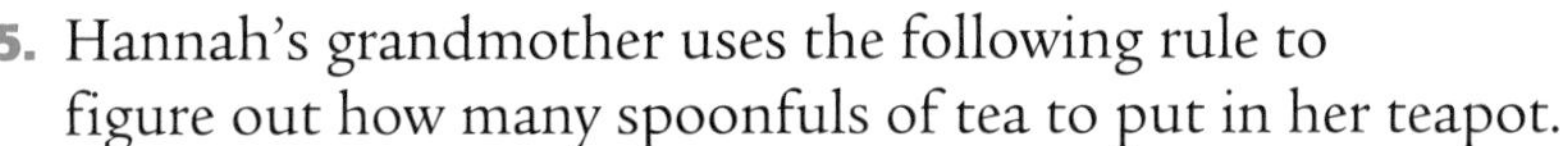

Real-World Link

It is believed that people first began drinking hot tea more than 5,000 years ago in China. Iced tea was not introduced until 1904 at the St. Louis World's Fair. An Englishman named Richard Blechynden added ice to the drink because no one was buying his hot tea.

5. Hannah's grandmother uses the following rule to figure out how many spoonfuls of tea to put in her teapot.

   Use one spoonful for each person and then add one extra spoonful.

   a. How much tea would Hannah's grandmother use for four people?

   b. Hannah's grandfather thinks this rule makes the tea too strong. Make up a rule that he might like better.

   c. Using your rule from Part b, how many spoonfuls of tea are needed for four people?

   d. Hannah's cousin, Amy, likes her tea much stronger than Hannah's grandmother does. Create a rule Amy might like. Use it to figure out how many spoonfuls of tea would be needed for four people.

6. A cookbook gives the following rule for roasting beef.

   Cook for 20 minutes at 475 °F. Then lower the heat to 375 °F and cook for 15 minutes per pound. If you like rare beef, remove the roast from the oven. If you like it medium, cook it an additional 7 minutes. If you like it well done, cook it an additional 14 minutes.

   What is the total cooking time for a four-pound beef roast if you like it medium?

The rules in Exercises 4–6 are fairly simple. Many rules involve more complicated calculations. If you do not need to find an exact value, you can sometimes use a simpler rule to find an approximation.

Develop & Understand: C

7. You can convert temperatures from degrees Celsius (°C) to degrees Fahrenheit (°F) using this rule.

   Multiply the degrees Celsius by 1.8 and add 32.

   Copy and complete the table to show some Celsius temperatures and their Fahrenheit equivalents.

Degrees Celsius	Degrees Fahrenheit
0	32
10	50
20	
30	
40	
50	

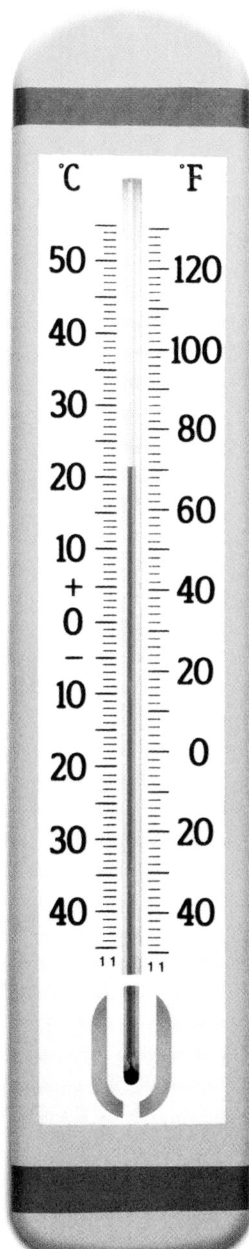

8. The Lopez family spent its summer vacation in Canada. Canadians use degrees Celsius to report temperatures. Ms. Lopez used the following rule to convert Celsius temperatures to Fahrenheit temperatures.

   Multiply the degrees Celsius by 2 and add 30.

   This rule makes it easy to do mental calculations, but it gives only an *approximation* of the actual Fahrenheit temperature.

   a. Complete this table to show the approximate Fahrenheit temperatures this rule gives for the listed Celsius temperatures.

Degrees Celsius	Approximate Degrees Fahrenheit
0	30
10	50
20	
30	
40	
50	

   b. For which Celsius temperature do the two rules above give the same result?

   c. For which Celsius temperatures in the table does Ms. Lopez's rule give a Fahrenheit temperature that is too high?

9. One day, the Lopez family flew from Toronto, where the temperature was 37°C, to Winnipeg, where the temperature was 23°C.

   a. Use the rule from Exercise 7 to find the exact Fahrenheit temperatures for the two cities.

   b. Use Ms. Lopez's rule from Exercise 8 to find the approximate Fahrenheit temperatures for the two cities.

   c. For which city did Ms. Lopez's rule give the more accurate temperature?

10. Look back at your answers to Exercises 8 and 9. What happens to the Fahrenheit approximation as the Celsius temperature increases?

Share & Summarize

1. Below are the first term and a rule for a sequence.

   First term: 20

   Rule: Write the number that is two units from the preceding number on the number line.

   a. Give the first few terms of two different sequences that fit the rule.

   b. Rewrite the rule so that only one of your sequences is correct.

2. At the corner market, bananas cost 49¢ a pound. Write a rule for calculating the cost of a bunch of bananas.

Real-World Link

About fifty percent of Toronto's two-and-a-half million population was born outside of Canada, making the city one of the most multicultural cities in the world. Toronto is the capital of the province of Ontario.

Investigation 2 Write Expressions

Vocabulary
variable

You have seen that algebraic expressions can be used to represent situations in which a quantity changes, or *varies*. In this investigation, you will write algebraic expressions to describe many other situations.

Develop & Understand: A

Jay, Lamar, Davina, Bart, Tara, and Freda live in a row of houses that are numbered 1 to 6. Use the clues below to help you solve Exercises 1–3.

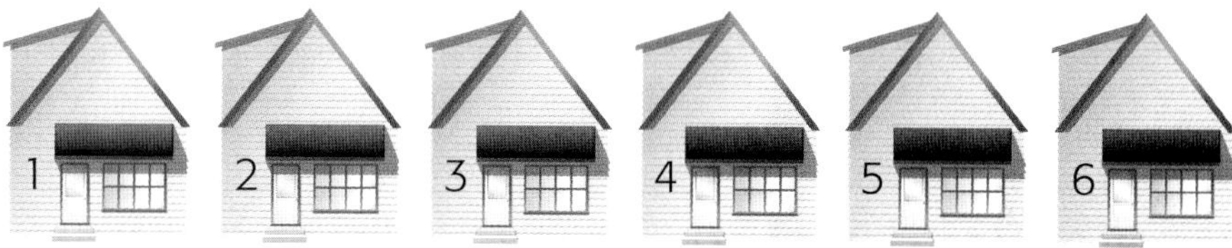

- Jay owns some DVDs.
- Lamar lives next to Jay and three houses from Freda. Lamar has 2 fewer DVDs than Jay.
- Davina lives on the other side of Jay and has three times as many DVDs as Lamar.
- Bart lives in house 2 and has 4 more DVDs than Jay.
- If Tara had 13 more DVDs, she would have four times as many as Jay.
- Jay acquired all of his DVDs from Freda, who lives in house 6. Freda had 17 DVDs before she gave p of them to Jay.
- The person in house 5 owns the most DVDs.

1. Determine who lives in each house. (Hint: Focus on where the people live. Ignore the other information.)

2. Let p stand for the number of DVDs that Jay has. Write an expression containing the variable p for the number of DVDs that each person has.

3. Use your expressions to help determine how many DVDs that Jay has. (Hint: Experiment with different values of p. Only one value of p gives the person in house 5 the most DVDs.)

In Exercises 1–3, the variable p could have only one value. The variable was used to represent the unknown quantity: Jay's number of DVDs. As you saw, it is helpful to be able to use a variable to represent an unknown quantity. Now you have two ways to think about variables. A **variable** is a quantity that can change or an unknown quantity.

Develop & Understand: B

For these situations, the variables represent unknown quantities rather than quantities that change.

4. Esteban bought N apples.
   a. Granny Smith bought four times as many apples as Esteban. Write an expression to show how many apples she bought.
   b. Esteban used 11 apples in a pie. How many does he have left?

5. Suppose a bag of potatoes weighs t pounds. In Parts a–d, write an expression for the number of pounds in each bag of vegetables below. Each expression should include the variable t.
   a. a bag of carrots weighing 3 pounds more than a bag of potatoes
   b. a bag of corn weighing 2 pounds less than a bag of potatoes
   c. a bag of broccoli weighing 5 pounds less than a bag of corn
   d. a bag of beans weighing 3 pounds more than a bag of broccoli
   e. Order the five bags from lightest to heaviest.

Real-World Link

The average newborn is about 20 inches long. It takes people about 4 years to double their length at birth and 8 more years to grow to three times their birth length. A person "only about 3.5 times the height of a newborn" is quite tall.

6. Baby Leanne is L inches tall. Write expressions, in terms of the variable L, to represent the heights in inches of the members of Leanne's family.
   a. her 4-year-old brother Tim, who is twice as tall as Leanne
   b. her 6-year-old sister Kerry, who is 5 inches taller than Tim
   c. her mother, who is 15 inches shorter than four times Leanne's height
   d. her father, who is 15 inches shorter than twice Kerry's height
   e. Baby Leanne is about 20 inches long. Check the expressions you wrote in Parts a–d by substituting this value and determining whether the heights are reasonable.

7. Sasha has k nickels in her pocket.

   a. Write an expression for the value of Sasha's nickels in cents. Find the value of the expression for $k = 4$.

   b. Sasha also has 60¢ in other coins. Write an expression for the total value, in cents, of all her change.

   c. Find the total value for $k = 5$ and $k = 12$.

   d. Yesterday, Sasha had k nickels in her pocket and no other money. She spent 70¢ on pencils. What was the total value of her coins after she bought the pencils? Could k be *any* whole number? Why?

8. Malik rides the city bus to school and back each weekday. The bus fare is F cents each way. He starts each week with $5 for bus fare. Write an expression for each of the following.

   a. the amount, in cents, Malik spends on bus fare in one day

   b. the amount he has left when he gets home on Monday

   c. the total bus fare he has spent by lunch time on Thursday

   d. the amount of bus fare he spends each week

   e. the amount he has left at the end of the week

   f. What would happen if F was 60?

9. You can sometimes make sense of an expression by inventing a meaning for its symbols. For example, for the expression $k - 4$, you can let k represent the number of kittens in a pet shop. Then $k - 4$ could stand for the number of kittens in the shop after four have been sold.

   Interpret each expression below by creating a meaning for the symbols.

   a. $d + 10$

   b. $3a$

   c. $f - 4$

Share & Summarize

Let h represent the number of hours Ella spent on homework last week.

1. Jack spent half as much time on his homework as Ella had on hers. Write an expression for the number of hours he spent.
2. Describe a situation that the expression $2h + 1$ might represent.
3. Write another algebraic expression containing the variable h. Describe a situation that your expression might represent.

Investigation 3 Evaluate Expressions

Vocabulary
exponent

When you study mathematics, it is important to know the shortcuts that are used for writing expressions. You already know that when you want to show a number times a variable, you can leave out the multiplication sign. Instead of $6 \times t$ or $6 \cdot t$, you can write $6t$. Note that the number is normally placed before the variable.

What if you want to show that the variable t is multiplied by itself? You could write $t \times t$, $t \cdot t$, or tt. However, an **exponent** is usually used to tell how many times a quantity is multiplied by itself. So, $t \cdot t$ is written t^2, and $t \cdot t \cdot t$ is written t^3.

Math Link
In the expression t^2, t is called the base and the ^{2} is the exponent.

Here are some other examples.

- The expression $2 \cdot s \cdot s \cdot s$ is written $2s^3$.
- The expression $m \cdot 5 \cdot m$ is written $5m^2$.
- The expression $2p \cdot p \cdot p$ is written $2p^3$.
- The expression $x \cdot x \cdot y \cdot 7$ is written $7x^2y$ or $7yx^2$.

It is also important to understand the rules for *evaluating,* or finding the value of, expressions. You learned some of these rules when you studied arithmetic.

Think & Discuss

Evaluate each expression without using a calculator.

$$7 + 8 \cdot 5 \qquad 7 - 6 \div 2$$

Now use your calculator to evaluate each expression. Did you get the same answers with the calculator as you did without?

Most calculators will multiply and divide before they add or subtract. This *order of operations* is a convention that everyone uses to avoid confusion. If you want to indicate that the operations should be done in a different order, you need to use parentheses.

Look at how the use of parentheses affects the value of the expressions below.

$$2 + 3 \cdot 5 = 2 + 15 = 17 \qquad 7 - 6 \div 2 = 7 - 3 = 4$$

$$(2 + 3) \cdot 5 = 5 \cdot 5 = 25 \qquad (7 - 6) \div 2 = 1 \div 2 = \frac{1}{2}$$

Which expression below means "add 4 and 6 and then multiply by n"? Which means "multiply 6 by n and then add 4"?

$$4 + 6n \qquad (4 + 6)n$$

There are also rules for evaluating expressions involving exponents.

Think & Discuss

Who is correct, Malik or Zach?

Does "find m^2 and multiply it by 4" give the same result as "find $4m$ and square it?" Try it for $m = 2$ to see.

Think again about who is correct, Malik or Zach. Then add parentheses to the expression $4m^2$ so it will give the other boy's calculation.

Develop & Understand: A

Rewrite each expression without using multiplication or addition signs.

1. $r + r$
2. $r \cdot r$
3. $t + t + t$
4. $5 \cdot g \cdot g \cdot g$
5. $5s + s$
6. $2.5m \cdot m$
7. Copy the expression $6 + 3 \cdot 4$.
   a. Insert parentheses, if necessary, so the resulting expression equals 18.
   b. Insert parentheses, if necessary, so the resulting expression equals 36.
8. Write an expression that means "take a number, multiply it by 6, and cube the result."

Evaluate each expression for $r = 3$.

9. $5r^2$
10. $(5r)^2$
11. $2r^4$
12. $(2r)^4$
13. When Kate, Jin Lee, Darnell, Zach, and Maya tried to find the value of $3t^2$ for $t = 7$, they came up with five different answers. Only one of their answers is correct.

    Kate: $t^2 = 49$, so $3t^2 = 349$

    Jin Lee: $3t^2 = 3 \cdot 7^2 = 42$

    Darnell: $3t^2 = 3 \cdot 7^2 = 3 \cdot 49 = 147$

    Zach: $3t^2 = 37^2 = 1{,}369$

    Maya: $3t^2 = 3 \cdot 7^2 = 21^2 = 441$

    a. Which student evaluated the expression correctly?
    b. What mistake did each of the other students make in thinking about the expression?

You have explored ways of expressing multiplication and repeated multiplication, like $3t$ and t^2. Now you will look at some ways to express division.

Example

Three friends won a prize of P dollars in the community talent show. They want to share the prize money equally. There are several ways to express the number of dollars each student should receive.

All three expressions in the example above are correct, but the forms $\frac{1}{3}P$ and $\frac{P}{3}$ are used more commonly in algebra than $P \div 3$.

Develop & Understand: B

14. In the prize-money example, P stands for the dollar amount of the prize, and $\frac{P}{3}$ shows each friend's share.

a. If the prize is \$30, how much will each friend get?

b. Suppose $\frac{P}{3} = \$25$. How much total prize money is there?

Find the value of each expression for $k = 5$.

15. $k^3 - 2$ **16.** $12 - k$ **17.** $\frac{k^2 - 1}{12}$ **18.** $\frac{7k}{5}$ **19.** $\frac{7}{5}k$

Develop & Understand: C

Kate, Zach, Maya, and Darnell earn money by selling greeting cards that they create on a computer.

20. In their first week in business, they sold 19 cards for B dollars each. They equally shared the money that they collected.

a. Write an expression that describes each friend's share.

b. What is the value of your expression if B is \$2.00? If B is \$2.60? If B is \$3.00?

c. The friends wrote the equation $\frac{19B}{4} = 7.60$ to describe what happened their first week in business. What does the equation indicate about the money that each friend made?

d. Starting with the equation in Part c, find how much the friends charged for each card.

21. This week, the friends received orders for 16 greeting cards.

a. If they sell the cards for B dollars each, how much will each friend earn?

b. The friends would like to earn \$12 apiece for selling the 16 cards. How much will they have to charge for each card?

22. In Exercises 20 and 21, the price of a greeting card was the only variable. Kate wants to be able to vary three amounts, the number of friends working on the cards, the number of cards, and the charge per card, and still be able to calculate how much each person would make.

Write an expression for how much each friend would earn in each situation.

a. Three friends sell 19 greeting cards for B dollars each

b. F friends sell 19 cards for B dollars each

c. Four friends sell P cards for B dollars each

d. F friends sell P cards for B dollars each

23. Suppose the price per card increases by \$2, from B to $(B + 2)$. Write an expression for how much F friends would each make after selling P cards at this new price.

24. Suppose the price per card increases by K dollars, from B to $(B + K)$. Write an expression for how much F friends would each make after selling P cards at this new price.

Share & Summarize

1. Write an algebraic expression that describes this set of calculations.

   Step 1. Take some number and square it.

   Step 2. Add 2 times the starting number.

   Step 3. Divide that result by your starting number.

   Step 4. Subtract your starting number.

2. Write an expression that describes this set of calculations.

   Step 1. Take some number and triple it.

   Step 2. Subtract your result from your starting number.

   Step 3. Divide that result by the square of your starting number.

   Step 4. Subtract the result from 10.

3. Find the value of $\frac{5 + 15x^2}{13}$ when $x = 2$.

Math Link

If you follow the steps in Question 1 carefully, your answer will be 2 when *n* is any positive number. Try it with your expression.

Investigation  4 Flowcharts

Vocabulary

backtracking

flowchart

In this investigation, you will use a tool that can help you evaluate expressions and solve equations.

In preparing for a party, Maya bought five bags of bagels and four extra bagels.

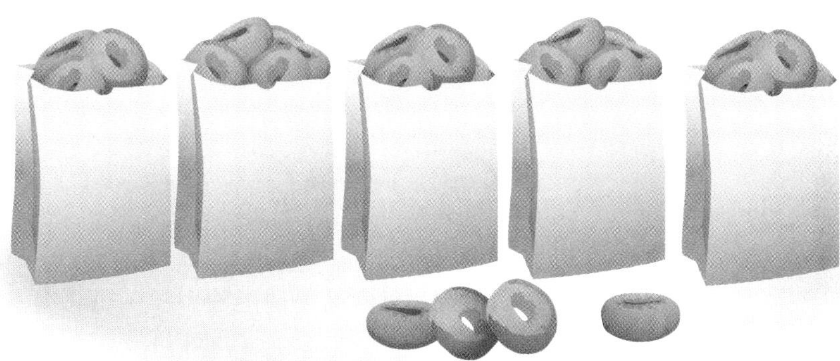

If n represents the number of bagels in each bag, the total number of bagels Maya purchased is $5n + 4$. To find the total number of bagels, you multiply the number in each bag by 5 and then add 4. You can make a diagram, called a **flowchart**, to show these steps.

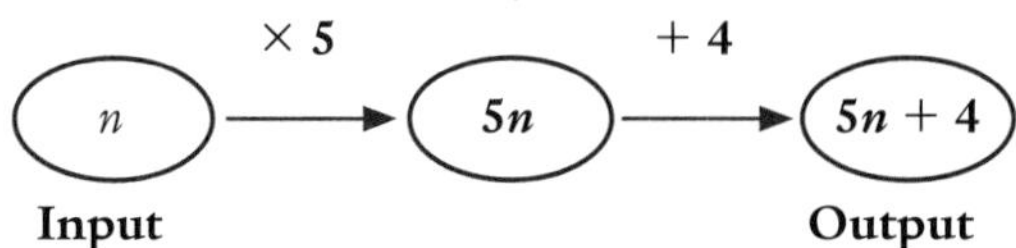

The oval at the left side of the flowchart represents the *input.* In this case, input is the number of bagels in each bag. Each arrow represents a *mathematical action.* The oval to the right of an arrow shows the result of a mathematical action. The oval at the far right represents the *output.* In this case, output is the total number of bagels.

To evaluate $5n + 4$ for a particular value of n, substitute that value for the input and follow the steps until you reach the output. Here is the same flowchart with an input, the value of n, of 3.

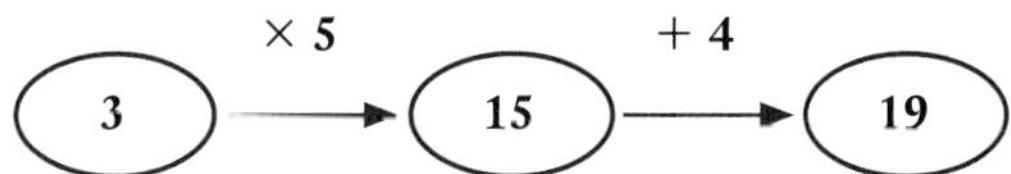

In Exercises 1–5, you will use flowcharts to find outputs for given inputs. You will also create flowcharts to match algebraic expressions.

Develop & Understand: A

Copy and complete each flowchart by filling the empty ovals.

1. 

2. 

3. × 3 − 1 ÷ 3
5

4. Consider the expressions $6y + 1$ and $6(y + 1)$.
   a. Make a flowchart for the expression $6y + 1$.
   b. Make a flowchart for the expression $6(y + 1)$. How is this flowchart different from the flowchart in Part a?

5. Consider the expressions $\frac{n + 2}{3}$ and $n + \frac{2}{3}$.
   a. Make a flowchart for each expression.
   b. Use your flowcharts to find outputs for three or four n values. Do both flowcharts give the same outputs? Explain why or why not.

Maya had $5n + 4$ bagels, where n is the number of bagels in each bag. Suppose she had a total of 79 bagels. How could you find the number of bagels in each bag? That is, how could you find the value of n for which $5n + 4 = 79$?

Leon used a flowchart to find this number by **backtracking**, or working backward. This is how he did it.

Since 79 is the output, I'll put it in the last oval.

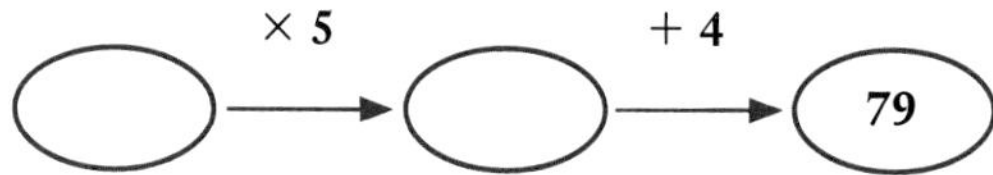

I add 4 to get 79, so the number in the second oval must be 75.

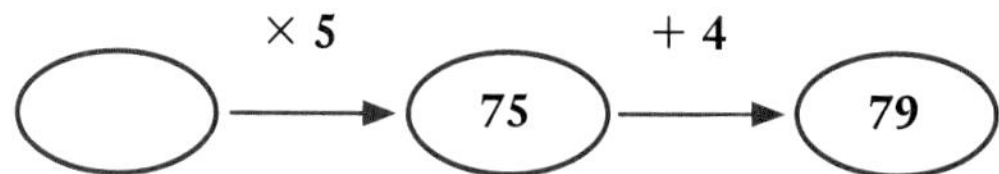

Since 5 times a number is 75, the number of bagels in each bag must be 15.

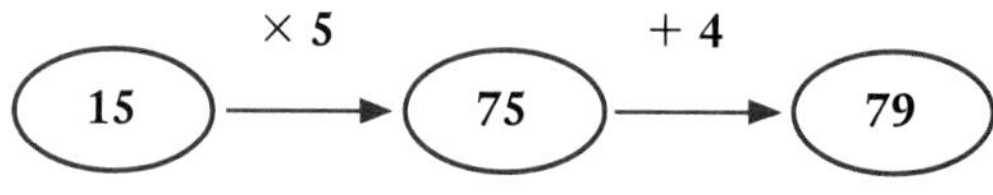

The input Leon found, 15, is the solution of the equation $5n + 4 = 79$. Now you will practice solving equations using the backtracking method.

Develop & Understand: B

6. Libby put an input into this flowchart and got the output 53. Use backtracking to find Libby's input.

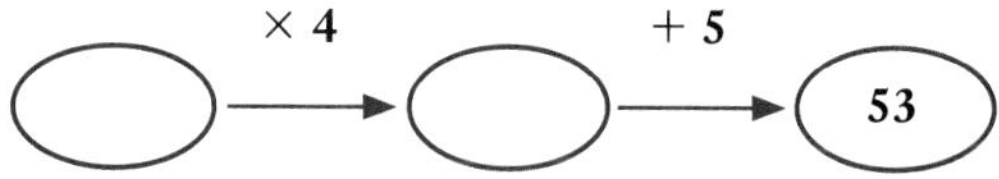

Leon solved some equations by backtracking. Exercises 7–9 show the flowcharts he used. Do Parts a and b for each flowchart.

a. Write the equation that Leon was trying to solve.

b. Backtrack to find the solution.

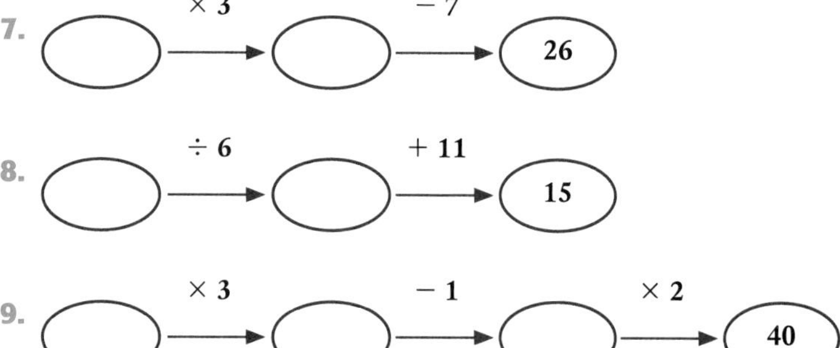

Now that you have some experience using backtracking to solve equations, you can use backtracking to solve more challenging equations.

Develop & Understand: C

10. Lakeisha and Mateo were playing a game called *Think of a Number.*

Lakeisha must use Mateo's output to figure out his starting number.

a. Draw a flowchart to represent this game.

b. What equation does your flowchart represent?

c. Use backtracking to solve your equation from Part b. Check your solution by following Lakeisha's steps.

11. Consider this expression.

$$\frac{2(n+1)}{3} - 1$$

a. Draw a flowchart for the expression.

b. Use backtracking to find the solution of $\frac{2(n+1)}{3} - 1 = 5$.

Share & Summarize

Create an equation that can be solved by backtracking. Write a paragraph explaining to someone who is not in your class how to use backtracking to solve your equation.

Practice & Apply

Use the first term or stage and the rule given to create a sequence. Tell whether your sequence is the only one possible. If it is not, give another sequence that fits the rule.

1. *First term*: 40
Rule: Divide the preceding term by 2.

2. *First stage*: △
Rule: Draw a shape with one more side than the preceding shape.

3. Starting with a closed geometric figure with straight sides of equal lengths, you can use the rule below to create a design.

Find the midpoint, or middle point, of each side of the figure. Connect the midpoints in order to make a new shape. It will be the same shape as the original but smaller.

a. Copy this square. Follow the rule three times, each time starting with the figure you drew the previous time.

b. Copy this triangle. Follow the rule three times, each time starting with the figure you drew the previous time.

c. Draw your own shape. Follow the rule three times to make a design.

4. Measurement Sergio is making a dessert that requires three eggs for each cup of flour.

a. How many eggs does he need for three cups of flour?

b. For a party, Sergio made a large batch of his dessert using a dozen eggs. How much flour did he use?

5. Economics Kendra uses this rule to figure out how much to charge for babysitting.

Charge $5 per hour for one child, plus $2 per hour for each additional child.

a. Last Saturday, Kendra watched the Newsome twins for 3 hours. How much money did she earn? Explain how you found your answer.

b. Mr. Foster hires Kendra to watch his three children for 2 hours. How much will she charge?

c. Does Kendra earn more for watching two children for 3 hours or three children for 2 hours?

d. Kendra hopes to earn $25 next weekend to buy her sister a birthday present. Describe two ways she could earn at least $25 babysitting.

6. Measurement You can convert speeds from kilometers per hour to miles per hour by using this rule.

Multiply the number of kilometers per hour by 0.62.

a. Convert each kilometers-per-hour value in the table below to a miles-per-hour value.

Kilometers per Hour	Miles per Hour
50	
60	
70	
80	
90	
100	
110	
120	

b. As part of his job, Mr. Lopez does a lot of driving in Canada. He uses the following rule to approximate the speed in miles per hour from a given speed in kilometers per hour.

Divide the number of kilometers per hour by 2 and add 10.

Use Mr. Lopez's rule to convert each kilometers-per-hour value in the table to an approximate miles-per-hour value.

c. For which kilometers-per-hour values from the tables are the results for the two rules closest?

d. For which kilometers-per-hour values in the table does Mr. Lopez's rule give a value that is too high?

7. Here are clues for a logic puzzle.

- Five friends, Alano, Bob, Carl, Dina, and Bonita, live in a row of houses numbered 1 to 5.
- They all keep tropical fish for pets.
- Dina got her fish from Bonita's next-door neighbor Alano. Alano had 21 fish before he gave f of them to Dina.
- Carl has twice as many fish as Dina, who lives in house 2. Carl does not live next to Dina.
- Bonita lives in house 4 and has 3 fewer fish than Carl.
- Bob has 4 more fish than Dina.

a. Who lives in which house?

b. Write expressions to describe the number of fish each person has.

c. If the five neighbors have a total of 57 fish, how many fish does Dina have?

8. Keshon has s stamps in his collection.

a. Keshon's older sister, Jamila, has three times as many stamps as Keshon. Write an expression for the number of stamps she has.

b. Jamila decides to give Keshon 13 of her stamps. How many stamps will she have left? How many stamps will Keshon have?

9. Franklin, a golden retriever, weighs p pounds. In Parts a–c, write an expression for the weight of the dog in pounds. Each expression should include the variable p.

a. Tatu, a pug, weighs 49 pounds less than Franklin.

b. Mia, a Chihuahua, weighs $\frac{1}{17}$ as much as Franklin.

c. Lucy, a Great Dane, weighs twice as much as Franklin, minus 15 pounds.

d. Franklin weighs 68 pounds. How much do Tatu, Mia, and Lucy weigh?

10. Mr. Karnowski has a stack of 125 sheets of graph paper. He is passing out g sheets to each of his students.

a. Write an expression for the number of sheets he will pass out to the first 5 students.

b. Write an expression for the number of sheets he will have left after he has given sheets to all 25 students in his class.

c. What would happen if g was 6?

11. The Marble Emporium sells individual marbles and bags of marbles. Each bag contains the same number of marbles.

a. Rita bought a bag of marbles, plus one extra marble. Is it possible that she bought an even number of marbles? Explain.

b. Lorena bought two bags of marbles, plus one extra marble. Is it possible that Lorena has an even number of marbles? Explain.

c. Challenge Let n stand for *any whole number*. Use what you have discovered in Parts a and b to write an algebraic expression containing n that means *any odd number*. Write another expression that means *any even number*.

Rewrite each expression without using multiplication or addition signs.

12. $t + t + t + t$

13. $4 \cdot xy \cdot xy$

14. $6g + g + g$

15. $8 \cdot c \cdot c \cdot d \cdot d$

16. Find the value of $9 - 3D^2$ for $D = 1.1$.

17. Without using a calculator, find the value of $9 - 3D^2$ for $D = \frac{1}{2}$.

Evaluate each expression for $h = 4$.

18. $\dfrac{h^2 - 3}{13}$

19. $\dfrac{3h}{4}$

20. $9h \div 2$

21. Luna, Mason, and Gabriel make and sell hand-painted t-shirts. They make a profit of D dollars for each shirt they sell.

a. If they sell 13 t-shirts, how much total profit will they make?

b. If the three friends equally divide the profit from selling the 13 shirts, how much will each receive?

c. Suppose each friend received $39 from selling the 13 shirts. How much profit did the students earn for each shirt?

d. Suppose F more friends join the business. If the group sells 20 t-shirts and divides the profit equally, how much will each friend receive?

e. If the new, larger group of friends sells X shirts, how much profit will each friend receive?

Copy and complete each flowchart.

22.

23.

24.

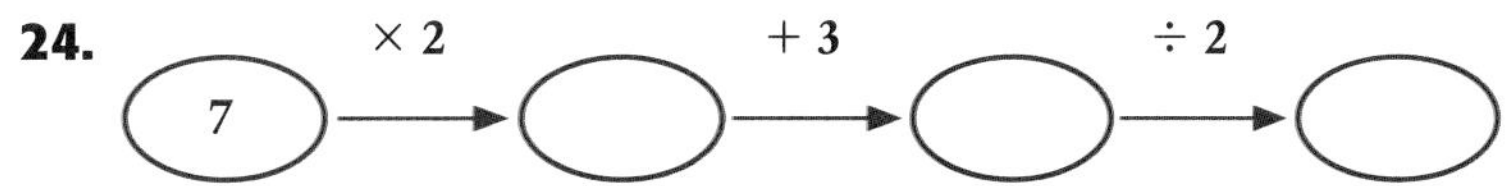

25.

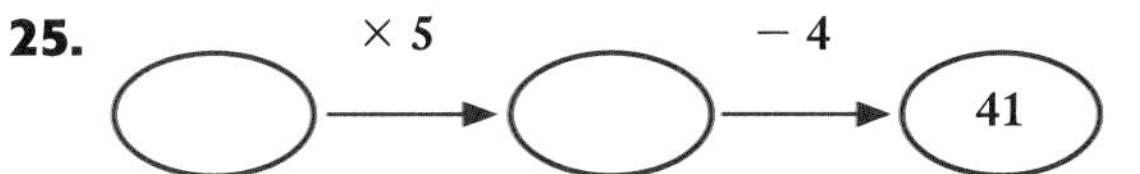

26. Consider the expression $3a \div 4$.

a. Draw a flowchart to represent the expression.

b. Use your flowchart to find the value of the expression for $a = 8$.

c. Use your flowchart to solve $3a \div 4 = 6$. Explain each step in your solution.

27. Lara drew this flowchart.

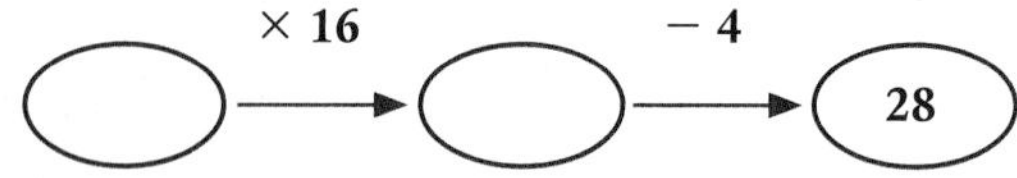

a. What equation was Lara trying to solve?

b. Copy and complete the flowchart.

28. In a game of *Think of a Number,* Jin Lee told the following to Darnell.

- *Think of a number.*
- *Subtract 1 from your number.*
- *Multiply the result by 2.*
- *Add 6.*

a. Draw a flowchart to represent this game.

b. Darnell said he came up with 10. Write an equation that you could solve to find Darnell's number.

c. Use backtracking to solve your equation. Check your solution by following Jin Lee's steps.

29. Luis wrote the expression $9(2x + 1) + 1$.

a. Draw a flowchart for Luis' expression.

b. Use your flowchart to solve the equation $9(2x + 1) + 1 = 46$.

30. Consider this equation.

$$\frac{3m \cdot 2}{6} = 1$$

a. Draw a flowchart to represent the equation.

b. Use backtracking to find the solution.

Connect & Extend

31. You can produce a sequence of numbers by applying the following rule to each term.

If the number is even, get the next number by dividing by 2. If the number is odd, get the next number by multiplying by 3 and adding 1.

a. Use this rule to produce a sequence with 1 as the first term. Describe the pattern in the sequence.

b. Now use the rule to produce a sequence with 8 as the first term. Keep finding new terms until you see a pattern in the sequence. Describe what happens.

c. Use the rule to generate two more sequences. Keep finding new terms until you see a pattern.

d. Using your calculator and the rule, generate a sequence with 331 as its first term. Again, keep finding new terms until you see a pattern.

e. Describe what you discovered in Parts a–d.

32. Measurement One mile is about 1.6 kilometers.

a. Which is the greater distance, 1 mile or 1 kilometer?

b. Los Angeles and New York City are about 2,460 miles apart. How many kilometers apart are the two cities?

c. If the speed limit on a road in Canada is 50 kilometers per hour, what is the speed limit in miles per hour?

d. Jin and Franklin traveled with their families by car to visit relatives during the summer. Jin traveled about 500 miles and Franklin traveled about 700 kilometers. Who traveled the greater distance? Explain your reasoning.

e. In Investigation 1, you learned that lightning is 1 mile away for every 5 seconds you count between the lightning and the following clap of thunder. About how many seconds would it take you to hear the thunder if the lightning was 1 kilometer away?

33. Tito's Taxi charges \$3.20 for the first mile plus \$2.40 for each additional mile.

a. Which expression gives the fare, in dollars, for a trip of d miles? Assume that the trip is at least one mile.

$d + 3.20 + 2.40 \qquad 3.20(d + 2.40)$

$2.40(d - 1) + 3.20 \qquad d(2.40 + 3.20)$

b. How much would it cost to travel one mile? Three miles?

c. Tito thought he might make more money if his drivers charged for every sixth of a mile, not every mile. The new rate is \$1.60 for the first $\frac{2}{6}$ mile, plus 40¢ for each additional $\frac{1}{6}$ mile.

If a person travels n sixths of a mile, which expression gives the correct fare? Assume that the person travels at least $\frac{2}{6}$ mile.

$1.60 + 40n \qquad 1.60 + 0.40n$

$1.60 + 40(n - 2) \qquad 1.60 + 0.40(n - 2)$

d. Use the formula you chose in Part c to calculate trip fares for one mile and three miles. Compare the results to those for Part b. Does Tito's Taxi make more money with this new rate plan?

Real-World Link

Profit is the amount of money a business earns. It is calculated by subtracting expenses, or what was spent, from income, or what was earned.

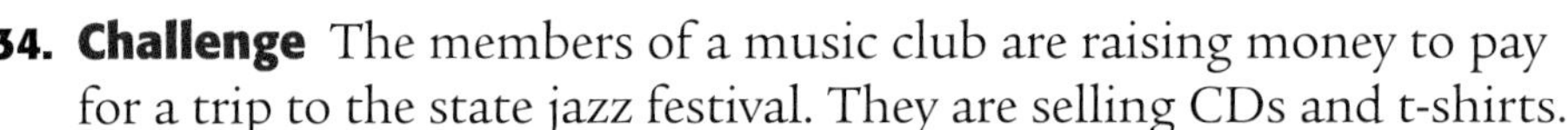

34. Challenge The members of a music club are raising money to pay for a trip to the state jazz festival. They are selling CDs and t-shirts.

a. Each CD costs the club \$2.40. The club is selling the CDs for \$12.00 apiece. Write an expression to represent how much profit the club will make for selling c CDs.

b. Each t-shirt costs the club \$3.50. The club is selling the t-shirts for \$14.50. Write an expression to represent how much profit the club will make for selling s t-shirts.

c. Now write an expression that gives the profit for selling c CDs and s t-shirts.

d. Ludwig sold 10 of each item. Johann sold 7 CDs and 12 t-shirts. Use your expression from Part c to calculate how much money each club member raised.

35. Preview Consider the expressions $x(x - 1)$ and $x^2 - x$.

a. Choose five values for x. Evaluate both expressions for those values. Organize your results in a table.

b. What appears to be the relationship between the two expressions?

36. Challenge In this exercise, you will compare x^2 and x for different values of x.

a. Find three values for x such that $x^2 > x$.

b. Find three values for x such that $x^2 < x$.

c. Find a value for x such that $x^2 = x$. Can you find more than one?

37. Number Sense In a game of *Think of a Number;* the input was 3 and the output was 2. Tell whether each of the following could have been the game's rule.

a. Think of a number. Double it. Add 3 to the result. Then divide by 5.

b. Think of a number. Subtract 2 from it. Multiply the result by 4. Then divide by 2.

c. Think of a number. Divide it by 3. Add 5 to the result. Then divide by 3 again.

38. In Your Own Words Describe a situation that can be represented by the expression $25n + 100$. Explain what the 25, the 100, and the n represent in your situation.

Mixed Review

39. The graphs give information about two video games. Use the graphs to determine whether each statement is true or false. Explain how you decided.

a. The more exciting game is less expensive.

b. The more difficult game has lower-quality graphics.

c. The less difficult game is more exciting.

d. The less expensive game has better graphics.

Expressions and Formulas

Inv 1	What's the Variable?	32
Inv 2	Formulas	35
Inv 3	Formulas and Spreadsheets	40

Before you write an expression to describe a situation, you need to figure out what is varying.

Zoe, Darnell, Maya, and Zach are members of a community group that raises money for various charities. Every year, the community group holds a calendar sale.

Last year's calendars pictured different animals. All the profits from the sale were given to a wildlife preservation fund. The students agreed to each donate \$2 of their own money to the fund in addition to the money that they collected from the calendar sale. They sold the calendars for \$12 each.

Think & Discuss

How much money did each student collect for the fund? Consider both the money they earned for selling the calendars and the money they donated themselves.

What is varying in this situation?

Write an expression for the total amount of money given to the fund by each student. Tell what your variable represents.

For its summer project, the community group hosted a walkathon to raise money for cancer research. A successful radio campaign prompted people to request more information. The community group responded by sending informational packets.

Zoe, Darnell, Maya, and Zach each spent time one weekend preparing packets, including addressing them. It takes about an hour for one student to prepare 80 packets. On Monday, after school, they each completed another 50 packets.

Zoe, Darnell, and Maya discussed how much time they spent on the project.

Think & Discuss

About how many packets did each of the three students prepare? Consider both the packets prepared over the weekend and those completed on Monday.

When you calculated how many packets each student prepared, what was varying?

Write an expression for the total number of packets prepared by a volunteer working for t hours.

Zach prepared 690 packets. How long did he spend preparing packets over the weekend?

Investigation  What's the Variable?

In this investigation, you will write algebraic expressions to match situations. You will figure out which of several situations match a given expression. You will also make up your own situations based on an expression and a given meaning for the variable.

Develop & Understand: A

1. A movie ticket costs \$6, and a box of popcorn costs \$2.

   a. Owen was meeting friends at the theater to see a movie. He arrived first and bought one ticket. He decided to buy a box of popcorn for each of his friends. Write an expression for the total amount Owen spent. Choose a letter for the variable in your expression, and tell what the variable represents.

   b. Another group of friends was meeting to see a movie. They planned to share one box of popcorn among them. Write an expression for the total amount the group spent, and tell what the variable represents.

2. A pancake recipe calls for flour. To make thicker pancakes, Mr. Gordon measures out an extra cup of flour. He then divides the flour into three equal portions so that he can make blueberry pancakes, banana pancakes, and plain pancakes. Write an expression for how much flour is in each portion, and tell what your variable represents.

3. Pablo made two round trips to the beach and then traveled another 600 yards to the concession stand.

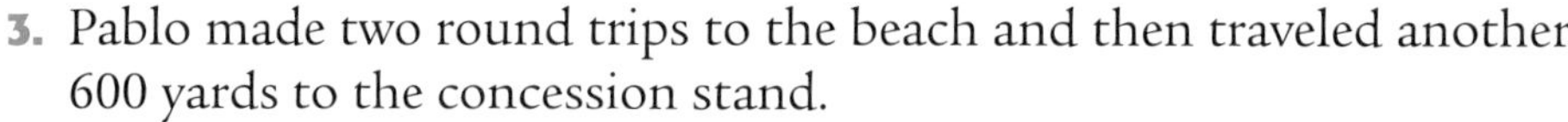

   a. Write an expression to represent the total distance Pablo traveled. Be sure to say what your variable represents.

   b. If Pablo walked a total of 1,584 yards, how far is it to the beach?

4. Ms. Franklin wants to till the ground to add new sections to her garden. She wants three square sections with the same side length and another section with an area of 7 square meters.

   a. Write an expression to represent the total area of the sections Ms. Franklin needs to till.

   b. Ms. Franklin's son, Justice, offers to help her. If they share the work equally, how many square meters will each have to till?

5. Holden received \$60 for his birthday. Each week, for x weeks, he spent \$4 of the money to go ice skating.

   a. How much money, in dollars, did Holden have left after x weeks?

   b. Does *any* number make sense for the value of x? If not, describe the values that do not make sense.

A single expression can represent numerous situations, depending on what the variable represents.

Example

Consider the expression $c + 10$.

- If c is the number of cents in Koto's piggy bank, $c + 10$ could represent the number of cents in the bank after she drops in another dime.
- If c is the number of gallons of gas left in Ms. Flores' gas tank, $c + 10$ could represent the number of gallons after she adds 10 gallons.
- If there are c members in the science club, $c + 10$ could represent the number of members after 10 new students join.

Think of some other situations $c + 10$ could represent.

Develop & Understand: B

In Exercises 6–11, decide whether the expression $2d + 5$ can represent the answer to the question. If it can, explain what d stands for in that situation.

6. Lucy bought two tickets to a symphony concert and five tickets to a movie. How much did she pay?

7. Sam bought several pens for \$2 each and a notebook for \$5. How much did he spend altogether?

8. A herd of zebras walked a certain distance to a watering hole. On the return trip, a detour added an extra 5 km. How many kilometers did the herd walk altogether?

9. There are two dogs in the house and five more in the yard. How many dogs are there altogether?

10. Because their parents were sick, Neva and Dario spent twice as many hours as usual doing housework during the week plus an extra 5 hours on the weekend. How many hours of housework did they do?

11. Blake gets paid \$5 for every hour he babysits plus a \$2 bonus on weekends. How much money does he earn if he babysits on Saturday?

Describe a situation that can be represented by each expression.

12. $4m - 3$, if m stands for the number of pages in a book

13. $4m - 3$, if m stands for the distance in kilometers from home to school

14. $4m - 3$, if m stands for the number of eggs in a waffle recipe

15. $4m - 3$, if m stands for the number of milliliters of water in a beaker of water

16. Describe two situations that can be represented by the expression $10 - x^2$. Discuss with your partner how your situations match the expression.

Share & Summarize

Describe two situations that can be represented by this expression.

$$3t + 7$$

Explain how your situations match the expression. Check by trying some values for the variable t and seeing whether the solutions make sense.

Investigation 2 Formulas

Vocabulary

formula

Materials

- graph paper

Weather reports in most countries give temperatures in degrees Celsius. If you are like most Americans, however, a temperature of 20°C is something you have heard, but you may not know whether it is very hot, pleasantly warm, or fairly chilly.

Think & Discuss

Test your sense of how warm these Celsius temperatures are.

- Which outdoor attire do you think is most appropriate when the temperature is 33°C? Is it a winter coat, jeans and a sweatshirt, or shorts and a t-shirt?
- Which outdoor activity do you think is most appropriate in 10°C weather? Is it swimming, soccer, or skiing?

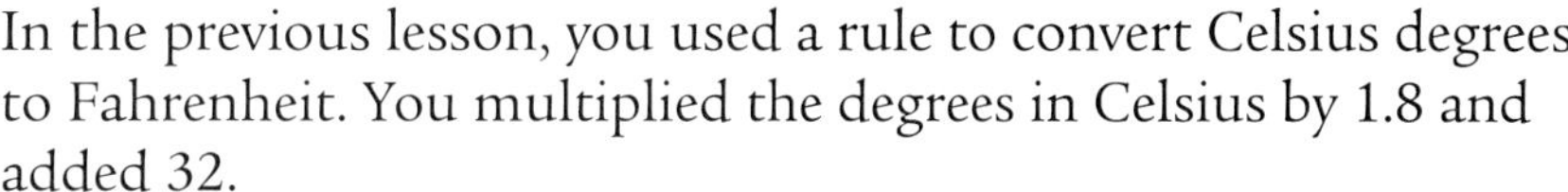

In the previous lesson, you used a rule to convert Celsius degrees to Fahrenheit. You multiplied the degrees in Celsius by 1.8 and added 32.

We often refer to this kind of rule as a **formula**, an algebraic "recipe."

The Celsius-Fahrenheit conversion formula is sometimes written as shown below.

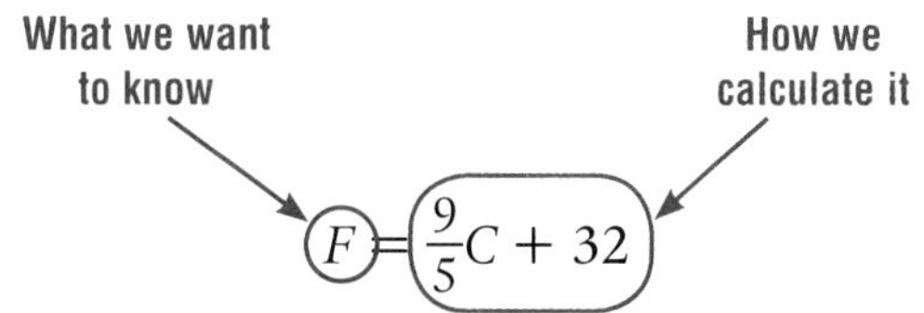

In the formula, the variable C represents the temperature in degrees Celsius. The variable F represents the temperature in degrees Fahrenheit.

This formula can also be written as shown below.

$$F = \frac{9C}{5} + 32$$

Example

Use the formula to find the Fahrenheit equivalent of 20°C.

$F = \frac{9}{5}C + 32$	Start with the formula.
$F = \frac{9}{5} \cdot 20 + 32$	Substitute 20 for C.
$F = 68$	Simplify.

So, 20°C is the same as 68°F.

Following the example above, find the Fahrenheit equivalents of 33°C and 15°C. Were your answers to the Think & Discuss questions on page 35 reasonable?

Develop & Understand: A

Real-World Link

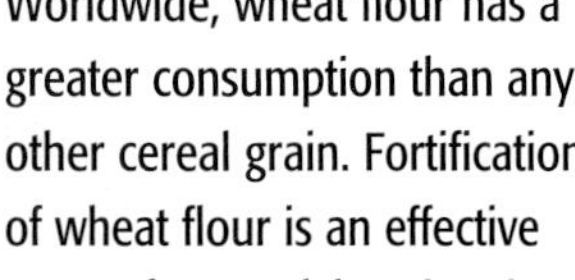

Worldwide, wheat flour has a greater consumption than any other cereal grain. Fortification of wheat flour is an effective means for supplying vitamins and minerals to the world population. The percentage of wheat flour fortification increased from 18% to 27% from 2004 to 2007.

1. Most pastry is made from flour, shortening, and water. There are different types of pastry. In *short pastry,* the relationship between the amount of flour F and the amount of shortening S is given by the following formula.

$$S = \frac{1}{2}F \qquad \text{or} \qquad S = \frac{F}{2}$$

   a. Alice wants to make lemon tarts with short pastry using exactly 800 grams of flour. How much shortening should she use?

   b. Daryl wants to make short pastry for cheese sticks, but he has only 250 grams of shortening. What is the maximum amount of flour he can use?

2. The relationship between flour and shortening in *flaky pastry* is given by the following formula.

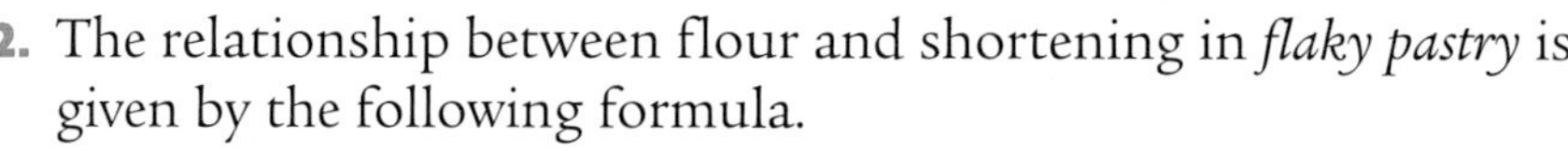

$$S = \frac{3}{4}F \qquad \text{or} \qquad S = \frac{3F}{4}$$

How much shortening would you need to add to 200 grams of flour to make flaky pastry?

3. Which type of pastry, short or flaky, needs more shortening for a given amount of flour? Check your answer by testing the amount of shortening for two amounts of flour.

4. You saw that the relationship between temperature in degrees Celsius and degrees Fahrenheit is given by this formula.

$$F = \frac{9}{5}C + 32$$

a. Convert 38°C to degrees Fahrenheit.

b. Water freezes at 0°C. How many degrees Fahrenheit is this?

c. Water boils at 100°C. How many degrees Fahrenheit is this?

d. Convert 50°F to degrees Celsius.

Math Link

The diameter is the longest measurement across a circle. It equals twice the radius.

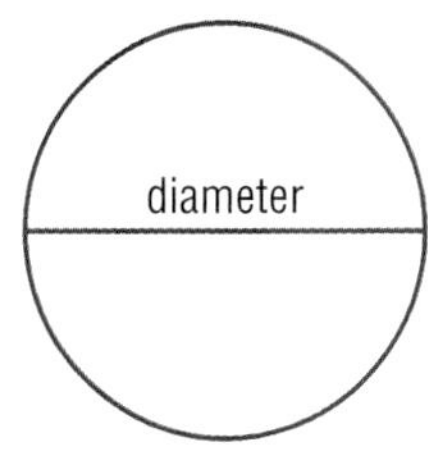

5. The formula for the area of a circle is $A = \pi r^2$, where r is the radius. The Greek letter π is not a variable. It is a number equal to the circumference of any circle divided by its diameter.

a. Your calculator probably has a button that automatically gives a good approximation for π. Do you recall some approximations that you have previously used?

b. What is the value of πr^2 for $r = 5.6$?

c. Why is the standard order of operations important when finding the area of a circle?

d. Tetromino's Pizza makes a pizza with a diameter of 12 inches. Use the formula to find the area of a 12-inch pizza.

e. Tetromino's 12-inch pizza costs $7.50. They also make a pizza with a diameter of 24 inches, which sells for $25. Which size is the better buy? Explain your answer.

f. Tetromino's smallest round pizza has an area of 78.5 in^2. Write the equation that you would need to solve to find the radius of the pizza. Solve your equation.

6. In Course 1, you found the measures of the interior angles of several polygons.

a. What is the sum of the interior angles of a square? Of a rectangle? Of any quadrilateral?

b. Write an expression for the sum of the interior angles of quadrilateral *ABCD*.

c. What is the measure of angle *A*?

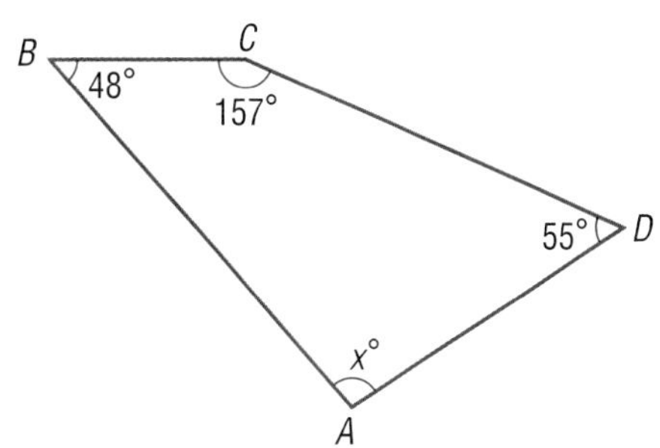

Now you will use formulas to help you analyze a race.

Develop & Understand: B

Three tortoises entered a ten-meter race.

Real-World Link

Tortoises live on land while their aquatic cousins, turtles, live in and around water. The limbs of tortoises and freshwater turtles have fingers and toes. Marine turtles have flat flippers.

Tortoise 1 was especially slow. He moved at only 0.9 meter per minute, so he was given a 3.1-meter head start.

Tortoise 2 roared along at 1.3 meters per minute and received no head start.

Tortoise 3 also received no head start. Her distance was equal to the square of the time, in minutes, since the race began, multiplied by 0.165.

7. It would be useful to have a formula that tells each tortoise's distance from the starting line at any time after the race began. For each tortoise, choose the formula below that describes its distance D from the starting point m minutes after the race began.

$D = 3.1m + 0.9$	$D = 1.3m$	$D = 0.165 + m^2$
$D = 3.1 + 0.9m$	$D = 0.165m^2$	$D = 10 - 1.3m$

Practice & Apply

1. A ticket to the symphony costs \$27 for adults and \$10.50 for children.
   a. Write an expression for the total ticket cost in dollars if A adults and three children attend the symphony.
   b. Write an expression for the total ticket cost if two adults and C children go to the symphony.

2. Anica and her friends are organizing a trip to the state skateboarding championships. The bus they will rent to take them there and back will cost \$500. Anica estimates that each person will spend \$40 for meals and souvenirs.
   a. If p people go on the trip, what will be the total cost for travel, meals, and souvenirs?
   b. What will be the cost per person if p people go on the trip?
   c. What will be the total cost per person if 25 people go on the trip?

3. A full storage tank contains 2,160 gallons of water. Every 24 hours, 4.5 gallons leak out of the tank.
   a. How many gallons of water will be in the tank H hours after it is filled?
   b. How many gallons of water will be in the tank after D days?
   c. How many gallons of water will be in the tank after W weeks?

In Exercises 4–6, determine whether the expression $9m - 4$ can be used to represent the answer to the question. If it can, explain what m represents.

4. The ski club bought nine lift tickets and received a \$4 discount on the total price. What was the total cost for the tickets?

5. Of the nine people on a baseball field, four are girls. How many are boys?

6. Last Saturday, Kenya listened to her new CD nine times. But the ninth time she played it, she skipped a 4-minute song she did not like. How much time did Kenya spend listening to her CD last Saturday?

In Exercises 7–9, describe a situation that the expression can represent.

7. $3p + 2$, if p is the price of a pepperoni pizza in dollars

8. $2x - 8$, if x is the number of ounces in a pitcher of lemonade

9. $\frac{L^2}{2}$, if L is the length of a square piece of paper

10. Three friends, Aisha, Mika, and Caitlin, have electronic robot toys that move at different speeds. They decide to have a ten-meter race.

- Aisha's robot travels 1 meter per second.
- Mika's robot moves 0.9 meter per second.
- Caitlin's robot's speed is 1.3 meters per second.

a. If they all start together, whose robot will win? How long will each robot take to reach the finish line?

b. To make the next race more interesting, the friends agree to give Mika's and Aisha's robots a head start.

- Mika's robot has a three-meter head start, so it has to travel only seven meters.
- Aisha's robot has a two-meter head start, so it has to travel only eight meters.

Whose robot wins this race? Whose comes in second? Whose comes in last?

c. The girls decide they would like the robots to finish as close to *exactly* together as possible. How much of a head start should Mika's and Aisha's robots have for this to happen?

Hint: Use your calculator. Caitlin's robot takes $10 \div 1.3 = 7.6923077$ seconds to travel ten meters. How can you position the other two robots so that they take this same amount of time?

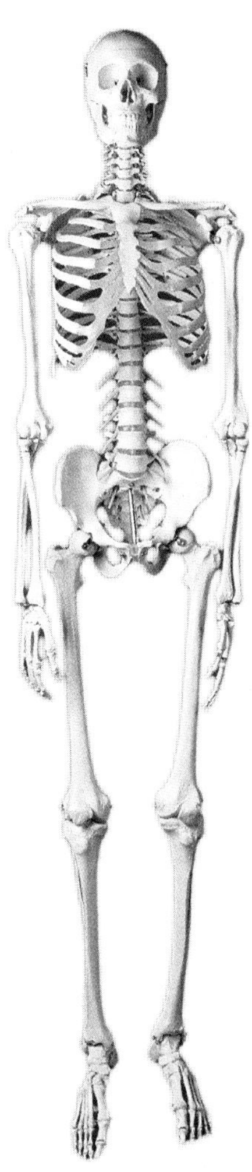

11. **Measurement** By measuring certain bones, forensic scientists can estimate a person's height. These formulas show the approximate relationship between the length of the tibia, or shin bone, t and height h for males and females. The measurements are in centimeters.

$$\text{males: } h = 81.688 + 2.392t$$

$$\text{females: } h = 72.572 + 2.533t$$

   a. How tall is a male if his tibia is 38 cm long? Give your answer in centimeters. Use the facts that 2.54 centimeters are about one inch and 12 inches equals one foot to find the height in feet.

   b. How tall is a female if her tibia is 38 centimeters long? Give your answer in centimeters and in feet.

   c. If a woman is 160 centimeters tall, or about 5.25 feet, how long is her tibia?

12. **Physics** The distance in meters an object falls in T seconds after it is dropped, not taking into account air resistance, is given by the formula below.

$$D = 4.9T^2$$

   a. A stone is dropped from a high cliff. How far will it have fallen after three seconds?

   b. After dropping from a plane, a parachutist counts slowly to ten before pulling the ripcord to unfold her parachute. How far does she fall while she is counting?

   c. Shelley dropped a pebble from a cliff and timed it as it fell. The pebble hit the water 3.7 seconds after she dropped it. How high is the cliff above the water?

   d. The watch Shelley used was accurate only to the nearest 0.1 second. This means that the actual time it took the pebble to fall may have been 0.1 second less or 0.1 second more than what Shelley timed.

   Find a range of heights for the cliff to allow for this error in timing the pebble's fall.

13. Geology Geologists estimate that Earth's temperature rises about 10°C for every kilometer k below Earth's surface.

a. Suppose it is 50°C on the surface of Earth. What is the formula for the temperature T in degrees Celsius at a depth of k km?

b. If it is 50°C on the surface, what is the temperature at a depth of 15 km?

Connect & Extend

14. A rental car costs $35 per day plus $0.10 per mile.

a. What is the cost, in dollars, to rent the car for five days if you drive a total of M miles?

b. What is the cost to rent the car for five days if you drive M miles each day?

c. What is the cost to rent the car for D days if you drive a total of 85 miles?

d. What is the cost to rent the car for D days if you drive 85 miles each day?

e. What would be the cost per person if three people share the cost of renting the car in Part d?

15. Geometry The diagram shows the square floor of a store. A square display case with sides of length three feet stands in a corner of the store. The manager wants the floor area painted. Assume the display case cannot be moved.

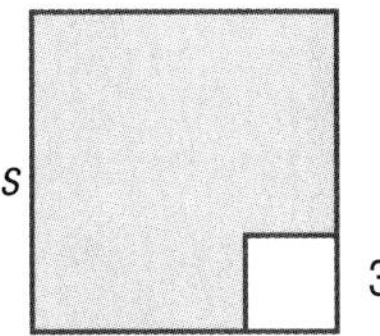

a. How many square feet need to be painted?

b. The manager hires two people to paint the floor. How many square feet would each person have to paint if they share the job equally?

c. Are there any restrictions on the value of s? If so, explain why and tell what the restrictions are. If not, explain why not.

Real-World Link

A *single* hit enables the batter to reach first base; a *double,* to reach second base; a *triple,* to reach third base; and a *home run,* to make a complete circuit of the bases and score a run.

A batting average and a slugging percentage are different. A *batting average* is a measure of the number of hits a player makes for every time at bat. A *slugging percentage* figures in how many bases the batter ran for every time at bat.

16. The formula below helps scuba divers figure out how long they can stay under water.

$$T = \frac{120V}{d}$$

T: approximate maximum time, in minutes, a diver can stay under water

V: volume of air in the diver's tank, in cubic meters, before compression

d: depth of the water, in meters

a. Antonio has one cubic meter of air compressed in his tank. He is 4 meters under water. How long can he stay down?

b. How long can Antonio stay eight meters under water with one cubic meter of air in his tank?

c. If Antonio wanted to stay four meters under water for one hour, how much air would he need?

17. Geometry The formula for the area of a trapezoid is $A = \frac{h(B + b)}{2}$, where B and b are the lengths of the two parallel sides and h is the height.

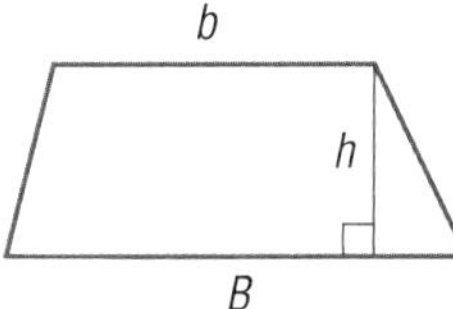

a. Find the area of a trapezoid with parallel sides of length 6 cm and 7 cm and height 5 cm.

b. The area of a trapezoid is 6 cm^2. What might be the values of h, B, and b?

18. A baseball player's slugging percentage P can be computed with the following formula.

$$P = \frac{S + 2D + 3T + 4H}{A}$$

S is the number of singles, D is the number of doubles, T is the number of triples, H is the number of home runs, and A is the number of official at bats.

a. Suppose a professional baseball player hit 112 singles, 20 doubles, 0 triples, and 66 home runs in 643 at bats. What was his slugging percentage?

b. What if the player hit 61 singles, 21 doubles, 0 triples, and 70 home runs in 509 at bats? What was his slugging percentage?

19. **In Your Own Words** Knowing that five pennies equal one nickel, Rob wrote the equation $5P = N$, where P is the number of pennies and N is the number of nickels. Explain why this equation does or does not represent this situation.

Mixed Review

Number Sense Fill in each ○ with >, <, or =.

20. $33\frac{1}{3}\%$ ○ $\frac{1}{3}$

21. $\frac{7}{8}$ ○ 85%

22. 0.398 ○ $\frac{2}{5}$

23. −5 ○ −1

24. $0.\overline{5}$ ○ $\frac{5}{9}$

25. $\frac{31}{40}$ ○ 75%

26. $\frac{347}{899}$ ○ $\frac{347}{900}$

27. $\frac{6}{7}$ ○ $\frac{7}{8}$

28. 80% ○ $\frac{45}{60}$

29. 0.01 ○ 0.1

Geometry Find each missing angle measure.

30.

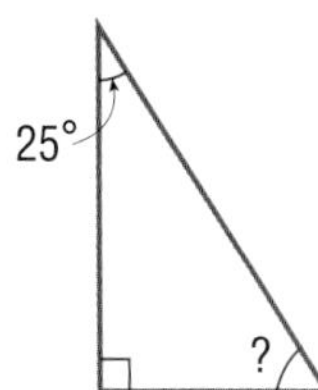

31.

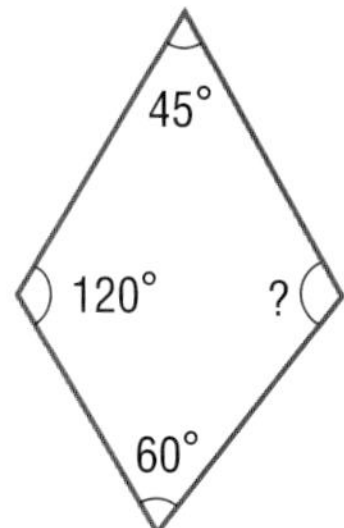

32. There are twenty-five students in Ms. Johnson's seventh grade class. Fifteen of the students are girls.

   a. Write a ratio that represents the number of girls to the number of boys.

   b. Write three equivalent ratios.

   c. What percent of the class is boys?

The Distributive Property

Inv 1	Bags and Blocks	51
Inv 2	The Same and Different	53
Inv 3	Use Parentheses	56
Inv 4	Combine Like Terms	60

Vocabulary

distributive property

If you have ever solved a multiplication problem like 4 • 24 by thinking, "It's 4 • 20 plus 4 • 4" or "It's 4 less than 4 • 25," you were using a mathematical property called the **distributive property**. Using this property, you can change the way you think about how numbers are grouped.

For example, rather than think about 4 groups of 24, you can think about 4 groups of 20 added to 4 groups of 4. As you will see, this property is helpful for more than just mental arithmetic.

You can use bags and blocks to help you think about algebraic expressions. Bags and blocks can help you look more closely at the distributive property, too.

Think & Discuss

Shaunda, Kate, and Malik are each holding one bag and two extra blocks.

Find the total number if each bag contains the following number of blocks.

- 5 blocks
- 20 blocks
- 100 blocks
- *b* blocks

How did you find your answers?

Here is how Sona and Omar found the total number of blocks when there were 20 in each bag.

Sona's method: "Each person has 22 blocks, so I multiplied 22 by 3 and got 66."

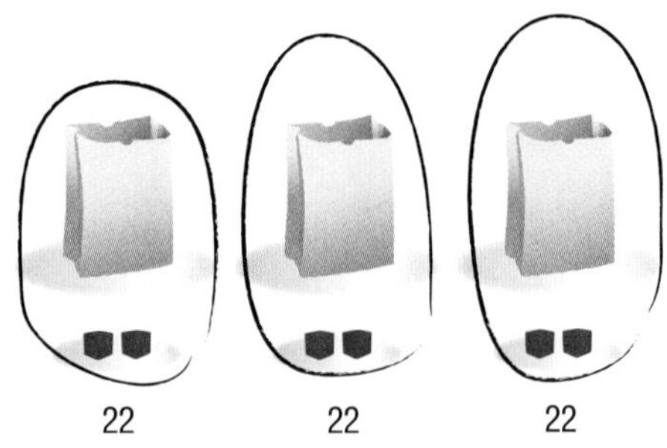

$$3(20 + 2) = 3 \cdot 22 = 66$$

Omar's method: "There are 3 bags, that's 60 blocks, and 3 sets of 2 leftover blocks, that's 6 more. So, there are 66 blocks."

Math Link

The following are four different ways to say the same thing.

3×22	$3(22)$
$3 \cdot 22$	$3 * 22$

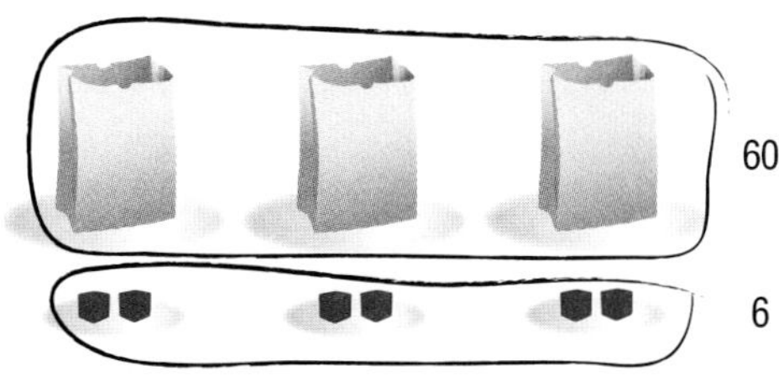

$$3 \cdot 20 + 3 \cdot 2 = 60 + 6 = 66$$

Sona's and Omar's methods both work no matter how many blocks are in each bag. You can express their methods in symbols for b blocks in each bag.

Sona's method	Omar's method
$3(b + 2)$	$3b + 3 \cdot 2 = 3b + 6$

Which way do *you* like to think about this situation? Or do you have another way?

Regardless of which method you prefer, the important thing to understand is that these two ways of looking at the situation give the same answer.

$$3(b + 2) = 3b + 6$$

Investigation 1 Bags and Blocks

One way to see how different expressions can describe the same situation is to consider different groupings of a given number of bags and blocks. Each grouping will have the same total number of blocks, so the expressions you create must represent the same quantity.

Develop & Understand: A

1. Brigitte placed three blocks in front of each of four bags. For each situation, show two ways of finding the total number of blocks that she has. If you need help, look back at Sona's and Omar's methods.

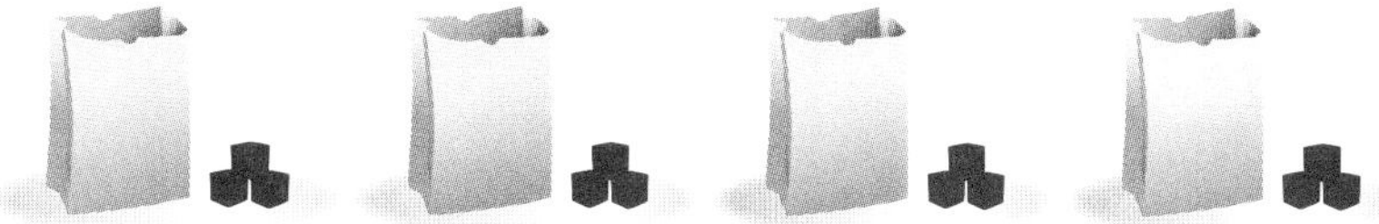

   a. 6 blocks in each bag
   b. 15 blocks in each bag
   c. 100 blocks in each bag
   d. b blocks in each bag

2. Keenan set four blocks in front of three sets of two bags. For each situation, show two ways of finding the total number of blocks that he has.

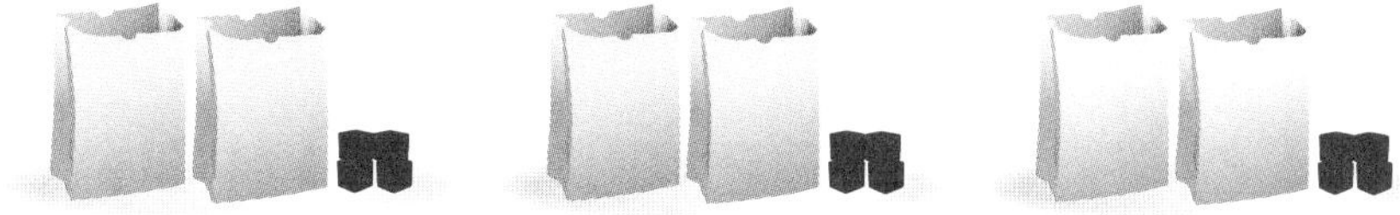

   a. 7 blocks in each bag
   b. 11 blocks in each bag
   c. 100 blocks in each bag
   d. b blocks in each bag

3. Flowcharts can also help you see different ways to express a quantity.
   a. Think about how you might create a flowchart to calculate the total number of blocks in the situation from Exercise 2. Draw and label a flowchart that has four ovals.
   b. Use an input of b for your flowchart. Find the output expression.
   c. Draw another flowchart to find the number of blocks in this situation. But this time, use only three ovals. Use an input of b and find the output expression.

4. Solana placed one block in front of each of four bags. Find two expressions for the total number of blocks that she has.

5. Simon and Zoe's teacher held up three bags and told the class that each contained the same number of blocks. She removed two blocks from each bag. How can you express the number of blocks still in the bags?

To answer this question, Simon and Zoe decided to experiment by starting with seven blocks in each bag.

a. If b blocks are in each bag, Zoe's reasoning can be expressed in symbols as $3b - 6$. Write an expression that fits Simon's reasoning.

b. Copy and complete the table to show the results of using Zoe's and Simon's methods for various numbers of blocks.

Number of Blocks	b	2	4	7	10	18	
Zoe's Method	$3b - 6$						60
Simon's Method				15			

Share & Summarize

Devon put five blocks in front of each of two bags.

1. Describe, in words, two ways of finding the total number of blocks if you know the number of blocks in each bag.

2. Write two rules, in symbols, for finding the total number of blocks if there are s blocks in each bag.

Investigation 2 The Same and Different

Vocabulary

equivalent expressions

In the last investigation, you discovered more than one way to find the total number of blocks in a bags-and-blocks situation. You can also use tables and flowcharts to help figure out why two expressions can *look* different but produce the same outputs.

Kaya, Maria, and Luis played the *What's My Rule?* game. Luis made up a rule for finding the output for any input. Next, Kaya and Maria gave Luis several inputs. Then, he told them the outputs his rule would produce. The girls organized the input/output values in a table.

Input	1	3	6	4	5
Output	8	12	18	14	16

Kaya and Maria used their data to guess Luis' rule. When they gave their rules, Luis had a dilemma.

Kaya and Maria described two rules that fit the table. If the input is represented by K, Kaya's rule is written in symbols as $2K + 6$. Maria's rule is written as $2(K + 3)$.

It is reasonable to believe that the calculation $2K + 6$ will always give the same result as the calculation $2(K + 3)$. To help you see *why* this works and why it *must* work no matter what K is, insert the missing multiplication signs.

Example

Start with $2(K + 3)$, and insert the missing multiplication sign.

$$2 \cdot (K + 3)$$

This is two groups of $(K + 3)$, which can be written as follows.

$$K + 3 + K + 3$$

This is the same as $K + K + 3 + 3$, or

$$2K + 6.$$

So, $2K + 6$ and $2(K + 3)$ are **equivalent expressions**. That means they must give the same result for every value of K.

Develop & Understand: A

1. Look more closely at the expressions $2K + 6$ and $2(K + 3)$.
   a. Draw a flowchart for each expression.
   b. For one of the expressions, you multiply first and then add. For the other, you add first and then multiply. Which expression is which?

2. You have seen that $2K + 6$ and $2(K + 3)$ are two ways of writing the same thing. And remember, Luis was thinking of a rule that was different from both of these but gave the same outputs.

   The rules below are also equivalent to $2K + 6$ and $2(K + 3)$. Using K for the input, write an expression for each rule. Show that your expression is equivalent to $2K + 6$.
   a. Add the input to 6 more than the input.
   b. Add 2 more than the input to 4 more than the input.
   c. Add 2 to the input, and double the sum. Add 2 to the result.
   d. Add 5 to the input, and double the sum. Subtract 4 from the result.

3. Write another rule, in words, that is equivalent to $2K + 6$.

You have seen that in the *What's My Rule?* game, the rule can often be written in more than one way. You will now look at input/output tables from the game and express each rule in two ways.

Develop & Understand: B

4. Here is an input/output table for a game of *What's My Rule.*

Input	0	1	2	3	4
Output	6	8	10	12	14

a. Use symbols to write two rules for this game. One rule should involve multiplying first and then adding. The other should involve adding first and then multiplying. Substitute some other inputs to check that the two rules give the same output.

b. Why do the two rules give the same output for a given input?

Use symbols to write two rules for the data in each table. One rule should involve multiplying first and then adding or subtracting. The other should involve adding or subtracting first and then multiplying.

5.

Input	2	3	4	5	6
Output	0	5	10	15	20

6.

Input	0	1	2	3	4
Output	4	4.5	5	5.5	6

Copy and complete each table. Then write another expression that gives the same outputs as the given expression. Check your expressions by substituting input values from your tables.

7.

j	0	1		8		20
$4(j + 2)$	8		20		76	

8.

x	2	3	4	5		18		
$5x + 5$	15				35		215	415

Share & Summarize

1. Write three expressions that are equivalent to $3P + 12$.

2. Use symbols to write two rules for the data in the table. One rule should involve multiplying first and then adding or subtracting. The other should involve adding or subtracting first and then multiplying.

Input, k	3	4	13	24	42	50
Output	0	4	40	84	156	188

Investigation 3 Use Parentheses

Vocabulary

expand

factor

The distributive property explains how you can write expressions in different ways. Using the distributive property to remove parentheses is called **expanding**. Using the distributive property to insert parentheses is called **factoring**.

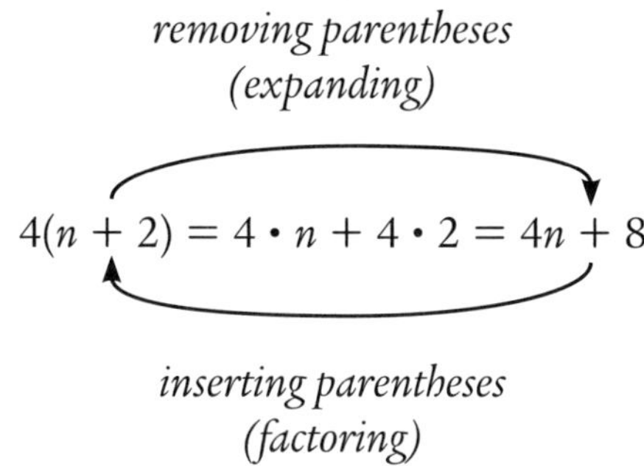

Expanding and factoring allow you to change the *form* of an expression, or what it looks like, without changing the output values it gives.

You have seen that rewriting an expression in a different form can be useful for simplifying calculations and comparing expressions. Later, you will see that rewriting expressions can help you solve equations.

In this investigation, you will practice using the distributive property to rewrite expressions. First, you will focus on expanding expressions, or removing parentheses.

Develop & Understand: A

1. Consider the equation $4(n + 2) = 4n + 8$.
   a. Is the equation true for $n = 2$?
   b. It is impossible to check that $4(n + 2) = 4n + 8$ for every value of n. How do you know that this equation is true for any value of n?

Decide whether each equation is true for all values of *n*. Explain how you decided.

2. $5(n + 6) = 5n + 30$
3. $7(n + 3) = 7n + 3$
4. $(n + 12) \cdot 9 = 9n + 108$
5. $1.5(n + 2.5) = 1.5n + 3.75$
6. $2(n + 4) = 2n + 4$
7. $(n + 9) \cdot 8 = n \cdot 8 + 72$

Develop & Understand: B

Determine whether there is a whole number or variable that divides both parts of each expression. If there is, use the distributive property to rewrite the expression using parentheses. Check the resulting expression by expanding it.

26. $4a + 8$
27. $4b + 12$
28. $4c + 17$
29. $3g - 15$
30. $5f + 13$
31. $8h - 24$

Factor each expression. Check the resulting expression by expanding it.

32. $22s + 33$
33. $34t - 4$
34. $45m + 25k$
35. $7j^2 + 3j$
36. $4t + 9t$
37. $8g^2 + 12g$
38. $10m + 15t + 25$
39. $8 - 16h^2 + 20h$

40. Every morning, Tonisha and her dog, Rex, run to the local park, around the park, and back home. It is 400 meters to the park and x meters around the park. Write two expressions for the total distance, in meters, they run in a week. One of your expressions should involve parentheses.

41. Explain why $4t + 9t$ and $6k + 9m$ can be factored. Explain why $4k + 9m$ cannot be factored.

42. **Prove It!** Show that $2k + 3k = 5k$ by first factoring $2k + 3k$. Then check that $2k + 3k = 5k$ for $k = 7$ and $k = 12$.

Share & Summarize

1. Create an expression in factored form. Give it to your partner to expand.
2. Create an expression that can be factored. Give it to your partner to factor.
3. Explain how expanding and factoring are related.

Investigation  Combine Like Terms

Vocabulary

like terms

monomial

polynomial

Real-World Link

In a recent year, 4,051,990 tax returns claimed clothing donations with a value of \$5,836,108,000.

Zach, Simon, and Maria collected clothes to donate to a local clothing drive. Each student brought the following items.

	Shirts	Pairs of Pants	Sweaters
Zach	3	5	1
Simon	4	4	2
Maria	5	2	3

Simon wrote a list of all of the clothes the three friends donated.

3 shirts + 4 shirts + 5 shirts + 5 pants + 4 pants + 2 pants +
1 sweater + 2 sweaters + 3 sweaters

Maria put all of the shirts in one bag. She put all of the pants in another bag and all of the sweaters in a third bag. She wrote a list of what was in each bag. Maria's list was used to label the bags.

Develop & Understand: A

1. Zach wondered how Simon's and Maria's lists could look so different. Explain how you could use Simon's list to create Maria's list.

2. Can you use Maria's list to create Simon's list? Why or why not?

3. At the shelter, the clothing was distributed among five people. Fill in the table to show what each person might have received.

	Shirts	Pairs of Pants	Sweaters
Jose	2	3	1
Lola			
Morris			
Cheyenne			
Ethan			

4. Maria put each person's clothing into a separate bag. She wrote a list of what was in each bag. For example, Maria's list for Jose's clothing was 2 shirts, 3 pants, and 1 sweater. Use your table entries to write a list like Maria's for Cheyenne's clothing.

5. Zach thought Maria's new list was too long. He thought she should label each bag with just one number, like she did in her first list. Explain whether or not it is possible to label the bags that way.

6. Think about the bags and blocks scenarios you used to explore the distributive property in Investigation 1. How do the bags and blocks apply to the donated clothing situations?

In this investigation, you will work with algebraic expressions called *polynomials.*

A **monomial** is a number or the product of a number and variables. The expressions x, $2y$, $\frac{1}{2}xy$, 7, and $45x^3$ are all examples of monomials.

A **polynomial** is an algebraic expression containing one or more terms. Each term must be a *monomial,* and the terms can be combined using only *addition* or *subtraction.*

Example

$x^2 + 3y$ is an example of a polynomial. It contains two terms: x^2 and $3y$. x^2 is a monomial because it is the product of a number, 1, and two variables, x, and x again. $3y$ is the product of 3 and y, so it is also a monomial.

$\frac{y}{8}$ is a monomial because it can be rewritten as $\frac{1}{8} \cdot y$, the product of a number and a variable.

Think & Discuss

Molly and Alonso each thought of an algebraic expression.

Molly wrote $3x \cdot x + 2$ on the board.

Alonso wrote $x^2y + 5$.

What makes each expression a polynomial?

Math Link

Some words use "mono" (meaning "one") and "poly" (meaning "more than one" or "extra"). A *polynomial* is an expression with several terms, while a *monomial* has only one term. A *monorail* has one of what item? A *polygon* has several of what mathematical figure?

Terms in an algebraic expression that have exactly the same variables and exponents are called **like terms**.

Example

$3x$ and $2x$ are like terms. $4y^2$ and $0.5y^2$ are like terms.

y^2 and $2x^2$ are not like terms because their variables are different. $3xy$, $5x$, and x^3 are not like terms either. Although they all have *x as a variable*, all of their variables and exponents do not match exactly.

When Maria put all of the shirts in one bag and added for a total, she was *combining like terms*. She found the total of each set of things that were all the same. You can do the same in a polynomial. Just as you factored expressions in Investigation 3, you can use the distributive property to combine like terms in a polynomial.

$$3x + 4x + 6 = x(3 + 4) + 6 = 7x + 6$$

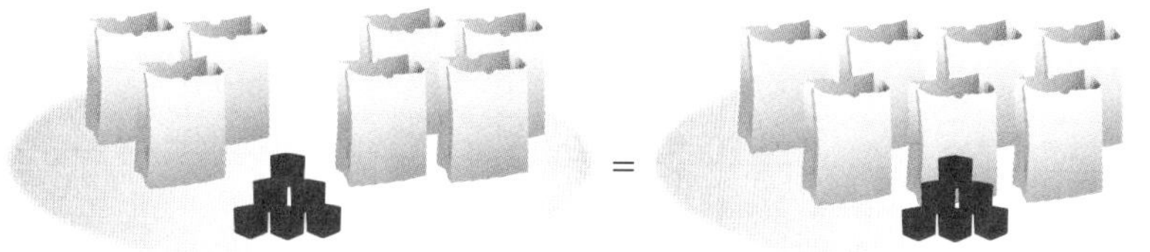

Develop & Understand: B

7. Use the distributive property to explain why $5b + 20b - b = 24b$.

In each exercise, combine like terms to simplify the polynomial as much possible.

8. $6x + 4x + 2x$
9. $20y - 8y + 4y$
10. $7x - 3y + 8x$
11. $3x + 2y + 5x + 18y + 1$
12. $x + x^2 + 3x + 2x^2$
13. $14x^2 + 4y^3 - 3x - 5 + 2y$
14. $5x + 8xy + 10x + 3xy$
15. $3x + 12xy + 6x + 7y^2 - 10xy$
16. Can you simplify the polynomial $x + x^2$ by combining like terms? Explain why or why not.

Share & Summarize

Draw a picture to represent the polynomial $3x + 5x + 3 + 10$. Draw a second picture to show how to simplify the polynomial by combining like terms.

Practice & Apply

1. Suppose there are five bags with n blocks in each bag and three extra blocks beside each bag.

a. Write one factored expression and one expanded expression for the total number of blocks as you did in Exercises 1–5 on pages 51 and 52.

b. Check that your two expressions give the same total for $n = 8$, $n = 12$, and $n = 25$.

c. Explain, using a picture if you like, why your two expressions must give the same total number of blocks for any value of n.

2. Suppose there are four sets of three bags with three blocks in front of each set. For each situation, show two ways to find the total number of blocks.

a. 5 blocks in each bag

b. 100 blocks in each bag

c. b blocks in each bag

3. Brandi has five bags with the same number of blocks in each. She removes two blocks from each bag. For each situation below, show two ways of finding the total number of blocks in the bags.

a. The bags start with 7 blocks in each.

b. The bags start with 50 blocks in each.

c. The bags start with p blocks in each.

4. Consider the expressions $3(T - 1)$ and $3T - 3$.

a. Copy the table, and complete the first four columns. In the last four columns, choose your own input values and calculate the output for both rules. Use fractions for at least two input values.

T	4		100					
$3T - 3$		30		87				
$3(T - 1)$	15	30						

b. Show that $3(T - 1)$ and $3T - 3$ are equivalent expressions.

c. Find two more expressions that are equivalent to $3(T - 1)$ and $3T - 3$.

Copy and complete each table. Then, for each table, write another expression that gives the same outputs. Check your new expression to make sure it generates the same values.

5.

h	0	1	2		42	
$2(h + 3)$	6			20		100

6.

r		5	10	11		23
$5(r - 2)$	5				100	

7.

n	4	5	6	8		
$\frac{n-4}{4}$	0				10	24

For each *What's My Rule?* table, find two ways to write the rule in symbols. Choose your own variable.

8.

Input	0	1	2	3	4
Output	8	10	12	14	16

9.

Input	0	1	2	3	4
Output	6	6.5	7	7.5	8

Math Link

Expanding an expression means rewriting it to remove the parentheses.

For Exercises 10–15, expand each expression. Check the resulting expression by making sure it gives the same values as the original expression for several values of the variable.

10. $5(3j + 4)$ **11.** $0.2(4k + 9)$ **12.** $2(n - 7)$

13. $n(n - 6)$ **14.** $3(3n + 3)$ **15.** $0(n + 875)$

Factor each expression, if possible.

16. $4a + 8$ **17.** $18 + 5b$ **18.** $3g - 15$

19. $18 - 2A$ **20.** $11v + 3v$ **21.** $5z^2 + 2z$

In each exercise, combine like terms to simplify the polynomial as much as possible.

22. $14x^2 + 5y - 6x + 23 + 7x - 3y$

23. $\frac{1}{3}xy + 13x + \frac{5}{6}xy - 4x$

24. $2y - 4x^2 + 2y^2 - 4x^2 + 14y - 3xy$

25. Thomas and Courtney are sorting footballs, basketballs, and tennis balls into different storage containers for gym classes. Here is how Thomas sorted the balls.

Container 1	Container 2	Container 3
9 footballs	12 basketballs	24 tennis balls

Courtney suggests sorting instead by gym class, A, B, and C. She puts some of each type of ball in each container.

a. Write a polynomial that represents the sorting that Thomas did.

b. Write one polynomial that represents how Courtney sorted the balls.

c. Which expression tells you more about the balls? Which tells you more about the gym classes?

d. If the gym classes need the same number of each type of ball, what polynomial would represent each container?

26. Logan's karate class is having a bake sale. In total, the class brought 96 cookies. Complete a table to show how many each person might have brought.

	Chocolate Chip	Sugar Cookies	Oatmeal Raisin	Peanut Butter
Aiden				
Hailey				
Logan				
Monica				

a. What is the shortest polynomial you could write? How many terms does it have?

b. How many terms would the longest possible polynomial have to represent the cookies brought? What situation causes the longest polynomial?

c. Hailey's mom takes all the cookies home to package them to sell. There are now six bags of equal numbers of cookies. Write a polynomial to represent the new bags.

Connect & Extend

27. Joshua has 4 bags of blocks. He removes 3 blocks from each bag.

a. Write two expressions for the total number of blocks in Joshua's bags now.

b. Explain, using a picture if you like, why your two expressions must give the same total number of blocks for any value of the variable.

28. The rock band "The Accidents" brought five boxes of its new CD, *Waiting to Happen,* on its tour. Each box holds the same number of CDs. The band hoped to sell all of the CDs, but it sold only 20 CDs from each box.

a. Write an expression that describes the total number of unsold CDs.

b. Write an expression that describes how much the band will earn, in dollars, by selling the remaining CDs for $15 each.

29. In this pattern of toothpicks, the width of the figure increases from one figure to the next. Using symbols, write two rules for finding the number of toothpicks in a figure. Use parentheses in only one of your rules. Use w to represent the width of a figure.

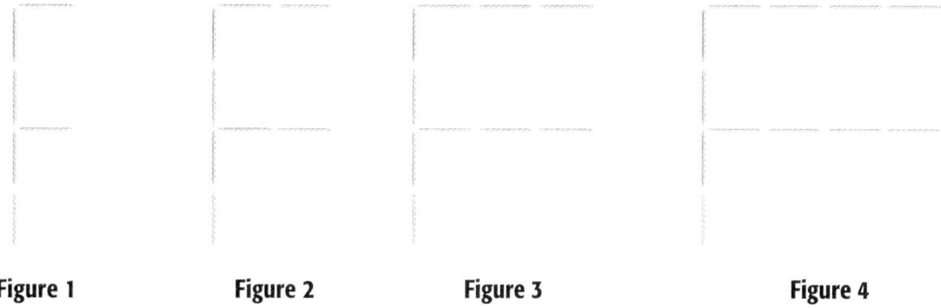

Figure 1 Figure 2 Figure 3 Figure 4

Expand each expression. Check your answer by substituting 1, 2, and 10 for the variable.

30. $7Q(Q + 8)$

31. $9L(M + 3L)$

32. $8R(\frac{1}{2} - R)$

33. $3(D^2 - \frac{D}{2})$

Factor each expression.

34. $3m^2 + 6m$

35. $2t + 4t^2$

36. $4L + 3L^2$

37. $3P + mP$

38. Challenge When you take any whole number, add $\frac{1}{2}$, and then multiply the sum by 2, the result is an odd number. Explain why.

39. Challenge This fraction contains variables.

$$\frac{8x + 4z + 2}{2}$$

To simplify this fraction, you might first think of it as

$$\frac{1}{2}(8x + 4z + 2).$$

Then you can apply the distributive property.

$$\frac{1}{2}(8x + 4z + 2) = \frac{1}{2}(8x) + \frac{1}{2}(4z) + \frac{1}{2}(2)$$

$$= 4x + 2z + 1$$

Use this method to simplify each fraction.

a. $\frac{12x + 6z + 3}{3}$ **b.** $\frac{10x^2 + 5x}{5}$

40. World Cultures In some countries, long multiplication is taught to students using a grid. Here is how to multiply 15 and 31 using a grid.

•	10	5
30	300	150
1	10	5

The product of 15 and 31 is 465, the total of the four numbers inside the grid. Use the grid method to find each product.

a. $14 \cdot 56$ **b.** $23 \cdot 23$ **c.** $45 \cdot 21$

41. Challenge You can use factoring to prove some interesting facts about numbers.

a. If you start with a whole number greater than 1 and add it to its square, the result will never be a prime number. Explain why not. (Hint: Use x to stand for the whole number. Then x^2 is its square, and $x^2 + x$ is the whole number added to its square.)

b. If you start with a whole number greater than 2 and subtract it from its square, the result can never be a prime number. Explain.

In each exercise, expand and combine like terms. Show that the original expression is equivalent to your final combined expression by substituting 2 for x and 3 for y.

42. $3x(4x + 6y^2) + y(y + 5x)$

43. $x(2y + 4) + 6y^2(2 + 3x) + 8x(y + y^2)$

44. Are the following expressions equivalent? Explain.

$$5xy(2x^2 + 3y + 4x) + x^2y(5 + 6y) + 3(x2y + 4xy^2) \text{ and}$$

$$10x^3y + 27xy^2 + 28x^2y + 6x^2y^2$$

45. In Your Own Words Suppose you are talking to a student two years younger than you are. Explain why $3(b + 4) = 3b + 12$, not $3b + 4$. You might want to draw a picture to help explain the idea.

46. Probability On a scale from 0 (impossible) to 100 (certain to happen), rate the chances of each event happening today.

a. You do homework.

b. Someone in the world has a baby.

c. Someone in your city has a baby.

d. You learn to drive.

e. It snows somewhere in your country.

Fill in the blanks.

47. _____ % of 152 = 114

48. 0.5% of 200 = _____

49. $33\frac{1}{3}$% of _____ = 15

50. _____ % of 20 = 60

51. 45% of 45 = _____

52. 80% of _____ = 320

53. In April 2008, you could have received about 40.125 Indian rupees for 1 U.S. dollar.

a. About how many U.S. dollars could you have received in exchange for 100 Indian rupees?

b. How many Indian rupees could you have received in exchange for 100 U.S. dollars?

CHAPTER 1

Review & Self-Assessment

Chapter Summary

In this chapter, you followed common rules and rules for creating sequences, and you wrote rules for others to follow. You learned that many situations can be described using algebraic expressions and *variables*. You evaluated expressions and used *flowcharts* to *backtrack* to solve equations.

By using different values for a variable, you explored what happens in a situation as a variable changes. You also worked with mathematical *formulas*.

You found that the same situation can often be described with several *equivalent expressions* and that sometimes one expression is more useful than another. You saw that you could change expressions into equivalent expressions by using the *distributive property* to *expand* and *factor* them.

Vocabulary

backtracking
distributive property
equivalent expressions
expand
exponent
factor
flowchart
formula
like terms
monomial
polynomial
variable

Strategies and Applications

The questions in this section will help you review and apply the important ideas and strategies developed in this chapter.

Using formulas and evaluating expressions

1. The formula for the circumference of a circle is $C = 2\pi r$, where r is the radius. If the radius of a circle is 4.3 cm, what is its circumference?

Evaluate each expression for $r = 2$.

2. $\frac{r}{2}$ **3.** $\frac{3}{2}r$ **4.** $3r^2$

Solving equations by backtracking

5. Consider the expression $4(2n - 1) + 5$.

a. Draw a flowchart to represent the expression.

b. Use backtracking to solve the equation $4(2n - 1) + 5 = 21$.

6. Find the length of a rectangle with width 4.2 cm and area 98 cm^2.

Using the distributive property

Use the distributive property to rewrite each expression.

7. $3(0.5r + 7)$ **8.** $\frac{10x - 6}{2}$

9. $3g + 9g^2$ **10.** $49s - 14s$

11. Find two rules, using symbols, that could produce this table.

Input	0	1	2	3	4
Output	4	8	12	16	20

Demonstrating Skills

Rewrite each expression without using multiplication or addition signs.

12. $s + s + s + s$ **13.** $7 \cdot b \cdot b \cdot b$ **14.** $7g + 3g$

15. Find the value of $6y^2$ for $y = 4$.

Insert parentheses to make each equation true.

16. $3 \cdot 5 - 2 = 9$ **17.** $4 + 7 \cdot 3 = 33$

Use the distributive property to expand each expression.

18. $5(x + 4)$ **19.** $3(a + 2b + 3c)$ **20.** $3y(y - \frac{1}{3})$

Factor each expression. Check the resulting expression by expanding it.

21. $25r - 50$ **22.** $16h - 4h^2$ **23.** $9x - 81y + 27z$

24. Julian's mother uses the following rule when making mashed potatoes.

Use two potatoes per person and add one for the pot.

a. How many potatoes would Julian's mother use for eight people?

b. Julian's mother used 23 potatoes to make her last batch of mashed potatoes. How many people did she serve?

c. Julian feels that his mother's rule makes too many mashed potatoes. Make up a rule that he might like better.

d. Using your rule from part c, how many potatoes are needed for five people?

25. Backpacks Unlimited has determined that students can safely carry up to 25 pounds in their backpacks without harming themselves. However, for children under ten years old, the weight allowance can be calculated by using the following rule.

Divide the child's age by 10 and multiple the result by 25.

a. How much does Backpacks Unlimited believe an eight-year-old child can safely carry?

b. According to this rule, how much can a 15-year-old safely carry?

c. A backpack weighs $12\frac{1}{2}$ pounds. What is the youngest age that can safely carry this backpack?

In each exercise, combine like terms and simplify the polynomial as much as possible.

26. $(7 \cdot 3) + (4 \cdot 3)$

27. $(6 \cdot 5) + (2 \cdot 8) + (6 \cdot 3)$

28. $7x - 4x + 3x$

29. $2x + 8x$

30. $21y - 18y + 5x$

31. $15x - 15y$

32. $4x^2 + 3x + 12x$

33. $16y^2 + 5x + 16x + 9y^2 + 7$

34. $7a + 11b + 13a - 5b$

35. $7x^2 + 11y^2 - 9x + 2$

36. $18y^3 - 4y^3 + 6y^3 - 7y$

37. $13x^2 + 2x + 5 + 11x + 8$

38. $21a^2 + 15b - 20a^2 + b$

39. $11x + 12y^2 + 13x - 4y^2 + 15x$

Test-Taking Practice

SHORT RESPONSE

1 Let n represent the cost of a notebook, p represent the cost of a package of pencils, and f represent the cost of a folder.

A Write an algebraic expression to represent the total cost C of 5 notebooks, 1 package of pencils, and 4 folders.

B A notebook costs \$4.85, a package of pencils costs \$2.35, and a folder costs \$0.55. Find the total cost of 5 notebooks, 1 package of pencils, and 4 folders.

Show your work.

Answer ______________________

MULTIPLE CHOICE

2 What is the value of $3m + 2n$ if $m = 5$ and $n = 3$?

A 8

B 16

C 19

D 21

3 Which one correctly combines like terms in $8 + 5x - 2x + 4$?

F $19x$

G $12 + 7x$

H $12 + 3x$

J $4 + 3x$

4 Find the width of a rectangle with length 6.4 mm and area 20.48 mm^2.

A 3.2 mm

B 3.84 mm

C 6.4 mm

D 14.08 mm

5 Which expression has the same value as $3(x + 4) - (y + 2)$?

F $2x + 10$

G $3x - y + 14$

H $3x - y + 10$

J $3x - 4y + 10$

CHAPTER 2

Exponents

Real-Life Math

Astronomical Figures The distances from the Sun to each of the planets in our solar system varies from about 35,980,000 miles to 2,796,000,000 miles. These distances are easier to write in shorthand, 3.598×10^7 miles and 2.796×10^9 miles. The distance from the Sun to the star nearest to it, Proxima Centauri, is about 25,000,000,000,000 miles. It would be much easier for an astronomer to write this distance as 2.5×10^{13} miles.

Mars, the fourth planet in the solar system, is 1.41×10^8 miles from the Sun. On July 4, 1997, the *Mars Pathfinder* from the National Aeronautics and Space Administration landed on Mars. The spacecraft put the surface rover *Sojourner* on the planet. *Sojourner* sent detailed photos back to Earth, giving us our first up-close views of our sister planet.

Think About It The Mars Pathfinder Mission was part of a NASA program with a spending limit of $150,000,000. Can you write this dollar amount without listing all of the zeros?

Contents in Brief

2.1 Factors and Multiples 74

2.2 Exponent Machines 92

2.3 More Exponent Machines 107

Review & Self-Assessment 120

Math Online

Take the **Chapter Readiness Quiz** at glencoe.com.

Dear Family,

The class is about to begin Chapter 2 about exponents, factors, multiples, and prime numbers. Exponents can be thought of as a shortcut method of expressing repeated multiplication. For example, $4 \cdot 4 \cdot 4$ is the same as 4^3. The base is 4, which is the number to be multiplied. The exponent is 3, which is the number of 4s that are multiplied together.

Key Concept–Exponents

Students will use a machine model to help learn about exponents. *Stretching machines* are a model of multiplication. They stretch any input by the number on the machine. This machine will stretch something 4 times. Suppose you put a one-inch piece of gum into the machine. How long will it be when it comes out the machine?

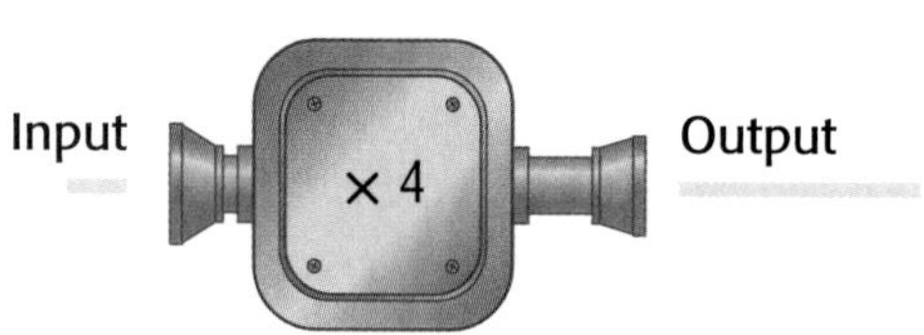

A *repeater machine* is a special type of stretching machine that models exponents. Look at the repeater machine at the right. It will stretch an input 4 times, then 4 times again, and then 4 times again. A one-inch piece of gum goes through the ×4 machine 3 times, for a total of 64 stretches, and comes out 64 inches long.

Chapter Vocabulary

base	**least common multiple**
common factor	**multiple**
common multiple	**order of operations**
composite number	**power**
exponent	**prime factorization**
factor	**prime number**
factor pair	**product law of exponents**
greatest common factor	**relatively prime**

Home Activities

Have your student think about common uses of exponents. Challenge them to use exponents to represent large numbers, such as the U.S. national debt.

Factors and Multiples

Inv 1 Factors 75

Inv 2 Prime Numbers 77

Inv 3 Common Factors 80

Inv 4 Multiples 83

Imagine that you work at a stretching factory. The factory has an incredible set of machines that will stretch almost anything.

To double the length of a stick of gum, for example, you can put it through a ×2 machine.

To stretch modeling clay to five times its length, you can use a ×5 machine.

The factory has machines for every whole-number stretch up to 200. This means there is a ×1 machine, a ×2 machine, a ×3 machine, and so on, up to ×200.

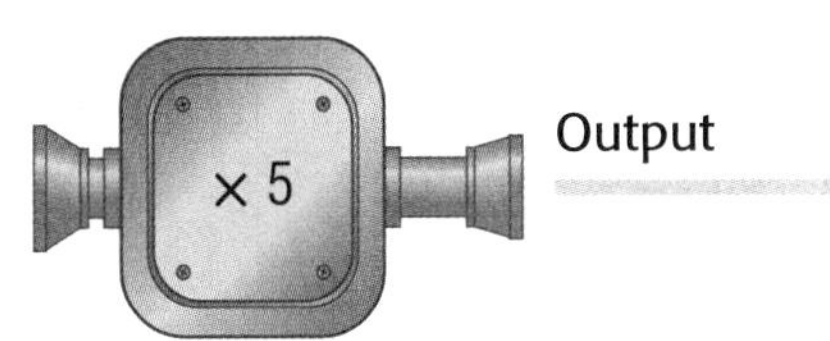

Real-World Link
Licorice candy is flavored with juice from the licorice herb. The herb is native to southern Europe and is often used to mask the taste of bitter medicines.

Think & Discuss

- Your supervisor has asked you to fill the following orders. Tell which machine you would use to perform each stretch.

  Order 1: Stretch a 1-foot chain into a 5-foot chain.

  Order 2: Stretch a 2-inch pencil into a 16-inch pencil.

  Order 3: Stretch a 3-inch wire into a 21-inch wire.

- Your co-worker, Winnie, sends a 1-foot piece of licorice through the ×2 machine five times. How does this change the length of the licorice?
- Another co-worker, Bill, puts a 1-foot piece of licorice through the ×1 machine five times. How does this change the length of the licorice?
- This afternoon, you will be using the ×2 machine and the ×3 machine to stretch 1-foot strands of spaghetti. Give ten different output lengths you can make using one or both of these machines.

Investigation 1 Factors

Vocabulary

factor

factor pair

Machines in the stretching factory can be hooked together. When machines are hooked together, the output of one machine becomes the input for the next. For example, when a 1-foot piece of licorice passes through the connection below, the ×2 machine stretches it into a 2-foot piece. Then the ×3 machine stretches the 2-foot piece into a 6-foot piece.

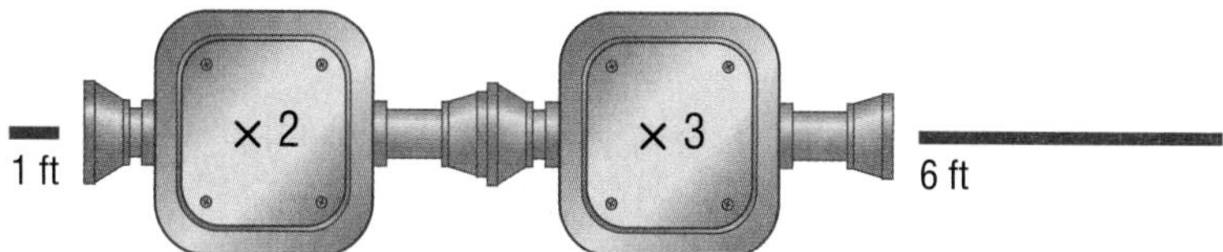

Develop & Understand: A

1. Winnie is using this connection to stretch rope.

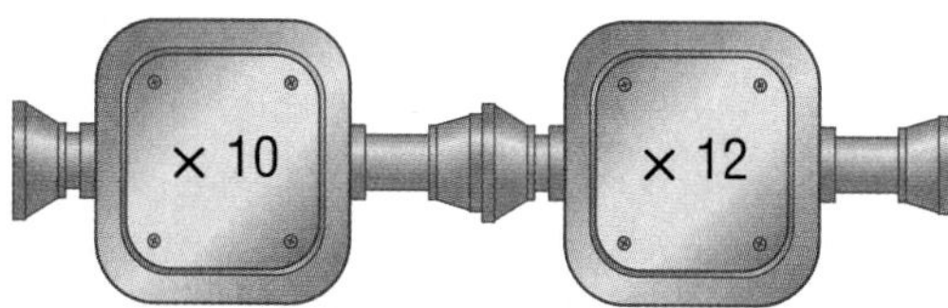

   a. What happens to a 1-foot length of rope put through this connection?

   b. What single machine would do the same stretch as Winnie's connection?

   c. Would Winnie get the same result if she used this connection? Explain.

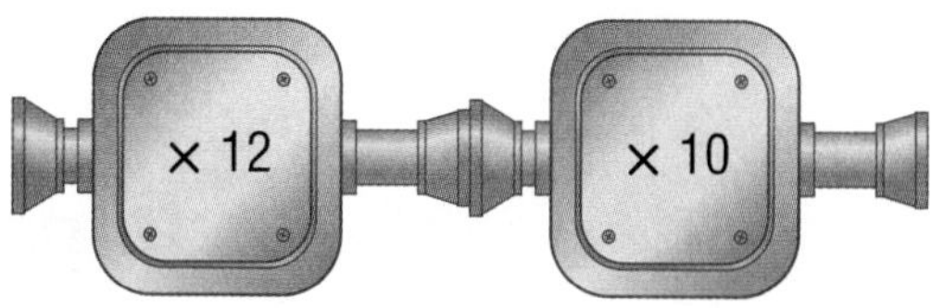

2. Bill wants to use a connection of five ×1 machines to stretch a rope to five times its length. Winnie says this will not work. Who is right? Why?

3. What happens when a 1-foot piece of rope is sent through a connection of two ×5 machines?

When a stretching machine breaks down, you can use your knowledge of connections to replace it.

Develop & Understand: B

4. If a ×12 machine breaks down, what machines can you connect to replace it?

5. If a ×51 machine breaks down, what machines can you connect to replace it?

6. Suppose a ×64 machine breaks down.

   a. What two-machine connections will replace it?

   b. What three-machine connections will replace it?

   c. What four-machine connections will replace it?

7. If a ×81 machine breaks down, what connections will replace it?

8. If a ×101 machine breaks down, what connections will replace it?

When you figure out what connections will replace a particular machine, you are thinking about *factors* of the machine's number. A **factor** of a whole number is another whole number that divides into it without a remainder. For example, 1, 2, 3, 4, 6, 8, 12, and 24 are factors of 24.

When you found two-machine connections, you were finding *factor pairs.* A **factor pair** for a number is two factors whose product equals that number. Here are the factor pairs for 24.

1 and 24　　2 and 12　　3 and 8　　4 and 6

Order does not matter in a factor pair. In fact, 2 and 12 is the same pair as 12 and 2.

Math Link

A number is divisible by:

- 2 if the ones digit is divisible by 2.
- 3 if the sum of the digits is divisible by 3.
- 4 if the number formed by the last two digits is divisible by 4.
- 5 if the ones digit is 0 or 5.
- 6 if the number is divisible by both 2 and 3.
- 8 if the number formed by the last three digits is divisible by 8.
- 9 if the sum of the digits is divisible by 9.
- 10 if the ones digit is 0.

Develop & Understand: C

List the factors of each number.

9. 21

10. 36

11. 37

12. 63

13. List all the factors of 150. Then list all the factor pairs for 150.

14. List all the factors of 93. Then list all the factor pairs for 93.

Share & Summarize

1. Explain how finding a connection to replace a broken machine involves finding factors.
2. Describe some methods you use for finding factors of a number.

Investigation 2 Prime Numbers

Vocabulary

composite number

prime factorization

prime number

Materials

- copy of the stretching-machine table

You, Winnie, and Bill plan to open your own stretching factory. Since stretching machines are expensive, you decide to buy only machines with stretches up to ×50. You also decide not to purchase any machine that can be replaced by connecting other machines together or by using the same machine more than once.

Develop & Understand: A

1. The table lists stretching machines from ×1 to×50. On a copy of the table, cross out all the machines that can be replaced by other machines. Circle the machines that cannot be replaced. When you finish, compare your table with your partner's.

×1	×2	×3	×4	×5	×6	×7	8	×9	×10
×11	×12	×13	×14	×15	×16	×17	×18	×19	×20
×21	×22	×23	×24	×25	×26	×27	×28	×29	×30
×31	×32	×33	×34	×35	×36	×37	×38	×39	×40
×41	×42	×43	×44	×45	×46	×47	×48	×49	×50

2. The machines that you crossed out can be replaced by using the machines that you circled.
   a. Explain how the ×24 machine can be replaced.
   b. Explain how the ×49 machine can be replaced.
   c. Choose three more crossed-out machines. Explain how each can be replaced by using the machines that you circled.

3. Use the idea of factors to explain why the circled machines cannot be replaced.

4. Why is $\times 2$ the only even-numbered machine that you need?

5. Winnie points out that you can save more money by not purchasing a $\times 1$ machine. Bill says, "There is no connection that can replace the $\times 1$ machine. We must buy it!"

   What do *you* think? Does your factory need the $\times 1$ machine? Explain. If you think it is not needed, cross it out in your table.

The numbers on the machines that cannot be replaced, excluding the $\times 1$ machine, are *prime numbers.* A **prime number** is a whole number greater than 1 with only two factors, itself and 1.

The numbers on the machines that can be replaced are *composite numbers.* A whole number greater than 1 with more than two factors is a **composite number**. The number 1 is neither prime nor composite.

Any composite number can be written as a product of prime numbers. Here are some examples.

$10 = 5 \cdot 2$ $\qquad 117 = 3 \cdot 3 \cdot 13$ $\qquad 693 = 3 \cdot 3 \cdot 7 \cdot 11$

When you write a composite number as a product of prime numbers, you are finding its **prime factorization**. You can use a *factor tree* to find the prime factorization of any number.

Example

Below are the steps for making a factor tree for the number 20. Two possible factor trees are shown.

- Find any factor pair for 20. Draw a "branch" leading to each factor.
- If a factor is prime, circle it. If a factor is not prime, find a factor pair for it.
- Continue until you cannot factor any further.

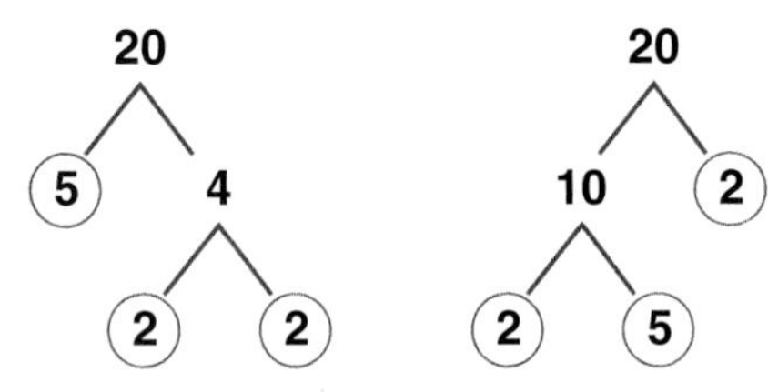

Both of these factor trees show that the prime factorization of 20 is $2 \cdot 2 \cdot 5$. Remember, when you multiply numbers, order does not matter.

Real-World Link

The greatest prime number discovered so far has 4,053,946 digits. This number can be written in a shorthand that uses an exponent.

$2^{13,466,917} - 1$

When a factor occurs more than once in a prime factorization, you can use *exponents* to write the factorization in a shorter form. Since $20 = 2 \cdot 2 \cdot 5$, the prime factorization of 20, using exponents, is $2^2 \cdot 5$.

Develop & Understand: B

6. Use a factor tree or another method to find the prime factorization of 24.

7. Use a factor tree or another method to find the prime factorization of 49.

8. Explain how your answers to Exercises 6 and 7 are related to the machines needed to replace a ×24 machine and a ×49 machine in your stretching factory.

Use a factor tree to find the prime factorization of each number. Write the prime factorization using exponents when appropriate.

9. 100

10. 99

11. 5,050

12. 111,111

Share & Summarize

1. Describe the difference between a prime number and a composite number. Give three examples of each.

2. Find the prime factorization of each composite number you listed in Question 1.

Investigation 3 Common Factors

Vocabulary

- common factor
- greatest common factor, GCF
- relatively prime

Business has been good at your factory. You have earned enough to buy one machine for every prime-number stretch from 2 to 101. However, there still do not seem to be enough machines to go around. You find that you and your co-workers often need to use the same machine at the same time.

Explore

Here are the orders that you and Winnie are scheduled to fill this morning.

Time	Your Orders	Winnie's Orders
10:00	stretch ×32	stretch ×16
10:30	stretch ×15	stretch ×21
11:00	stretch ×99	stretch ×6
11:30	stretch ×49	stretch ×14

Which machines do you and Winnie need to use at the same time?

Rearrange the schedule so both you and Winnie can fill your orders.

The scheduling exercises in Explore occurred when the stretch numbers had a *common factor*. A **common factor** of two or more numbers is a number that is a factor of all of the numbers.

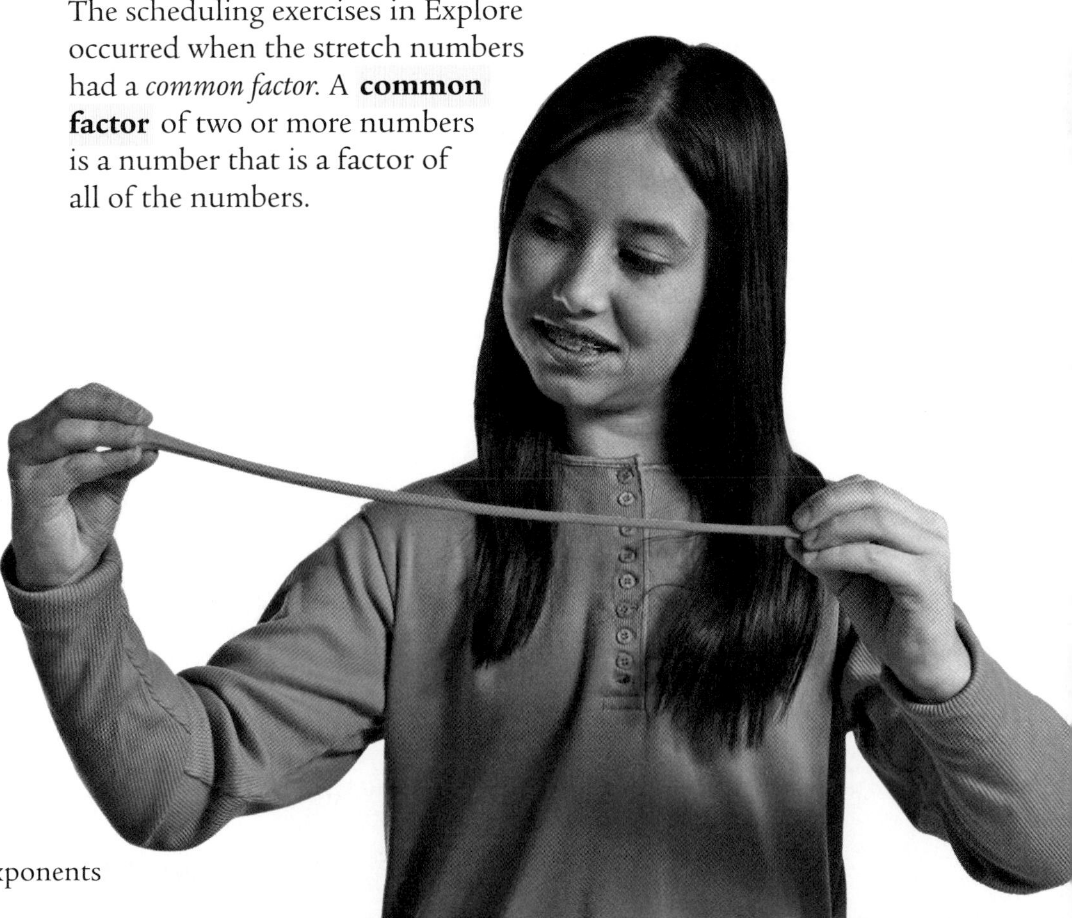

Develop & Understand: A

In Exercises 1–3, list the factors of each number. Then find the common factors of the numbers.

1. 100 and 75
2. 33 and 132
3. 36, 84, and 112
4. Is it possible for two numbers to have 8 as a common factor but not 2? If so, give an example. If not, explain why not.
5. Is it possible for two numbers to have no common factors? If so, give an example. If not, explain why not.
6. Is it possible for two numbers to have only one common factor? If so, give an example. If not, explain why not.

The **greatest common factor** of two or more numbers is the greatest of their common factors. The abbreviation **GCF** is often used for greatest common factor. Two or more numbers are **relatively prime** if their only common factor is 1.

Example

Look at the common factors of 12 and 16.

Factors of 12: **1**, **2**, 3, **4**, 6, 12 Factors of 16: **1**, **2**, **4**, 8, 16

The common factors of 12 and 16 are 1, 2, and 4. Therefore, 12 and 16 are *not* relatively prime. The GCF of 12 and 16 is 4.

Now consider the common factors of 8 and 15.

Factors of 8: **1**, 2, 4, 8 Factors of 15: **1**, 3, 5, 15

The only common factor of 8 and 15 is 1. Therefore, 8 and 15 are relatively prime. The GCF of 8 and 15 is 1.

Develop & Understand: B

The *GCF Game* will give you practice finding greatest common factors and determining whether numbers are relatively prime.

GCF Game Rules

For the game board, create a 4 × 4 grid of consecutive numbers. One possible board is shown here.

12	13	14	15
16	17	18	19
20	21	22	23
24	25	26	27

Then follow this procedure.

- Player A circles a number.
- Player B circles a second number.
  Player B's score for this turn is the GCF of the two circled numbers. Cross out the circled numbers.
- Any number that is not crossed out is circled by Player B.
- Another number that has not been crossed out is circled by Player A. Player A's score for the turn is the GCF of the two numbers. Cross out the circled numbers.
- Until all numbers have been crossed out, continue to take turns. The player with the higher score at the end of the game wins.

1. Play *GCF Game* four times with your partner. As you play, think about strategies that you can use to score the most points.
2. Describe the strategies that you used when playing *GCF Game.* Use the terms *relatively prime* and *greatest common factor* in your explanation.

Share & Summarize

1. Here is the board for a round of *GCF Game*. Your partner has just circled 15. Which number would you circle? Why?

12	13	14	15
16	17	18	19
20	21	22	23
24	25	26	27

2. You make the first move in this round of *GCF Game.* Which number would you circle? Why?

6	7	8	9
10	11	12	13
14	15	16	17
18	19	20	21

Investigation 4 Multiples

Vocabulary

common multiple

multiple

least common multiple (LCM)

Materials

- 3 strips of paper labeled 1–40

In the previous investigations, you found factors of a given number. Now you will find numbers that have a given number as a factor.

For example, think about numbers that have 20 as a factor. You can easily generate such numbers by multiplying 20 by whole numbers.

$$20 \cdot 1 = 20$$
$$20 \cdot 2 = 40$$
$$20 \cdot 3 = 60$$
$$20 \cdot 4 = 80$$

The products 20, 40, 60, and 80 are *multiples* of 20. A **multiple** of a whole number is the product of that number and another whole number.

Develop & Understand: A

1. List four multiples of 20 that are greater than 1,000.
2. Is there a limit to how many multiples a whole number greater than 0 can have? Explain.
3. Is there a limit to how many factors a whole number greater than 0 can have? Explain.
4. Is a number its own factor? Is a number its own multiple? Explain.
5. How can you determine whether 1,234 is a multiple of 7?

A **common multiple** of a set of numbers is a multiple of all the numbers. The **least common multiple** is the smallest of these. The abbreviation **LCM** is often used for least common multiple.

Example

Consider the multiples of 6 and 9.

Multiples of 6: 6, 12, **18**, 24, 30, **36**, 42, 48, **54**, 60, 66, **72**, ...

Multiples of 9: 9, **18**, 27, **36**, 45, **54**, 63, **72**, 81, ...

The numbers 18, 36, 54, and 72 are common multiples of 6 and 9. The LCM is 18.

Develop & Understand: B

6. Use a strip of paper labeled with the whole numbers from 1 to 40.

1 2 3 4 5 6 7 8 9 10 11 12 13 14 15 16 17 18 19 20 21 22 23 24 25 26 27 28 29 30 31 32 33 34 35 36 37 38 39 40

   a. Circle all the multiples of 2 between 1 and 40. Draw a square around all the multiples of 3.

   b. List the common multiples of 2 and 3 between 1 and 40. What do these numbers have in common?

   c. What are the next three common multiples of 2 and 3?

   d. What is the LCM of 2 and 3?

7. Use a new strip of paper labeled from 1 to 40.

1 2 3 4 5 6 7 8 9 10 11 12 13 14 15 16 17 18 19 20 21 22 23 24 25 26 27 28 29 30 31 32 33 34 35 36 37 38 39 40

   a. Circle all the multiples of 4 between 1 and 40. Draw a square around all the multiples of 6.

   b. List the common multiples of 4 and 6 between 1 and 40. What do these numbers have in common?

   c. What are the next three common multiples of 4 and 6?

   d. What is the LCM of 4 and 6?

8. Look at your third numbered strip.

1 2 3 4 5 6 7 8 9 10 11 12 13 14 15 16 17 18 19 20 21 22 23 24 25 26 27 28 29 30 31 32 33 34 35 36 37 38 39 40

   a. Which two numbers greater than 1 have the most common multiples between 1 and 40?

   b. What is the LCM of these two numbers?

   c. How do you know these two numbers have the most common multiples of all the pairs you could choose?

Use the idea of LCMs to help solve the next set of exercises.

Develop & Understand: C

Jada often attends family reunions in the summer.

- Her mother's family, the Coles, has a reunion every third summer.
- Her father's family, the Brooks, has a reunion every fourth summer.
- Her stepfather's family, the Turners, has a reunion every sixth summer.

This summer, Jada attended all three reunions.

9. Think about the Cole and Brooks reunions.

a. How many years will it be until the Cole and Brooks reunions occur in the same summer again? Explain how you found your answer.

b. How many years will it be until the second time both reunions happen again?

c. Will both reunions be held 48 years from now? Will both be held 70 years from now? How do you know?

10. Now think about the Cole and Turner reunions.

a. How many years will it be until the Cole and Turner reunions occur in the same summer again? Explain how you found your answer.

b. How many years will it be until the second time both reunions happen again?

c. Will both reunions be held 40 years from now? How do you know?

11. Now consider the Turner and Brooks reunions.

a. How many years will it be until the Turner and Brooks reunions are held in the same summer again? Explain how you found your answer.

b. How many years will it be until the second time both reunions happen again?

c. Will both reunions be held 72 years from now? How do you know?

12. How many years will it be until all three reunions occur in the same summer again?

Share & Summarize

Explain how your answers to the Exercises 9–12 are related to the ideas of common multiples and least common multiples.

On Your Own Exercises

Lesson 2.1

Practice & Apply

1. The ×18 machine at the stretching factory breaks down one day.
   a. What two-machine connections could you use to replace this machine?
   b. What three-machine connections could you use?
   c. What four-machine connections could you use?
2. What connections can be used to replace a ×37 machine?
3. List all the factors of 28. Then list all the factor pairs for 28.
4. List all the factors of 56. Then list all the factor pairs for 56.

Describe how you could fill the order using only prime number stretching machines.

5. Stretch a piece of rope to 12 times its original length.
6. Stretch a piece of twine to 21 times its original length.
7. Stretch a crayon to 72 times its original length.

Use a factor tree or another method to find the prime factorization of the given number. Use exponents in your answer when appropriate.

8. 84   9. 64   10. 1,272

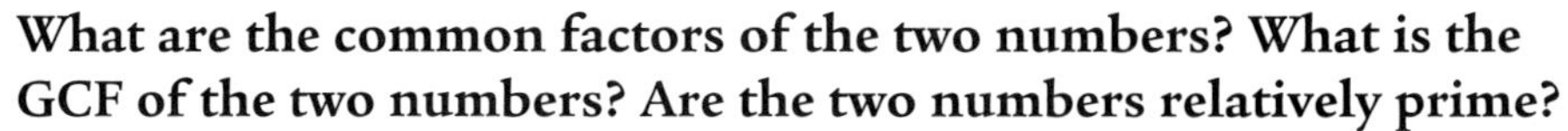

What are the common factors of the two numbers? What is the GCF of the two numbers? Are the two numbers relatively prime?

11. 13 and 24
12. 50 and 75
13. 144 and 54
14. 24 and 35

15. Suppose you are playing the *GCF Game* with your partner. Below is the board so far. Your partner has just circled 24. Which number would you circle? Why?

12	13	14	15
16	17	18	19
20	21	22	23
24	25	26	27

16. Suppose you are playing the *GCF Game* with your partner. Below is the board so far. You circle the first number in this round. Which number should you circle? Why?

16	17	18	19
20	21	22	23
24	25	26	27
28	29	30	31

List five multiples of each number.

17. 14

18. 66

19. 25

20. Is 3,269,283 a multiple of 3? How do you know?

21. Is 1,000 a multiple of 3? How do you know?

22. List all the common multiples of 3 and 5 that are less than 60.

23. List all the common multiples of 7 and 14 that are less than 60.

24. List all the common multiples of 2, 3, and 15 that are less than 100.

25. Adriana earns money by walking dogs after school and on weekends. She walks Madeline every other day, Buddy every fourth day, and Ernie every fifth day. Today, she walked all three dogs.

a. How many days will it be before Adriana walks Madeline and Buddy on the same day again?

b. How many days will it be before she walks Buddy and Ernie on the same day again?

c. How many days will it be before she walks Madeline and Ernie on the same day again?

d. How many days will it be before she walks all three dogs on the same day again?

e. Which dogs will Adriana walk 62 days from today?

Connect & Extend

26. Geometry Simone wants to form a rectangle from 12 square tiles.

a. Sketch every possible rectangle Simone could make.

b. What are the factor pairs of 12? How are the factor pairs related to the rectangles you sketched in Part a?

c. Simone wants to form a rectangle from 20 square tiles. Give the dimensions of each possible rectangle she could make. Tell how you know you have found all the possibilities.

27. Daryll said, "I am thinking of a number. The number is less than 50, and 2 and 4 are two of its nine factors. What is my number?"

28. Emma said, "I am thinking of a number. The number is less than 100, 19 is one of its four factors, and the sum of its digits is 12. What is my number?"

Math Link

The next greatest perfect number is 496. As of 1999, there were only 38 known perfect numbers, all of which are even. No one knows how many perfect numbers there are or whether there are any odd perfect numbers.

29. Jacinta said, "One of the factors of my number is 24. What other numbers must also be factors of my number?"

30. The *proper factors* of a number are all of its factors except the number itself. For example, the proper factors of 6 are 1, 2, and 3. A *perfect number* is a number whose proper factors add to that number. For example, 6 is a perfect number because $1 + 2 + 3 = 6$.

There is one perfect number between 20 and 30. What is it?

31. *Goldbach's conjecture* states that every even number except 2 can be written as the sum of two prime numbers. For example, $10 = 3 + 7$ and $24 = 19 + 5$. Although the conjecture appears to be true, no one has ever been able to prove it.

a. Write each of the numbers 4, 36, and 100 as the sum of two prime numbers.

b. Choose two even numbers of your own. Write each as the sum of two prime numbers.

32. A *reversible prime* is a prime number whose digits can be reversed to form another prime number. For example, 13 is a reversible prime because reversing its digits gives 31, another prime number. There are nine reversible primes between 10 and 100. List them all.

33. *Twin primes* are prime numbers that differ by 2. For example, 11 and 13 are twin primes. Find three more pairs of twin primes.

34. History The Greek mathematician and astronomer Eratosthenes created a technique for finding prime numbers. The technique is known as the *Sieve of Eratosthenes* because nonprime numbers are sifted out, like sand through a sieve, leaving the prime numbers.

a. To create the Sieve of Eratosthenes, start with a grid of numbers from 1 to 100 and follow these steps.

1	2	3	4	5	6	7	8	9	10
11	12	13	14	15	16	17	18	19	20
21	22	23	24	25	26	27	28	29	30
31	32	33	34	35	36	37	38	39	40
41	42	43	44	45	46	47	48	49	50
51	52	53	54	55	56	57	58	59	60
61	62	63	64	65	66	67	68	69	70
71	72	73	74	75	76	77	78	79	80
81	82	83	84	85	86	87	88	89	90
91	92	93	94	95	96	97	98	99	100

Step 1. Cross out 1.

Step 2. Circle 2. Then cross out every multiple of 2 after 2, that is, 4, 6, 8, 10, and so on.

Step 3. Circle 3. Cross out every multiple of 3, that is, 6, 9, 12, 15, and so on, after 3 that has not already been crossed out.

Step 4. Circle 5, the next number that is not crossed out. Cross out every multiple of 5, that is, 10, 15, 20, 25, and so on, after 5 that has not already been crossed out.

Continue in this manner, circling the next available number and crossing out its multiples, until every number is either crossed out or circled. When you are finished, all the prime numbers between 1 and 100 will be circled.

b. Explain why this technique guarantees that only the prime numbers will be circled.

Real-World Link

Using only a primitive tool and shadows to measure the angle of the Sun's rays, Eratosthenes, who lived more than 2,200 years ago, was able to accurately calculate the circumference of Earth.

35. A kindergarten teacher passed out 30 large stickers and 90 small stickers to his students. Each child received the same number of large stickers and the same number of small stickers. No stickers were left.

a. How many children could have been in the class? List all of the possibilities.

b. How is the greatest number of children that could have been in the class related to the GCF of 30 and 90?

Math Link

Two whole numbers are consecutive if their difference is 1.

36. In Parts a–d, determine whether each pair of consecutive numbers is relatively prime.

a. 7 and 8 **b.** 14 and 15 **c.** 20 and 21 **d.** 55 and 56

e. Choose two more pairs of consecutive numbers. Determine whether they are relatively prime.

f. Use evidence from Parts a–e to make a *conjecture*, an educated guess, about any pair of consecutive numbers.

Real-World Link

The Constitution states that a senator must be at least 30 years old, have been a U.S. citizen for nine years, and, when elected, be a resident of the state from which he or she is chosen.

37. Social Studies In the United States, senators are elected for six-year terms and presidents are elected for four-year terms. One of the Senate seats in Nebraska was up for election in the year 2000, the same year a president was elected.

a. What is the next year the same Senate seat will be up for election and a presidential election will take place? Assume the seat is not vacated early.

b. List three other years when this will happen. What do these years have in common?

c. When was the last year *before* 2000 that this Senate seat was up for election in the same year as a presidential election?

38. Tai found that the LCM of 3 and 5 is 15 and that the LCM of 2 and 3 is 6. He concluded that he could find the LCM of any two numbers by multiplying them.

a. Copy and complete this table to show the product and the LCM of each pair of numbers.

b. Is the LCM of two numbers always equal to their product? If not, use the table to predict when the LCM of a pair of numbers is the product.

Numbers	Product	LCM
2 and 3	6	6
2 and 4		
2 and 5		
3 and 4		
3 and 6		
6 and 9		
4 and 10		

39. To find the LCM of two numbers, Ivan first writes their prime factorizations. He then decides where each prime number appears the most. For example, 3 appears twice for 252, so he writes down 3 two times. If the most a prime appears is once, then he writes that number only once. For example, he wrote the prime factorizations of 210 and 252.

$$210 = 2 \cdot 3 \cdot 5 \cdot 7 \qquad 252 = 2 \cdot 2 \cdot 3 \cdot 3 \cdot 7$$

2 appears twice for 252, 3 appears twice for 252, 5 appears once for 210, 7 appears once for both 210 and 252. So, the LCM is $2 \cdot 2 \cdot 3 \cdot 3 \cdot 5 \cdot 7$, or 1,260.

Try Ivan's method on at least two more pairs of numbers.

40. In Your Own Words Choose a composite number between 20 and 100. Use the ideas from the lesson to write a short report about your number. Your report should include a discussion about the factors and multiples of your number and give your number's prime factorization.

Mixed Review

41. Probability Sophia bought two dozen superballs to give out at her birthday party. She put them in a bag. A third of the superballs were blue, 25% were striped, an eighth were hot pink, and the rest were made to glow in the dark.

a. Rohan arrived first. Sophia asked him to reach into the bag without looking and take a ball. What is the probability that Rohan's ball will glow in the dark?

b. Rohan picked a blue ball. Lucita arrived next and said she hoped to pick hot pink. What is the probability she will get her wish?

c. Lucita did not get what she wanted. But Emilio, who picked next, did get hot pink. What was the probability of this happening?

d. Emilio traded with Lucita for her striped superball. Tadi picked next, hoping for a glow-in-the-dark ball. What is the probability of him choosing one?

Use the distributive property to rewrite each expression with parentheses.

42. $18 - 4f$

43. $13 - 26m$

44. $45x + 15$

45. $16x^2 + 48x$

Exponent Machines

Business is good at your stretching factory. In fact, you have received more orders than you can handle. Winnie has noticed that many of the orders that you receive can be filled by using just one machine.

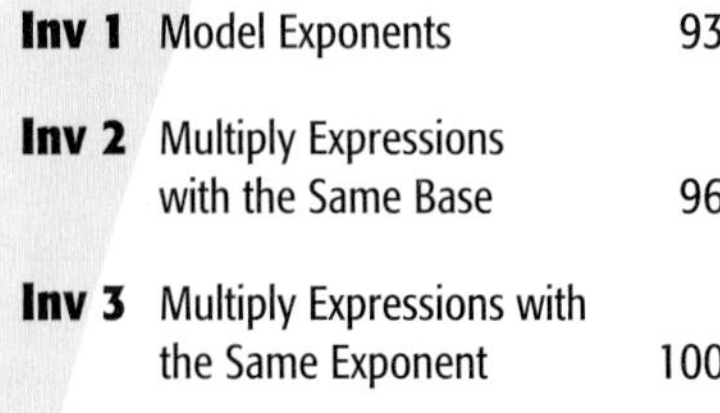

Inv 1 Model Exponents 93

Inv 2 Multiply Expressions with the Same Base 96

Inv 3 Multiply Expressions with the Same Exponent 100

Winnie thinks that this new machine can be used instead of connecting two of the same machines together. If you specialize in single-machine orders, you can save time. No one will have to run from one machine to another, and you will never have to worry about two people needing the same machine at the same time. You decide to give Winnie's plan a try.

Think & Discuss

If you start with a one-inch piece of taffy, what lengths between 1 and 30 inches can you make by using a single machine twice?

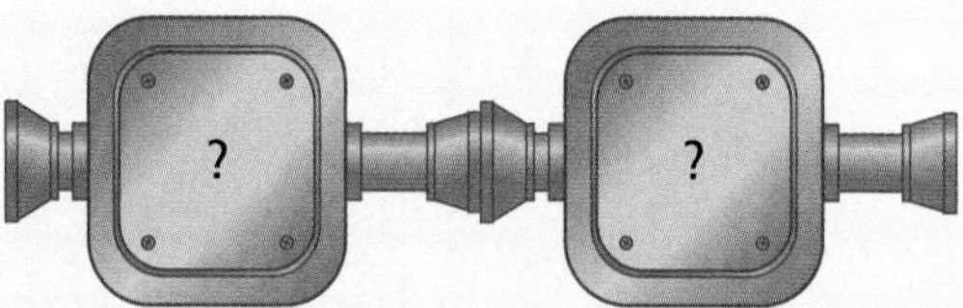

What lengths can you make by using a single machine three times?

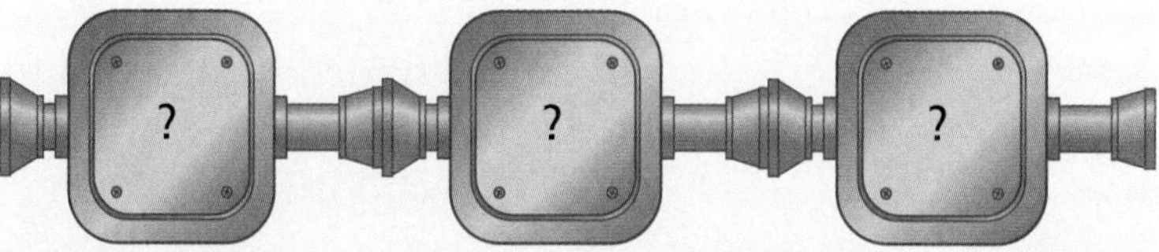

Winnie filled an order that stretched a one-inch piece of taffy to a length less than 30 inches, using a single machine four times. What machine might she have used?

Will your new plan of using one machine per order mean that there are lengths that your factory cannot produce?

Investigation  Model Exponents

Vocabulary

base

power

The stretching machines occasionally break down at your factory. Your employees usually find a way to work around this problem.

Develop & Understand: A

1. One of the $\times 10$ machines broke down. Annie connected two machines together to do the same work as a $\times 10$ machine. The output from the first machine becomes the input for the second.

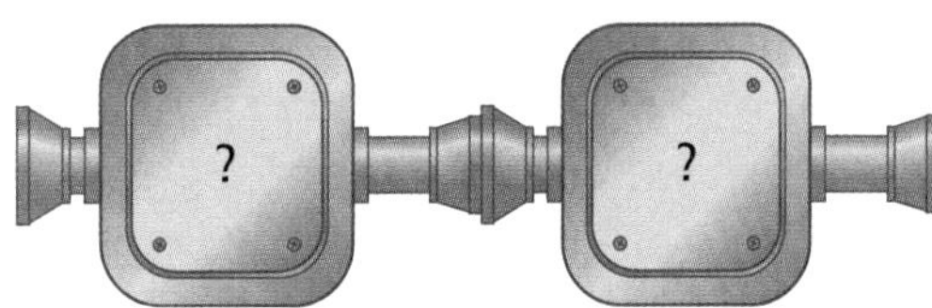

What two machines connected together do the same work as a $\times 10$ machine? Is there more than one arrangement that will work?

2. What stretching machine does the same work as two $\times 2$ machines connected together?

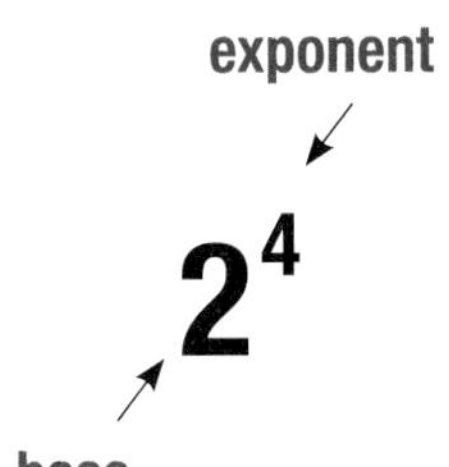

If you send a 1-inch piece of taffy through a $\times 2$ machine four times, its length becomes $1 \cdot 2 \cdot 2 \cdot 2 \cdot 2 = 16$ inches. You learned that $2 \cdot 2 \cdot 2 \cdot 2$ can be written 2^4. The expression 2^4 is read "two to the fourth **power**." The 2 is called the **base**, and the raised 4 is called the exponent.

Suppose some of the machines automatically feed a piece of taffy through several times. You write an exponent to indicate the number of times each machine is applied. For example, sending taffy through a $\times 2^4$ machine is the same as putting it through a $\times 2$ machine four times.

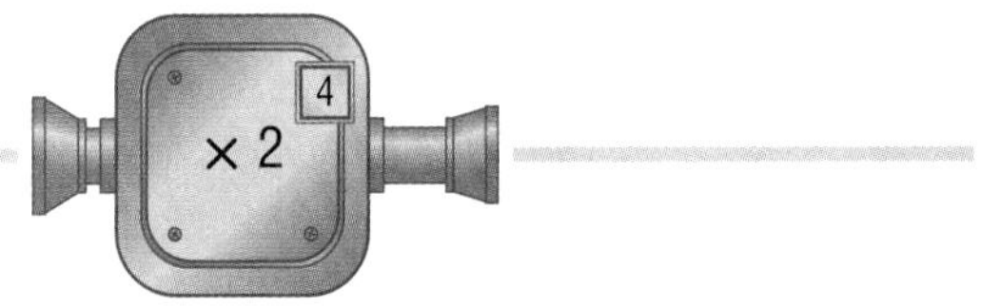

These adjusted machines are *repeater* machines because they repeatedly stretch the input by the same factor. The original machine, or *base* machine, is a $\times 2$ machine.

Develop & Understand: B

For each repeater machine, tell how many times the base machine is applied and how much the total stretch is.

3. **4.** **5.**

6. Find three repeater machines that will do the same work as a $\times 64$ machine. Draw or describe them using exponents.

7. In a repeater machine with 0 as an exponent, the base machine is applied 0 times.

   a. What do these machines do to a piece of chalk?

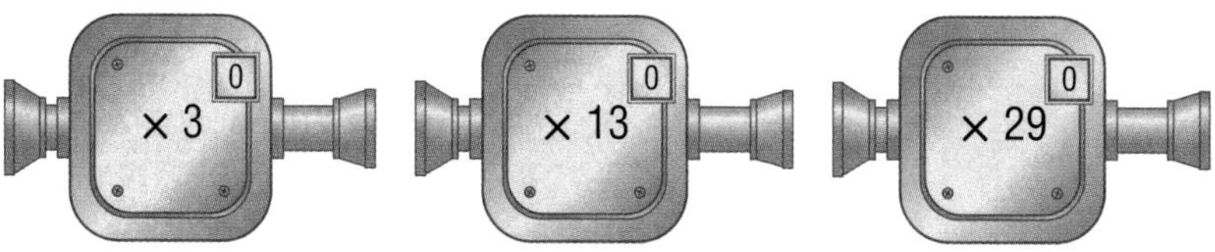

   b. What do you think the value of 6^0 is?

Evaluate each expression without using a calculator.

8. 7^0

9. 2^5

10. 4^2

11. 3^3

Peter found a $\times\frac{1}{2}$ machine in a corner of the factory. He was not sure what it would do, so he experimented with some licorice. This is what he found.

Each piece of licorice was compressed to half its original length. Peter decided this was a *shrinking* machine.

Develop & Understand: C

Try a shrinking machine for yourself.

12. If a foot-long sandwich is put into the machine to the right, how many inches long will it be when it emerges?

Like stretching machines, shrinking machines can be used in connections and repeater machines.

13. What happens when one-inch gum is sent through these connections?

a.

b.

14. Evan put a one-inch stick of gum through a $\times\left(\frac{1}{3}\right)^2$ machine. How long was the stick when it came out the machine?

15. This stick of gum came out of a $\times\left(\frac{1}{2}\right)^2$ machine. Estimate the length of the input stick in inches. Explain how you found your answer.

16. Armando put a one-inch piece of gum through a $\times\frac{1}{10}$ repeater machine. It came out $\frac{1}{100{,}000}$ in. long. What is the missing exponent?

Find a single machine that will do the same job as the connection given below.

17. a $\times 2^3$ machine followed by a $\times\left(\frac{1}{2}\right)^2$ machine

18. a $\times 5^{99}$ machine followed by a $\times\left(\frac{1}{5}\right)^{100}$ machine

Evaluate each expression without using a calculator.

19. $\left(\frac{1}{2}\right)^5$

20. $\left(\frac{1}{4}\right)^2$

21. $\left(\frac{1}{3}\right)^3$

22. $\left(\frac{1}{7}\right)^0$

Real-World Link

For centuries, Native Americans chewed the sap, called *chicle*, of the sapodilla tree. Settlers took up the habit. In the late 1800s, chewing gum began to be made commercially from chicle, sugar, and flavorings.

Share & Summarize

1. If you put a two-inch toothpick through this machine, how long will it be when it comes out the machine?

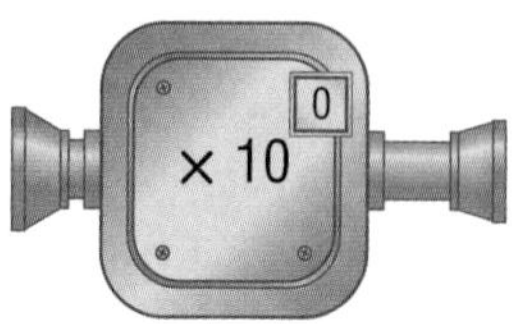

2. A one-inch-long beetle crawled into a repeater machine and emerged $\frac{1}{16}$ in. long. Through what repeater machine did it go? Is there more than one possibility?

3. What single repeater machine does the same work as a $\times 3^5$ machine followed by a $\times\left(\frac{1}{3}\right)^3$ machine?

Investigation 2 Multiply Expressions with the Same Base

Vocabulary

order of operations

product law of exponents

Evan's supervisor asked him to stretch a one-inch noodle into a 32-inch noodle. Evan intended to send the noodle through a $\times 2^5$ machine, but he accidentally put it through a $\times 2^3$ machine. His co-workers had several suggestions for how he could fix his mistake.

Will all of the suggestions work? Why or why not?

Evan's mistake was not so terrible after all. He figured out how to find single repeater machines that do the same work as some connections.

Example

Evan found a single machine for this connection.

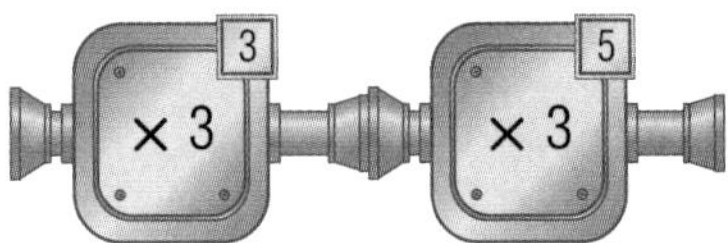

He reasoned, "The first machine tripled the noodle's length three times. The second machine took that output and tripled *its* length five times. The original noodle's length was tripled $3 + 5$ times, or eight times in all. That means a $\times 3^8$ machine would do the same thing."

Develop & Understand: A

Follow Evan's reasoning to find a single repeater machine that will do the same work as each connection.

1. 

2. 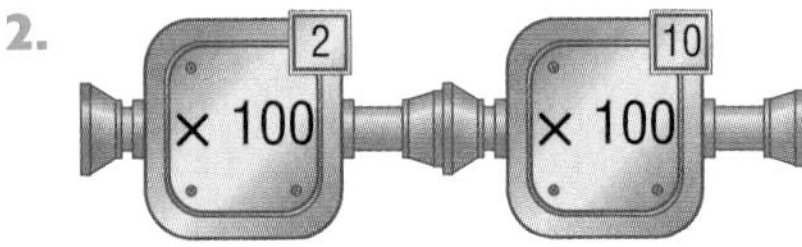

3. 

4. 

You have seen that a connection of repeater machines with the same base can be replaced by a single repeater machine. When you multiply exponential expressions with the same base, you can replace them with a single expression.

Example

Malik thought about how he could rewrite the expression $2^{20} \cdot 2^{5}$.

Malik's idea is one of the **product laws of exponents**, which can be expressed like the following.

Multiplying Expressions with the Same Base
$a^{b} \cdot a^{c} = a^{b+c}$

Actually, this law can be used with more than two expressions. As long as the bases are the same, to find the product you can add the exponents and use the same base.

$$\begin{aligned} 3^{2} \cdot 3^{3} \cdot 3^{10} &= 3^{2+3+10} \\ &= 3^{15} \end{aligned}$$

Develop & Understand: B

Rewrite each expression as a power of 2. It may help to think of the expressions as connections of $\times 2$ machines.

5. $2 \cdot 2$
6. $2^5 \cdot 2^4$
7. $2^{10} \cdot 2$
8. $2^{10} \cdot 2^{10}$
9. $2^0 \cdot 2^3$
10. $2^m \cdot 2^n$

For each connection, determine whether there is a single repeater machine that will do the same work. If so, describe or draw it.

11. 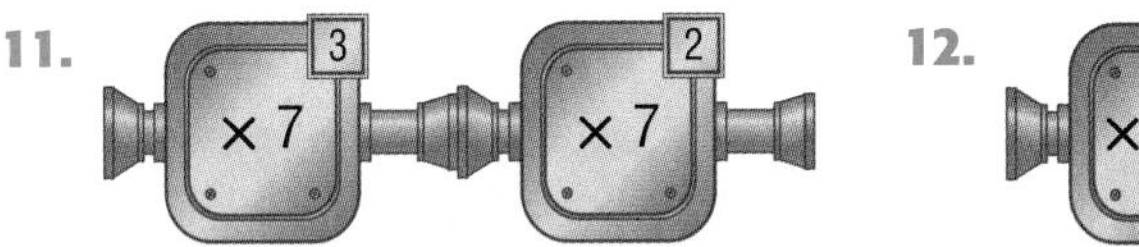

12. 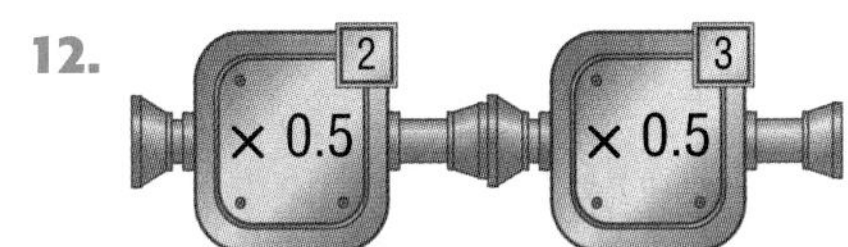

13. 

14. 

15. 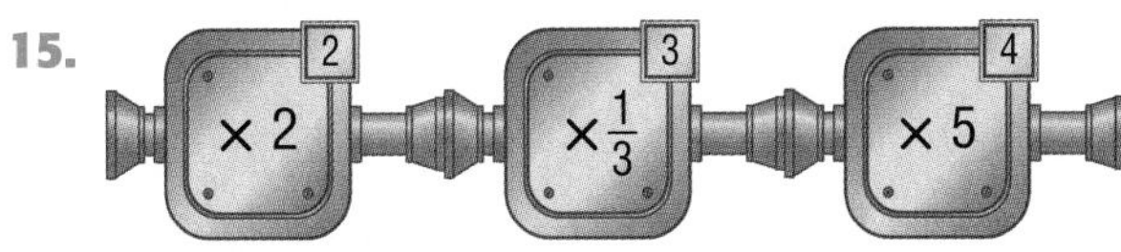

Tell what number should replace the question mark to make each statement true.

16. $5 \cdot 5 \cdot 5 = ?^3$
17. $2.7^{10} \cdot 2.7^? = 2.7^{12}$
18. $\left(\frac{1}{5}\right)^4 \cdot \left(\frac{1}{5}\right)^6 = \left(\frac{1}{5}\right)^?$
19. $1.2^? \cdot 1.2^2 = 1.2^2$
20. $\frac{2}{3} \cdot ? = \left(\frac{2}{3}\right)^2$
21. $a^{20} \cdot a^5 = a^?$

If possible, rewrite each expression using a single base. For example, $6^2 \cdot 6^3$ can be rewritten 6^5.

22. $3^3 \cdot 3^3 \cdot 3^4$
23. $2^5 \cdot 5^2$
24. $4 \cdot 4 \cdot 4$
25. $4x^3 \cdot y^2$
26. $4^4 \cdot 2^2$
27. $a^n \cdot a^m$

When you simplify algebraic expressions involving exponents, it is important to keep the **order of the operations** in mind.

Math Link

By the order of operations, $2b^3$ means $2 \cdot b^3$, not $(2 \cdot b)^3$.

Example

$$\begin{aligned} 2b^3 \cdot 5b^2 &= 2 \cdot b^3 \cdot 5 \cdot b^2 \\ &= 2 \cdot 5 \cdot b^3 \cdot b^2 \\ &= 10b^5 \end{aligned}$$

Develop & Understand: C

Simplify each expression.

28. $2b \cdot 2b$

29. $5z^2 \cdot 6z^2$

30. $5a^2 \cdot 3a^4$

31. $z^2 \cdot 2z^3 \cdot 6z^2$

Share & Summarize

1. Find a single repeater machine to do the same work as this connection.

2. In your own words, write the rule that lets you figure out the missing exponent in equations like $3^3 \cdot 3^{16} = 3^?$ and $r^{15} \cdot r^? = r^{23}$.

Investigation 3 Multiply Expressions with the Same Exponent

Combining expressions that have the same base can make products much easier to understand. Do you think you can simplify products that do not have the same base?

Develop & Understand: A

1. Maxine has an order from a golf course designer to put palm trees through a $\times 2^3$ machine and then through a $\times 3^3$ machine. She thinks she can do the job with a single repeater machine. What single repeater machine should she use?

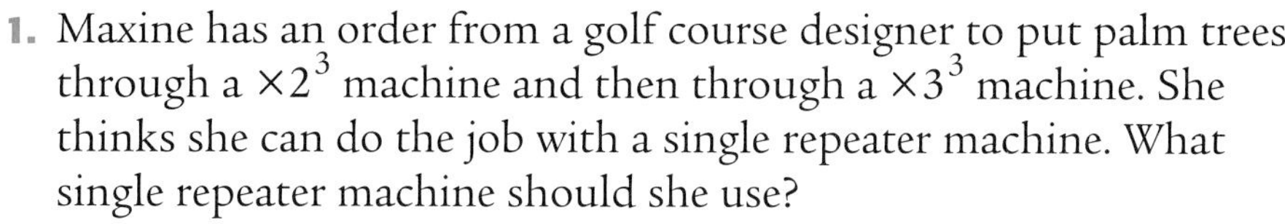

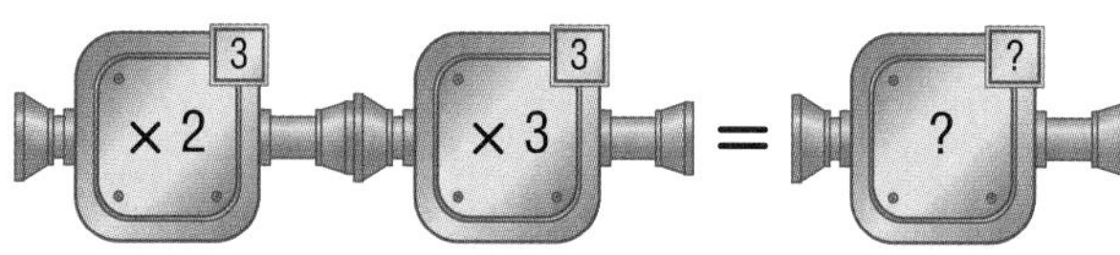

2. Maxine also needs to stretch some pine trees to 10^2 times their original lengths. Her $\times 10$ machine is still broken, and someone is using the $\times 10^2$ machine. Find a connection of two repeater machines that will do the same work as a $\times 10^2$ machine. To get started, think about the connection that you could use to replace the $\times 10$ machine.

For each connection, find a single repeater machine to do the same work.

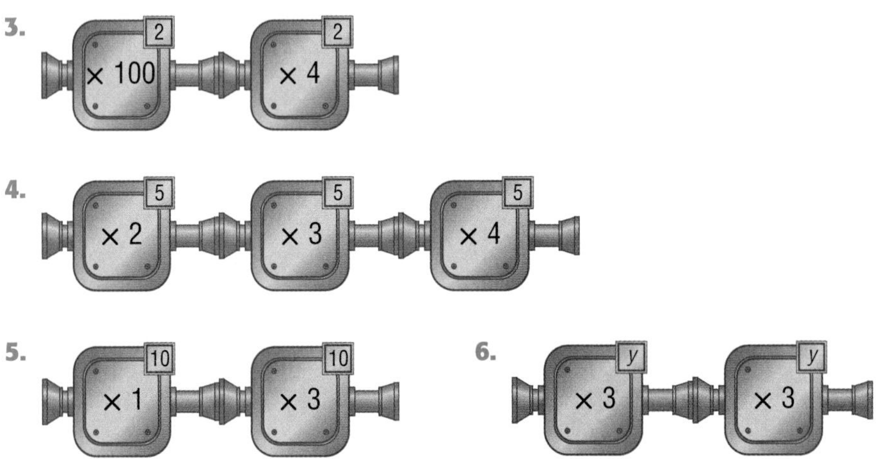

You can use the same kind of thinking that you used in the above exercises to multiply expressions with the same exponent.

Example

Maya multiplied $4^2 \cdot 3^2$ by thinking about stretching machines.

Use Maya's idea to multiply $5^3 \cdot 2^3$. Use your calculator to check your answer.

Maya's idea is another *product law of exponents.*

Multiplying Expressions with the Same Exponents
$a^c \cdot b^c = (a \cdot b)^c$

You can use this law with more than two expressions. If the exponents are the same, multiply the expressions by multiplying the bases and using the same exponent. For example, $2^8 \cdot 3^8 \cdot 7^8 = (2 \cdot 3 \cdot 7)^8 = 42^8$.

Develop & Understand: B

Rewrite each expression using a single exponent. For example, $2^3 \cdot 3^3$ can be rewritten 6^3.

7. $100^2 \cdot 3^2$
8. $10^{20} \cdot 25^{20}$
9. $5^{100} \cdot 5^{100} \cdot 2^{100}$
10. $\left(\frac{1}{3}\right)^a \cdot \left(\frac{1}{5}\right)^a$
11. $1{,}000^5 \cdot 3^5$
12. $x^2 \cdot y^2$

For each connection, determine whether there is a single repeater machine that will do the same work. If so, describe it.

13.

14.

15.
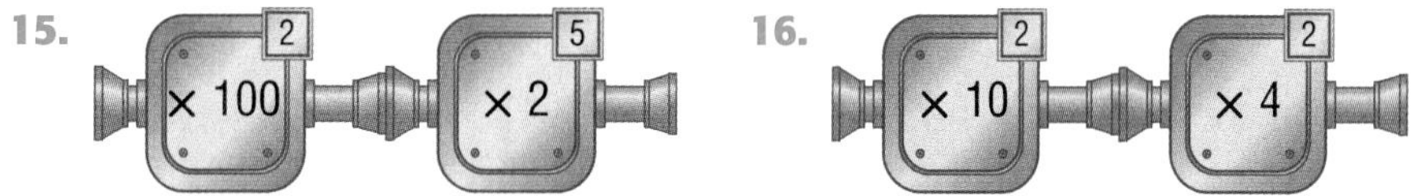

16.

17.

Tell what number should replace the question mark to make each statement true.

18. $100^? \cdot 3^2 = 300^2$
19. $4^{20} \cdot 25^{20} = ?^{20}$
20. $5^{100} \cdot \left(\frac{1}{2}\right)^{100} = ?^{100}$
21. $\left(\frac{1}{3}\right)^3 \cdot \left(\frac{1}{5}\right)^3 = \left(\frac{1}{15}\right)^?$

You have worked with the product laws of exponents. Do you think there are similar laws for sums of exponential expressions? You will explore this question in the next exercise set.

Develop & Understand: C

Determine whether each statement is true or false. Explain how you decided.

22. $2^3 + 2^4 = 2^7$

23. $3^2 + 5^2 = 8^2$

24. From your answers to Exercises 22 and 23, do you think $2^a + 2^b = 2^{a+b}$ for any numbers a and b?

25. From your answers to Exercises 22 and 23, do you think $a^2 + b^2 = (a + b)^2$ for any numbers a and b?

Share & Summarize

1. In your own words, write the rule that lets you find the missing base in equations like $3^{30} \cdot 2^{30} = ?^{30}$ and $5^{15} \cdot ?^{15} = 100^{15}$.

Find the missing numbers.

2. $15^3 \cdot 4^3 = ?^3$

3. $3^7 \cdot 3^? = 3^{12}$

4. $\left(\frac{2}{5}\right)^{12} \cdot \left(\frac{2}{5}\right)^3 = \left(\frac{2}{5}\right)^?$

5. $2^2 \cdot ?^2 = 18^2$

6. You have seen two product laws for working with exponents. Explain how you know when to use each rule.

Product Laws

$$a^b \cdot a^c = a^{b+c}$$

$$a^c \cdot b^c = (a \cdot b)^c$$

On Your Own Exercises
Lesson 2.2

Practice & Apply

Supply the missing information for each diagram.

1.

2.

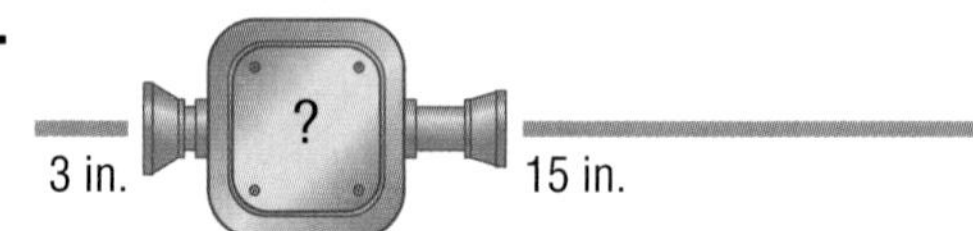

3.

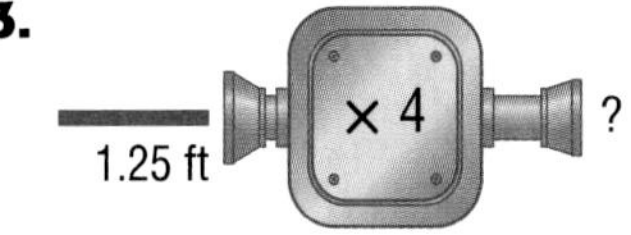

4.

If possible, find a connection that will do the same work as the given stretching machine. Do not use ×1 machines.

5.

6.

7.

8.

9. Find two repeater machines that will do the same work as a ×81 machine.

10. Find a repeater machine that will do the same work as a $\times\frac{1}{8}$ machine.

11. This stick of gum exited a $\times 3^2$ machine. Without measuring, estimate the length of the input stick in centimeters. Explain how you found your answer.

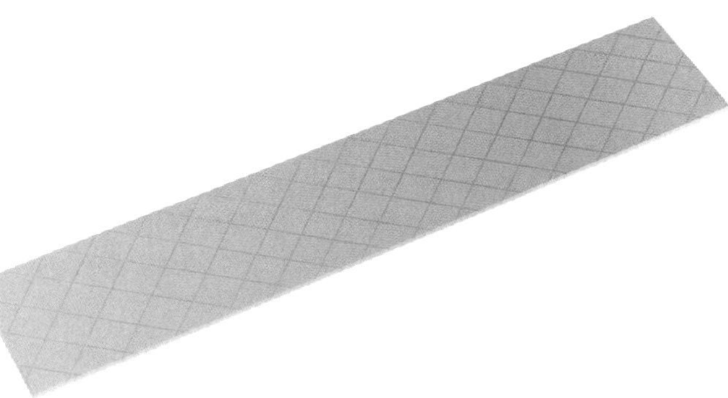

Evaluate each expression without using your calculator.

12. 6^2

13. 9^2

14. $\left(\frac{1}{3}\right)^3$

15. $3 \cdot 2^3$

16. 8^1

17. x^1

Do each calculation in your head. (Hint: Think about stretching and shrinking machines.)

18. $\left(\frac{1}{3}\right)^5(3)^5$

19. $(4)^6\left(\frac{1}{4}\right)^6$

20. $\left(\frac{1}{10}\right)^7(10)^8$

If possible, find a repeater machine to do the same work as each connection.

21. × 2 [2] × 3 [2] × 5 [1]

22. × 5 [2] × 5 [2]

If possible, rewrite each expression using a single base.

23. $6^4 \cdot 6^2$ **24.** $4^{10} \cdot 5^3$ **25.** $1.1^{100} \cdot 1.2^{101}$

26. $7^{10} \cdot 7^k$ **27.** $4^n \cdot 4^m$ **28.** $a^{20} \cdot a^5$

Simplify each expression.

29. $3c \cdot 5c$ **30.** $10z^2 \cdot z^2$ **31.** $2a^3 \cdot 10a$

If possible, find a repeater machine to do the same work as each connection.

32. **33.**

34. **35.**

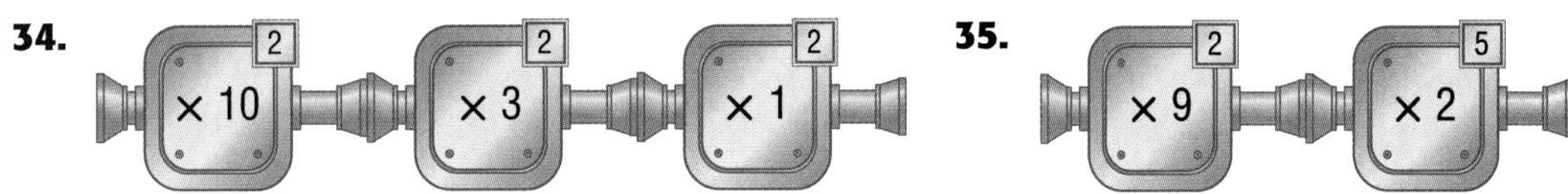

If possible, rewrite each expression using a single base and a single exponent.

36. $2^3 \cdot \left(\frac{1}{3}\right)^3$ **37.** $4^0 \cdot 5^2$ **38.** $1^{100} \cdot 5^2$

39. $3^9 \cdot \left(\frac{1}{3}\right)^7$ **40.** $\left(\frac{1}{2}\right)^2 \cdot \left(\frac{1}{5}\right)^2$ **41.** $6^4 \cdot 2^4$

Connect & Extend

42. Patrick has a one-inch sewing needle. He says that putting the needle through a $\times 3$ machine five times will have the same effect as putting it through a $\times 5$ machine three times. Jane says the needles will turn out different lengths.

a. Who is correct? Explain how you know.

b. In Part a you compared 3^5 and 5^3. Are there *any* two different numbers, a and b, for which $a^b = b^a$? Your calculator may help you.

43. The left column of the chart lists the lengths of input pieces of ribbon. Stretching machines are listed across the top. The other entries are the outputs for sending the input ribbon from that row through the machine from that column. Copy and complete the chart.

Input Length	Machine			
	×2	?	?	?
	1	5		
3				15
	14		7	

44. The left column of the chart lists the lengths of input chains of gold. Repeater machines are listed across the top. The other entries are the outputs you get when you send the input chain from that row through the repeater machine from that column. Copy and complete the chart.

Input Length	Repeater Machine			
	$\times 2^3$	?	?	?
	40		640	
2				
		96		768

Math Link

The only factors of a prime number are the number itself and 1. The *prime factorization* of a number expresses the number as a product of prime numbers, $12 = 2 \cdot 2 \cdot 3$.

45. Write the prime factorization of each number using exponents.

a. 27 **b.** 12 **c.** 100 **d.** 999

46. In Your Own Words Choose one of the product laws of exponents and explain it so a fifth grader could understand. Use stretching and shrinking machines in your explanation if you wish.

Mixed Review

Fill in the blanks to make each a true statement.

47. $\frac{1}{8} + \frac{1}{3} + \_\_\_\_ = 1$

48. $\frac{2}{7} + \frac{3}{10} + \_\_\_\_ = 1$

Sketch a graph to match each story.

49. Paul charted his height from when he was age 3 until he turned age 12.

50. The space shuttle was launched on Monday, orbited the planet for ten days, and then returned to Earth.

51. Darius threw the ball into the air and watched as it rose, fell, and then hit the ground.

More Exponent Machines

Inv 1	Division Machines with Exponents	107
Inv 2	Divide Expressions with Exponents	109
Inv 3	Power Law of Exponents	111
Inv 4	The Tower of Hanoi	114

Peter arrived at the resizing factory one day to find a whole new set of machines. Instead of × symbols, the machines had ÷ symbols. Peter's supervisor explained that the factory was getting more orders for shrinking things. The factory purchased a new type of shrinking machine that uses whole numbers instead of fractions. For example, a ÷2 machine divides the length of whatever enters the machine by 2.

Think & Discuss

How long will a meterstick be after traveling through each machine?

1. 

2. 

3. 

4. 

Investigation 1 Division Machines with Exponents

The new shrinking machines can be connected to other shrinking machines and to stretching machines.

Develop & Understand: A

If a yardstick is put into each machine, how many inches long will it be when it comes out of the machine?

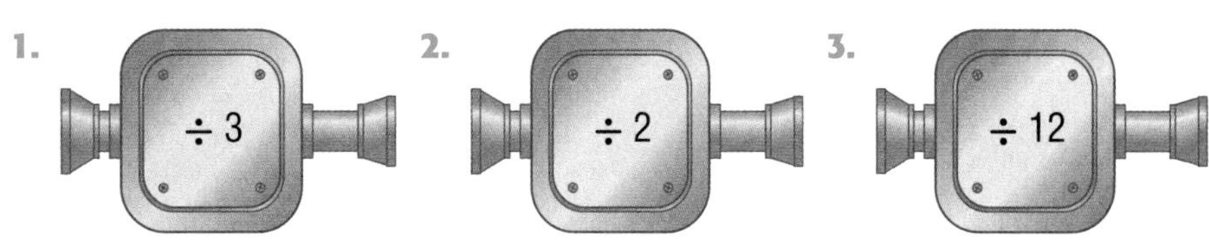

4. 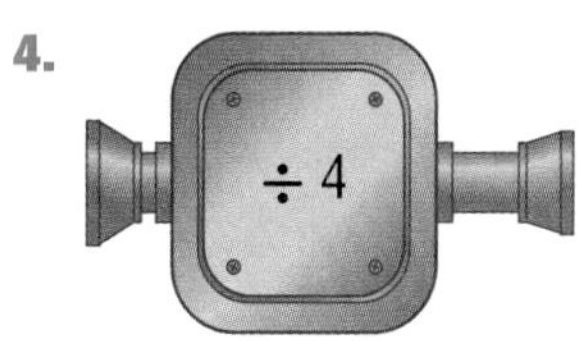

5. 

Some of the new shrinking machines are designed to make repeater machines. If a yardstick is put into each repeater machine, how many inches long will it be when it exits?

6. 

7. 

8. 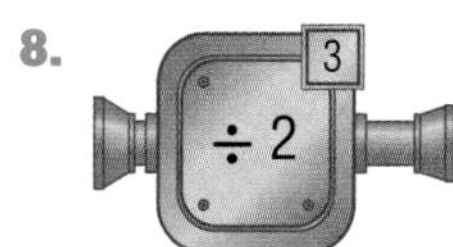

9. If a one-inch paper clip is put through this connection, what will be its final length?

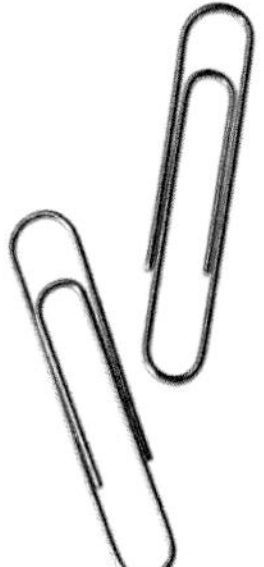

Find a single repeater machine to do the same work as each connection.

10. 

11. 

12. 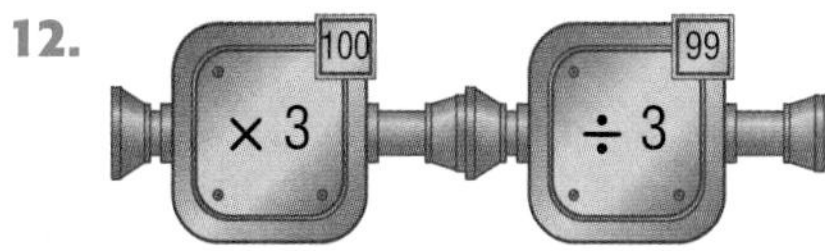

Share & Summarize

1. Find two repeater machines that will shrink a one-inch gummy worm to $\frac{1}{16}$ of an inch.
2. A one-inch gummy worm was sent through this connection and emerged 25 inches long. What is the first machine in the connection?

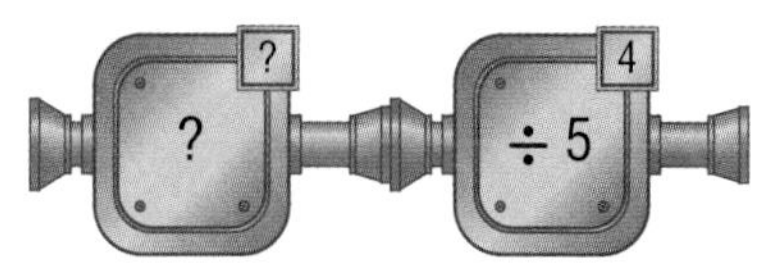

Investigation 2 Divide Expressions with Exponents

The machine model can help you divide exponential expressions.

Example

Here is how Shaunda thought about $3^5 \div 3^3$ connection.

×3 5
÷3 3
I think about a connection of a ×3 5 and a ÷3 3 machine. When something goes through the connection, it is stretched 5 times and shrunk 3 times.

×3 5
÷3 3
×3 2
Each ÷3 shrink cancels a ×3 stretch. So the three shrinks cancel three of the stretches...leaving two stretches. The connection does the same work as a ×3 2 machine.

When I calculate 3⁵ ÷ 3³, I can just subtract the number of shrinks from the number of stretches. This is the same as subtracting exponents,
3⁵ ÷ 3³ = 3⁵⁻³ = 3².

Shaunda had discovered one of the *quotient laws of exponents.*

Dividing Expressions with the Same Base
$a^b \div a^c = a^{b-c}$ $\qquad \dfrac{a^b}{a^c} = a^{b-c}$

Develop & Understand: A

Write each expression as a power of 2.9.

1. $2.9^5 \div 2.9^3$
2. $2.9^{10} \div 2.9^6$
3. $2.9^{13} \div 2.9^{12}$
4. $2.9^{13} \div 2.9^{11}$
5. $2.9^{25} \div 2.9^{25}$
6. $2.9^8 \div 2.9^3$

Evaluate each expression without using your calculator.

7. $2^3 \div 2^3$
8. $2^{10} \div 2^{10}$
9. $2^{15} \div 2^{15}$

If possible, rewrite each expression using a single base.

10. $5^5 \div 5^2$
11. $3^4 \div 3^2$
12. $2^5 \div 3^4$
13. $3^2 \div 3^0$
14. $2^8 \div 3^2$
15. $a^{10} \div a$
16. $a^n \div a^m$
17. $a^n \div b^m$

Though you may not be able to rewrite an expression using a single base, there may still be a way to simplify the expression.

Example

Kate wondered if it would be possible to simplify an expression like $2^2 \div 3^2$. Simon and Jin Lee tried to help her decide.

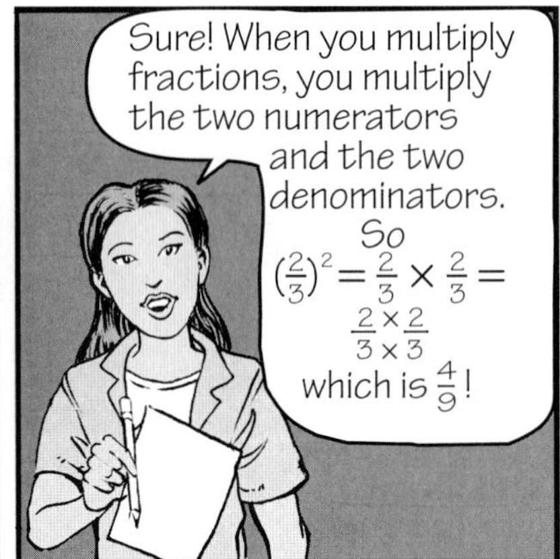

This led the class to discuss another quotient law of exponents.

Dividing Expressions with the Same Exponent
$a^c \div b^c = (a \div b)^c$
$\frac{a^c}{b^c} = \left(\frac{a}{b}\right)^c$

Develop & Understand: B

Write each expression in the form c^m. The variable c can be a fraction or a whole number.

18. $2^2 \div 5^2$

19. $5^6 \div 5^2$

20. $a^2 \div 2^2$

21. $a^9 \div b^9$

22. $3^7 \div 3^4$

23. $x^n \div y^n$

24. Prove It! Show that $\frac{a^3}{b^3} = \left(\frac{a}{b}\right)^3$, for $b \neq 0$. (Hint: Write a^3 and b^3 as products of a's and b's. Regroup the quotient into a string of $\left(\frac{a}{b}\right)$'s.)

25. Do you think $\frac{a^c}{b^c} = \left(\frac{a}{b}\right)^c$ for $b \neq 0$ and any value of c greater than or equal to 0? Explain.

Share & Summarize

In your own words, write the rule that lets you find the missing exponent in equations like $5^{16} \div 5^8 = 5^?$ and $r^? \div r^{10} = r^5$.

Investigation 3 Power Law of Exponents

Your resizing factory is a success. You decide to order a shipment of *super machines.* Here is one of the new machines.

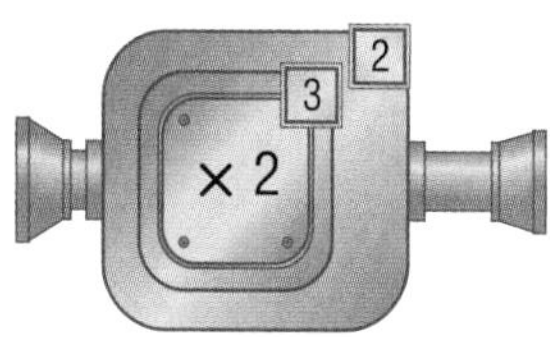

Peter was excited to try the machine, but he was not sure what it would do. He found this diagram in the manufacturer's instructions.

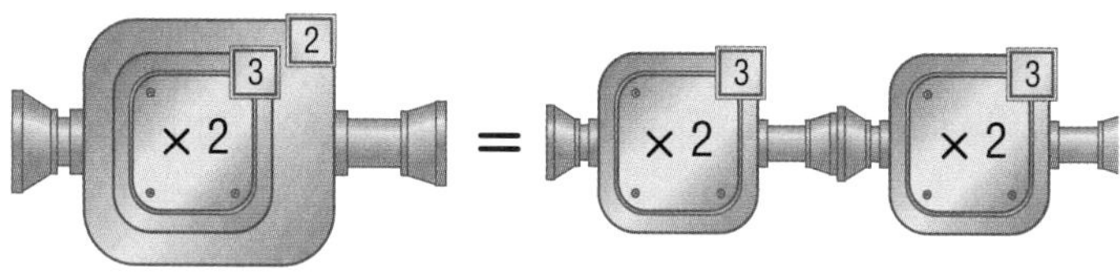

Real-World Link

Improperly discarded chewing gum can be a litter nightmare. Schools and businesses spend thousands of dollars every year cleaning chewing gum from floors, carpets, and furniture.

Think & Discuss

If Peter puts a one-inch stick of gum through the above super machine, how long will it be when it comes out the machine?

How many times would Peter need to apply a ×2 machine to do the same work as this new super machine?

Peter used the shorthand notation $\times(2^3)^2$ to describe this super machine. Why is this shorthand better than $\times 2^{3\,2}$?

Describe or draw a connection of three repeater machines that will do the same work as a $\times(4^2)^3$ machine.

Develop & Understand: A

Careta has an order from a kite maker who wants to stretch different types of string. If a one-inch piece of kite string is sent through each super machine, how long will it be when it exits?

1. 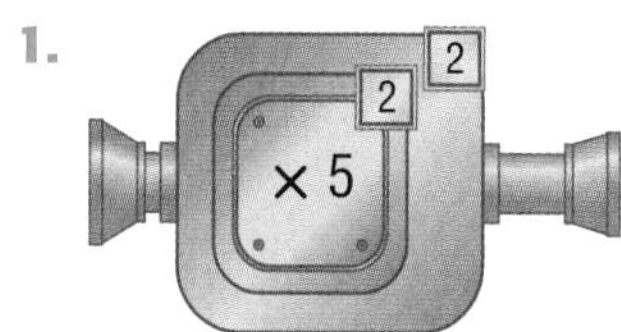

2. 

3. 

4. 

5. For each super machine in Exercises 1–4, describe or draw a connection of two or more repeater machines that would do the same work.

6. Describe a super machine that will stretch a one-inch strand of kite string to 81 inches. Do not use 1 for either exponent.

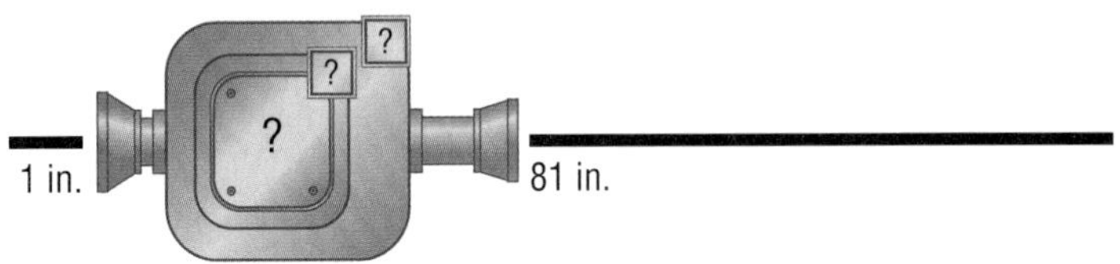

7. Consider the expression $(4^3)^2$.

   a. Draw a super machine that represents this expression.

   b. How many times would you need to apply a $\times 4^3$ machine to do the same work as your super machine?

   c. How many times would you need to apply a $\times 4$ machine to do the same work as your super machine?

The expressions represented by the super machines can be simplified.

Example

In Exercise 7, Zoe added the exponents and decided it would take five $\times 4$ machines to do the same work as the $\times(4^3)^2$ super machine. Maya thought the answer was six. She drew a diagram to explain her thinking to Zoe.

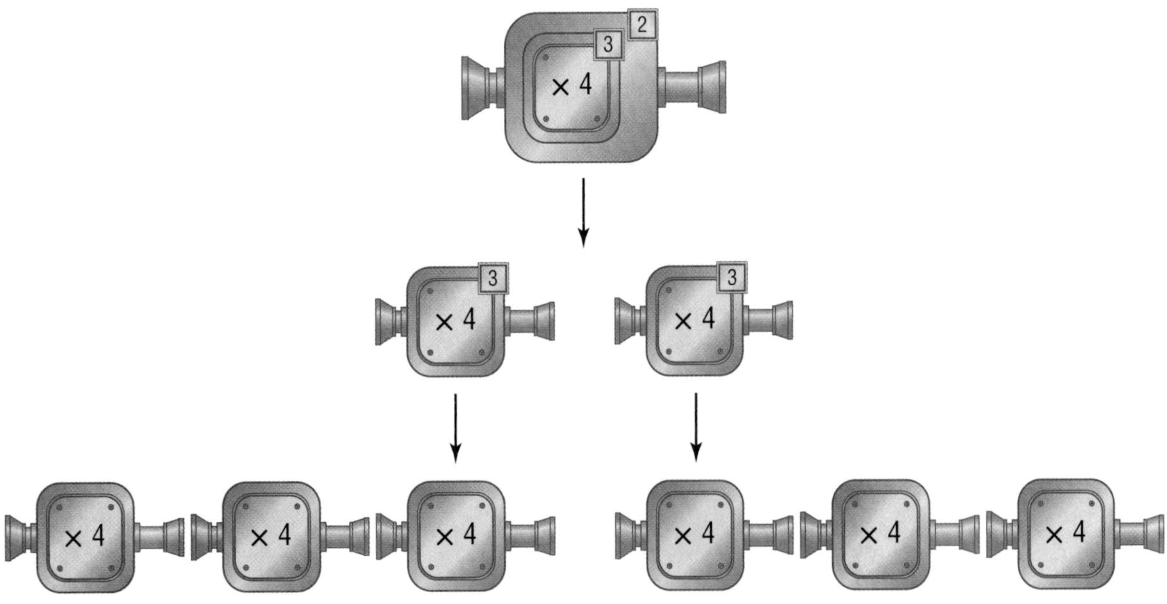

Maya's diagram convinced Zoe that she should have multiplied the exponents rather than adding them. The diagram shows that $(4^3)^2 = 4^{3 \cdot 2} = 4^6$.

The diagram Maya used demonstrates an example of the *power of a power law of exponents.*

Raising a Power to a Power

$$(a^b)^c = a^{b \cdot c}$$

Develop & Understand: B

Write each expression as a power of 2.

8. $(2^4)^2$
9. $(2^2)^3$
10. $(2^4)^3$
11. $(2^2)^2$
12. $(2^1)^7$
13. $(2^0)^5$

Tell what number should replace the question mark to make each statement true.

14. $4 = 2^?$
15. $4^? = (2^2)^3$
16. $4^4 = (?^2)^4$
17. $2^3 \cdot 4^? = 2^{13}$
18. $2^7 \div 4^3 = 2^?$
19. $?^3 \div 2^3 = 2^3$
20. $8^2 = 2^?$
21. $2^7 + 2^7 = 2^?$

22. Use a calculator to check your answers for Exercises 8–21. To evaluate an expression like $(2^4)^2$, push the exponent key twice. To find $(2^4)^2$, you could use the following keystrokes.

2 [^] 4 [^] 2

or

2 [y^x] 4 [y^x] 2

Share & Summarize

Tell whether each equation is true or false. If an equation is false, rewrite the expression to the right of the equal sign to make it true.

1. $3^5 \cdot 3^3 = 3^{15}$
2. $(2^5)^0 = 1$
3. $(a^3)^4 = a^7$
4. $a^{30} \div a^2 = a^{15}$
5. $(a^5)^3 = a^{15}$
6. $a^4 \cdot a^3 = a^7$
7. Explain the difference between $(a^b)^c$ and $a^b \cdot a^c$. Refer to stretching and shrinking machines if they help you explain.

Inquiry

Investigation  The Tower of Hanoi

Materials

- 6 blocks, labeled 1, 2, 3, 4, 5, and 6

The Tower of Hanoi is a famous puzzle invented in 1883 by French mathematician Edouard Lucas. Lucas based the puzzle on the following legend.

At the beginning of time, the priests in a temple were given three golden spikes. On one of the spikes, 64 golden disks were stacked, each one slightly smaller than the one below it.

The priests were assigned the task of moving all the disks to one of the other spikes while being careful to follow these rules.

- *Move only one disk at a time.*
- *Never put a larger disk on top of a smaller disk.*

When they completed the task, the temple would crumble and the world would vanish.

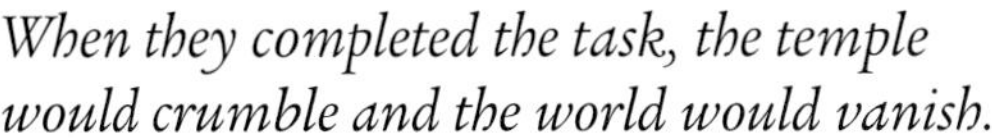

In this investigation, you will figure out how long it would take to move all the disks from one spike to another. We will start by assuming the disks are so large and heavy that the priests can move only one disk per minute.

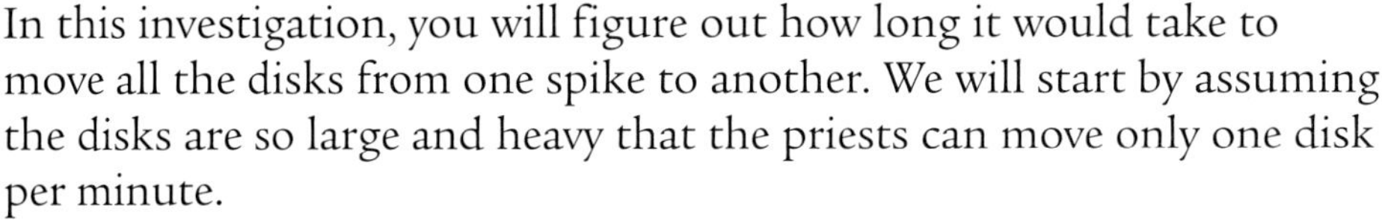

Make a Prediction

Imagine that the spikes are labeled A, B, and C and that the disks start out on spike A. Since it can be overwhelming to think about moving all 64 disks, it may help to first consider a much simpler puzzle. Remember, a larger disk cannot be placed on top of a smaller disk.

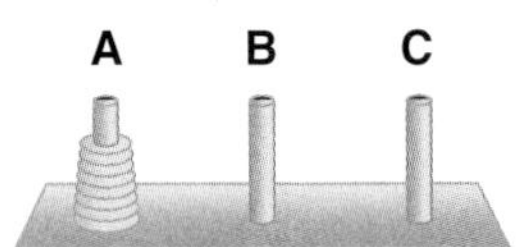

1. Suppose the puzzle started with only one disk on spike A. How long would it take to move the disk to spike B?
2. Suppose the puzzle started with two disks on spike A. How long would it take to move both disks to spike B? What would be the moves?
3. Try again with three disks. How long would it take? What would be the moves?

4. Predict how the total time required to solve the puzzle will change each time you increase the number of disks by one.

5. Predict how long it would take to move all 64 disks. Write down your prediction so you can refer to it later.

Try It Out

Luckily, you do not need 64 golden disks to try the Tower of Hanoi puzzle. You can model it with some simple equipment. Your puzzle will have five "disks," rather than 64. You will need a blank sheet of paper and five blocks, labeled 1, 2, 3, 4, and 5.

Label your paper with the letters A, B, and C as shown. Stack the blocks in numerical order with 5 on the bottom, next to the A.

To solve the puzzle, you need to move all the blocks to another position, either B or C, following these rules.

- Move only one block at a time.
- Never put a larger number on a smaller number.

This is not an easy puzzle. To solve it, you might want to start with a puzzle using only two or three blocks. As you explore, look for a *systematic* way to move all the blocks to a new position.

Try It Again

6. Solve the puzzle again for towers of 1, 2, 3, 4, and 5 blocks. This time, count the number of moves it takes you to solve each puzzle. Record your data in a table.

Tower Height	1	2	3	4	5
Number of Moves					

7. Describe any patterns you see that might help you make predictions about how many moves it would take for larger towers.

8. Use your pattern to fill in a table like the one below. Then use a sixth cube to test your prediction for a tower of height 6.

Tower Height	6	7	8	9	10
Number of Moves					

Go on

9. Write an expression for the number of moves it would take to solve the puzzle for a tower of height t. Add 1 to each entry in the second row of your table for Question 6 and then look at the pattern again.

Back to the Legend

10. Assume that one disk is moved per minute. Figure out how long it would take to solve the puzzle for the heights shown in the table below. Report the times in appropriate units. After a while, minutes are not very useful.

Tower Height	1	2	3	4	5	6	7	8	9	10
Number of Moves										
Time										

11. How long would it take to move all 64 disks? Give your answer in years. How does your answer compare to your prediction in Question 5?

12. You are able to move blocks at a much faster pace than one per minute. What if the disks the priests used were smaller and lighter, so they could also work faster?

    a. If one disk is moved per second, how long would it take to finish the puzzle?

    b. If ten disks are moved per second, how long would it take to finish?

What Did You Learn?

13. When you move a piece in the Tower of Hanoi puzzle, you often have two choices of where to place it. Explain how you decide which move to make.

14. Suppose the legend is true and the priests can move pieces at the incredible rate of ten per second. Do you think they are likely to finish the puzzle in your lifetime? Explain.

15. Write a newspaper article about the Tower of Hanoi puzzle. You might mention the legend and the time it takes to move the disks for towers of different heights.

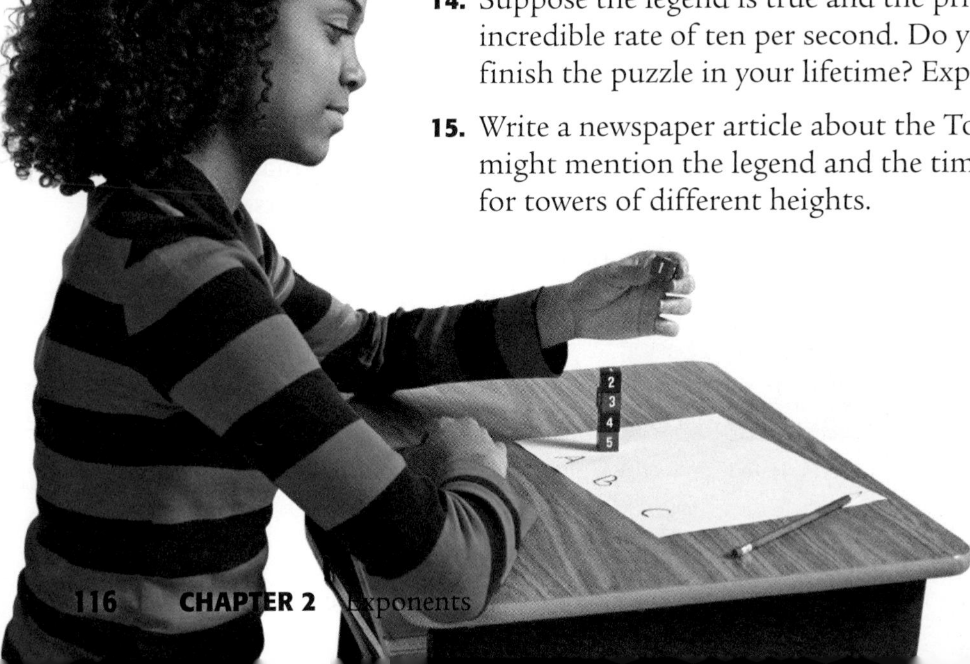

Practice & Apply

A 12-inch ruler is put through each machine. Find its exit length.

1. 

2. ÷ 2

3. 

4. ÷ 3

5. 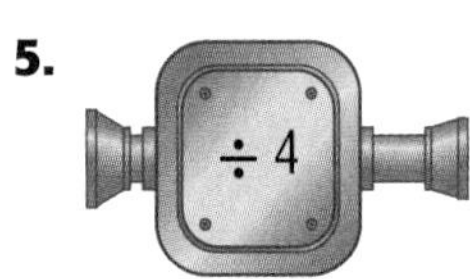

Write each expression using a single base.

6. $2.7^{10} \div 2.7$

7. $0.2^{10} \div 0.2^{10}$

8. $a^{20} \div a^{5}$

9. $\frac{s^{100}}{s}$

10. $\frac{a^{n}}{a^{m}}$

11. $5^{4} \div 5^{2}$

12. Ana was working with this super machine.

a. How many times would Ana need to apply a $\times 3^3$ machine to do the same work as this super machine?

b. How many times would she need to apply a $\times 3$ machine to do the same work as this machine?

c. If Ana puts a one-inch bar of silver through this machine, how long will it be when it exits?

13. Here is Evan's favorite super machine.

a. How many times would Evan need to apply a $\times 10^3$ machine to do the same work as this super machine?

b. How many times would he need to apply a $\times 10$ machine to do the same work?

c. If Evan puts a one-inch bar of gold in this machine, how long will it be when it comes out the machine?

Write each expression as a power of 2.

14. $(4^2)^3$

15. $2^4 \div 4^2$

16. $2^3 \div 2^3$

Write each expression using a single base and a single exponent.

17. $(0.5^2)^2$

18. $(2.5^9)^0$

19. $(1.9^5)^3$

20. $(p^{25})^4$

21. $(p^4)^{25}$

22. $(5^n)^2$

23. $(5^2)^n$

24. $(4n)^3$

25. $(3^2)^3$

26. $(5^2)^2$

27. $(99^1)^5$

28. $(100^0)^5$

Connect & Extend

29. **Preview** Describe two super machines that would stretch a bar of gold to one million times its original length.

Math Link
5,280 feet = 1 mile

Preview A one-inch bar of titanium is put through each super machine. How long will it be when it comes out of the machine? Give your answers in inches, feet, and miles.

30. 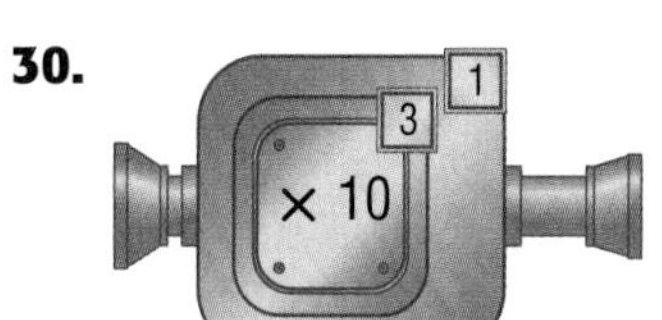

31. 

32. 

33. 

34. How long will the bar of copper be when it exits this connection?

For Exercises 35–49, rewrite each expression using a single base or a single exponent, if possible. For example, $3^7 \div 5^7$ can be rewritten $\left(\frac{3}{5}\right)^7$. You will have to decide which exponent laws to apply. Here are the laws you have seen so far.

Product Laws	Quotient Laws	Power of a Power Law
$a^b \cdot a^c = a^{b+c}$	$a^b \div a^c = a^{b-c}$	$(a^b)^c = a^{b \cdot c}$
$a^c \cdot b^c = (a \cdot b)^c$	$a^c \div b^c = (a \div b)^c$	

35. $7^5 \div 3^5$
36. $7^5 \cdot 5^7$
37. $s^{100} \cdot s$
38. $1{,}000^{14} \div 2^{14}$
39. $6^5 \div 6^2$
40. $100^{100} \div 25^{25}$
41. $(2^0)^3$
42. $10^2 \cdot \left(\frac{1}{2}\right)^2$
43. $6^b \cdot 6^b$
44. $(x^2)^x$
45. $d^2 \cdot d^0$
46. $10^p \div 5^p$
47. $a^n \cdot a^m$
48. $(4x)^2 \div x^2$
49. $(10^2)^3 \div 10^2$

50. In Your Own Words Choose one of the quotient laws of exponents and explain the law so a fifth grader could understand it. You can refer to stretching and shrinking machines if they help you explain.

Challenge Use what you know about exponents to write each expression in a simpler form.

51. $m^3n^5 \cdot m^5n^2$

52. $a^3b^4c \cdot 2ab^3c^2$

53. $14^p \div 7^{2p}$

54. $2y^3z \cdot (6yz^4)^2$

55. $\dfrac{x^2yz^3}{x^3y^2z}$

56. $\dfrac{15m^3n^2}{60k^3m^4n^2}$

Mixed Review

57. Geometry A rectangle has a perimeter of 12 feet. Its length is 4 feet. Find the width.

58. The circumference of a circle is 35 centimeters. Find the diameter. State your answers in terms of π.

59. Statistics Karina kept a record in minutes of the length of each phone call she made last week. Here are the results.

7	37	3	24	29	54	12	18	25
15	19	22	32	35	18	21	15	22

a. Make a histogram of the phone call times. Use ten-minute intervals on the horizontal axis.

b. Describe the distribution of phone call times.

CHAPTER 2

Review & Self-Assessment

Vocabulary

base

common factor

common multiple

composite number

exponent

factor

factor pair

greatest common factor, GCF

least common multiple, LCM

multiple

order of operations

power

prime number

product law of exponents

relatively prime

Chapter Summary

In this chapter, you used the idea of stretching machines to help think about *factors, multiples,* and *prime* and *composite numbers.* You learned to find the prime factorization of a number and to find the *greatest common factor* or *least common multiple* of a group of numbers.

Stretching and shrinking machines also helped you understand the laws of exponents.

Product Laws	Quotient Laws	Power of a Power Law
$a^b \cdot a^c = a^{b+c}$	$a^b \div a^c = a^{b-c}$	$(a^b)^c = a^{b \cdot c}$
$a^c \cdot b^c = (a \cdot b)^c$	$a^c \div b^c = (a \div b)^c$	

Strategies and Applications

The questions in this section will help you review and apply the important ideas and strategies developed in this chapter.

Understanding and applying concepts related to factors and multiples

1. Describe the difference between a prime number and a composite number. Give two examples of each.
2. Explain how to use a factor tree to find the prime factorization of a number. Illustrate your explanation with an example.
3. How can you determine if a number is a multiple of another number? Explain.
4. Lisa and Claudio are discussing whether or not two numbers will always have a common multiple. Lisa says that two numbers always have a common multiple while Claudio says that it is possible for two numbers not to have a common multiple. Who is correct? Explain your reasoning.
5. Is it possible for a number to have fewer multiples than factors? Discuss.
6. Kamilah set her computer's clock to ding every 20 minutes, quack every 30 minutes, and chime every 45 minutes. At noon, her computer dinged, quacked, and chimed at the same time.
   - **a.** In how many minutes will Kamilah's computer ding and quack at the same time?
   - **b.** In how many minutes will her computer quack and chime at the same time?

c. In how many minutes will her computer ding, chime, and quack at the same time?

d. What sounds will the computer make at 4:30 P.M.?

e. How many times per hour will Kamilah's computer clock ding? Quack? Chime?

Strategies and Applications

The questions in this section will help you review and apply the important ideas and strategies developed in this chapter.

Working with stretching and shrinking machines

If a 1-cm stick of gum is sent through each super machine, how long will it be when it exits?

7. 

8. 

Find a single repeater machine to do the same work as each connection.

9. 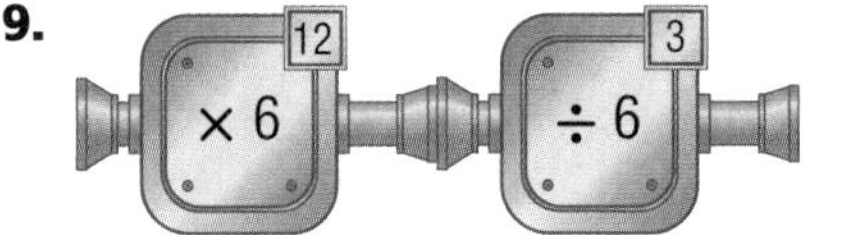

10. 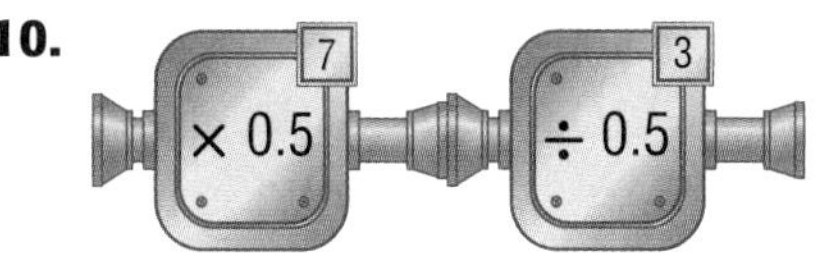

Understanding the laws of exponents

11. Explain why $a^b + a^c = a^{b-c}$. You can refer to stretching and shrinking machines if they help you explain.

12. Explain why $2^0 = 1$. You can refer to stretching and shrinking machines if they help you explain.

13. Explain why $(5^2)^2$ and $5^2 \cdot 5^2$ both equal 5^4 when they each follow a different law of exponents.

Demonstrating Skills

Simplify each expression.

14. $2^3 \cdot 2^4$ **15.** $(7^6)^2$ **16.** $4^2 \cdot 4^2$ **17.** $11^7 + 11^4$ **18.** $\frac{x^{12}}{x^9}$

Tell whether each equation is true or false. If an equation is false, rewrite the expression to the right of the equal sign to make it true.

19. $2^4 \cdot 2^3 = 2^7$ **20.** $(3^3)^3 = 3^6$ **21.** $5^{20} \div 5^4 = 5^{24}$

22. $(7^3)^0 = 7$ **23.** $a^{10} \div a^5 = a^5$ **24.** $3^4 \cdot 3^2 \cdot 3^4 = 3^{32}$

Tell what number should replace each question mark to make each statement true.

25. $4^2 \cdot ?^7 = 4^9$ **26.** $(7^3)^? = 1$ **27.** $\frac{10^5}{10^2} = 10^?$

28. $\left(\frac{1}{4}\right)^2 \cdot ? = \left(\frac{1}{4}\right)^3$ **29.** $(1.2^7)^? = 1.2^{10}$

Rewrite each expression using a single exponent.

30. $15^5 \times 10^5$ **31.** $2^8 \cdot 9^8$ **32.** $\left(\frac{1}{5}\right)^7 \cdot \left(\frac{1}{3}\right)^7$

33. $2{,}000^2 \cdot 4^2$ **34.** $\left(\frac{2}{3}\right)^9 \cdot \left(\frac{2}{5}\right)^9$

Use what you know about exponents to write each expression in simplest form.

35. $a^2bc^3 \cdot ab^4c^2$ **36.** $(2xy^3)^2 \cdot \frac{2}{y^2}$ **37.** $3mn \cdot lm^5n$

38. $\frac{6a^4b^8}{2ab^3}$ **39.** $\frac{7m^4n^4}{21l^2m^3n}$

Determine whether each pair of numbers is relatively prime. For those that are not, give one common factor of the pair.

40. 37 and 42 **41.** 8 and 9 **42.** 51 and 119

43. 18 and 21 **44.** 100 and 101

List the factors of each number.

45. 18 **46.** 64 **47.** 43 **48.** 32 **49.** 121

List three multiples of each number.

50. 7 **51.** 11 **52.** 42 **53.** 80 **54.** 104

Give the prime factorization of each number. Use exponents when appropriate.

55. 246 **56.** 56 **57.** 81 **58.** 150 **59.** 71

Find the greatest common factor and the least common multiple of each pair of numbers.

60. 9 and 12 **61.** 11 and 23 **62.** 14 and 28

63. 6 and 54 **64.** 8 and 20

65. Is 1,392 a multiple of 8? Explain how you know.

66. Is 7 a factor of 227? Explain how you know.

67. List all the common multiples of 3 and 5 which are less than 50.

68. Dalton said, "One of the factors of my number is 18. What other numbers must also be factors of my number?

69. Diana said, "I am thinking of a number. It is less than 50. 9 is one of its 9 factors and the sum of its digits is nine. What is my number?

Tower of Hanoi

70. How might the use of technology make the moving of the tower of disks from one rod to the next become easier to complete?

71. Would the Tower of Hanoi puzzle be possible for you to solve in your lifetime if you moved one disk per minute? Explain your reasoning.

72. Why is the Tower of Hanoi puzzle considered a mathematical exercise?

Test-Taking Practice

SHORT RESPONSE

1 A large tour group is waiting at a bus station for two buses to visit New York City. One bus comes every 12 minutes and a second bus comes every 10 minutes. At the earliest, how many minutes will the group have to wait for both buses to be at the station?

Show your work.

Answer ____________________

MULTIPLE CHOICE

2 Find the GCF of 18 and 24.
A 2
B 3
C 6
D 72

3 Find the LCM of 4, 8, and 16.
F 24
G 16
H 48
J 64

4 Which list contains only prime numbers?
A 5, 21, 33
B 7, 13, 15
C 11, 37, 79
D 27, 31, 63

5 Find the prime factorization of 45.
F $2 \cdot 3 \cdot 5$
G $2 \cdot 2 \cdot 5$
H $3 \cdot 3 \cdot 5$
J $5 \cdot 5 \cdot 3 \cdot 3$

6 Which factorization is a prime factorization?
A $2 \cdot 42 \cdot 5$
B $2 \cdot 3 \cdot 5$
C $3 \cdot 5 \cdot 6$
D $3 \cdot 3 \cdot 9$

7 Find the GCF of 2, 3, and 6.
F 1
G 2
H 3
J 6

CHAPTER

3

Signed Numbers

Real-Life Math

Soaring to New Heights How high is the highest mountain? How deep is the deepest part of the ocean? How are height and depth measured?

We call height and depth *elevation.* We measure elevation from sea level, the average level of the ocean. Parts of the world that are at sea level have an elevation of 0. Denver, the "mile-high" city, is approximately one mile above sea level. Its elevation is approximately one mile. Death Valley is 282 feet below sea level. We say that its elevation is −282 feet.

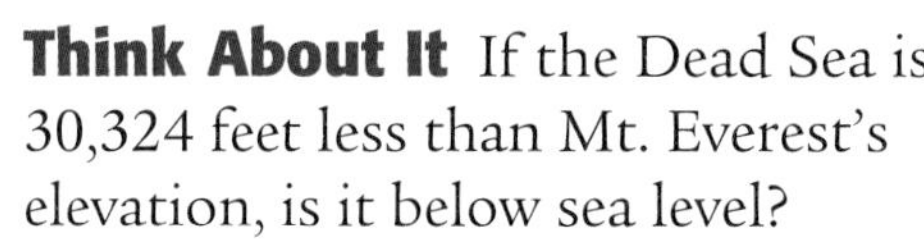

Think About It If the Dead Sea is 30,324 feet less than Mt. Everest's elevation, is it below sea level?

Contents in Brief

3.1 Add and Subtract with Negative Numbers 126

3.2 Multiply and Divide with Negative Numbers 154

Review & Self-Assessment 169

Math Online

Take the **Chapter Readiness Quiz** at glencoe.com.

Dear Family,

In Chapter 3, the class will be working with negative numbers. Students will use numbers that have negative signs in sums, differences, products, quotients, exponents, and graphs.

Key Concept–Addition of Signed Numbers

The class will use a variety of models to represent the addition of signed numbers. Two models are shown below for finding the sum of 3 and -2.

Two-Color Chip Model

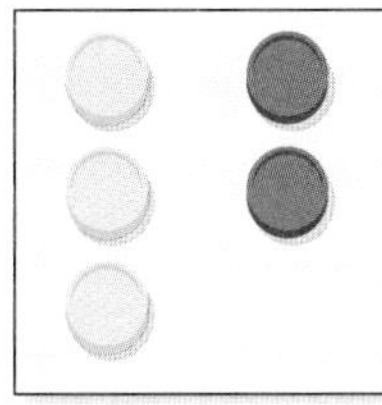

The numbers 3 and -2 are modeled.

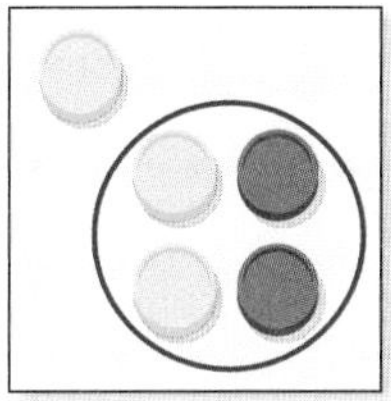

The grouping of 2 and -2 forms a zero pair.

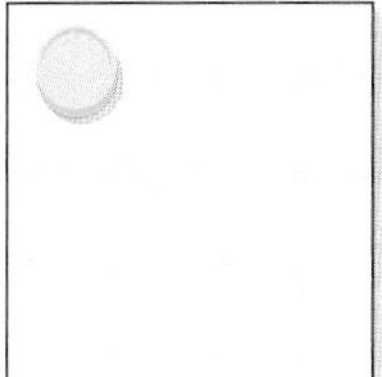

The sum equals 1.

Number-Line Model

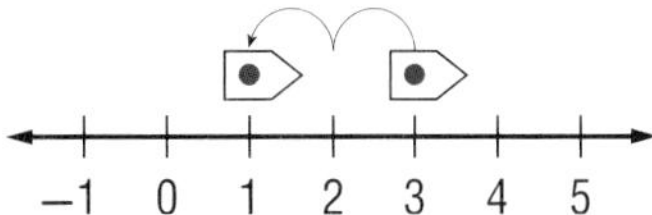

Your student will also use models to simplify expressions involving subtraction, multiplication, and division. In the final investigation, the class will apply these skills to analyze data sets involving signed numbers.

Home Activities

Look for occurrences of negative numbers in real-world situations.

- Elevations below sea level
- The national debt
- Temperatures below zero
- Sports-related situations, such as below par golf scores and football carries resulting in loss of yardage
- Balances of overdrawn bank accounts

Add and Subtract with Negative Numbers

You have probably seen negative numbers used in many situations. Below par golf scores, elevations below sea level, temperatures below 0°, and balances of overdrawn bank accounts can all be described with negative numbers.

Inv 1 The Two-Color Chip Model 128
Inv 2 A Number Line Model 131
Inv 3 Add and Subtract on the Number Line 134
Inv 4 Equivalent Operations 140
Inv 5 Inequalities and Negative Numbers 143
Inv 6 Predict Signs of Sums and Differences 146

Vocabulary

absolute value

Zoe and Darnell have different ways of thinking about negative numbers.

Zoe and Darnell both used positive numbers when they thought about negative numbers. Zoe actually used the *absolute value* of −4. Since −4 and 4 are both four units from 0, both have absolute value 4.

To show the absolute value of a number, draw a vertical line on each side of the number.

$$|4| = 4$$

$$|-4| = 4$$

Math Link

The *absolute value* of a number is its distance from 0 on the number line.

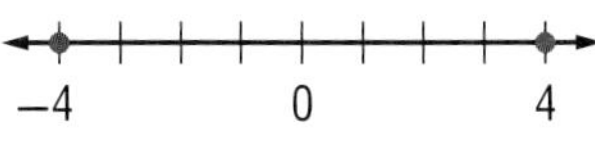

Think & Discuss

These questions will help you review some important ideas about negative numbers.

- Order these numbers from least to greatest.

$$4 \quad -5 \quad 0 \quad -3.5 \quad 4.2 \quad -0.25 \quad 1.75$$

- Express each number using the fewest negative signs possible.

$$-(-2)$$

$$-(-(-(-(7))))$$

$$-(-(-5.5))$$

- How can you tell whether a number is positive or negative by counting how many negative signs it has?
- How can you tell whether $-x$ is positive or negative?
- Explain why the following is true.

$$\text{If } x \text{ is positive, then } |x| = x.$$

- Tell for what numbers the following is true.

$$|x| = -x$$

In this lesson, you will expand your understanding of negative numbers by learning how to add and subtract negative and positive numbers.

Throughout the lesson, do your calculations without a calculator. This will help you better understand why the operations work the way they do.

Investigation 1 The Two-Color Chip Model

Materials

- two-color counters (or dry beans)
- mat

Math Link

Integers are natural numbers, their opposites, and zero.

{..., −3, −2, −1, 0, 1, 2, 3, ...}

Learning the "rules" for adding and subtracting integers is easier when you have a model for what is happening. The model in this investigation involves chips, or counters, that represent the numbers.

Think carefully about how you are using the chips to model each situation. Later, you will not use the chips, but the chip model can become a mental model for helping you think about integers.

Think & Discuss

In any situation, if you add an amount to an existing quantity and then subtract the same amount, you end up with the original quantity. Also, if you subtract an amount and then add the same amount, you end up with the original quantity.

Suppose you have $20. You earn $10 doing yard work. You spend $10 at the movies. How much money do you have? Explain how you know.

Discuss with a partner a situation that involves subtracting an amount and then adding the same amount.

Develop & Understand: A

It is important to be able to make zero when using the chip model to add and subtract positive and negative numbers. One positive chip (yellow) and one negative chip (red) combine to make zero.

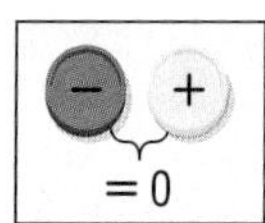

For Exercises 1–6, use a–f below.

a. 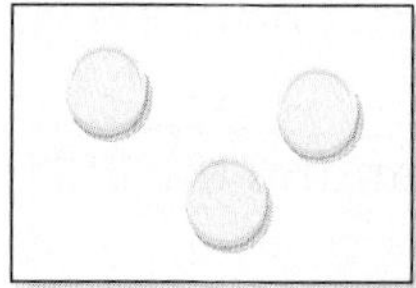

b. 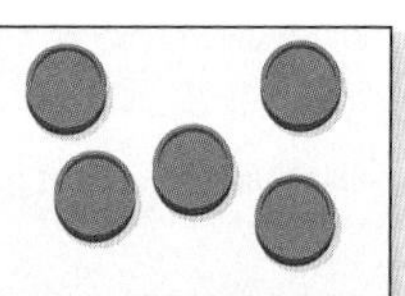

c. 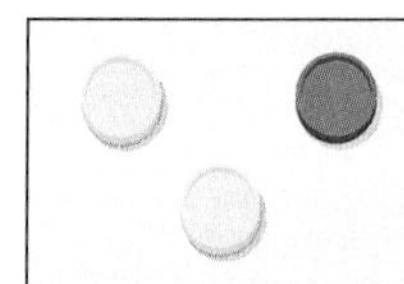

d. 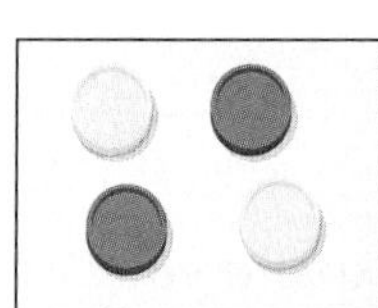

e. 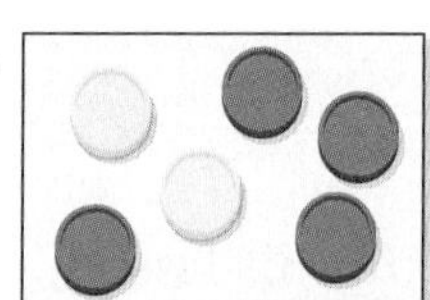

f. 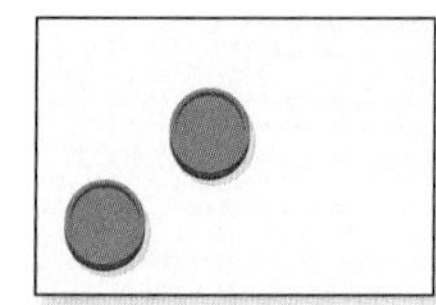

1. Draw each set on your paper. Circle chip pairs that form zeros. Count the remaining chips to find the value of the set.

2. What chips would you add to each set to make the set equal to zero?

3. What chips would you add to each set to write an expression that is equal to 2? For example, for A, $3 + (-1) = 2$. Write an expression for each set.

4. What chips would you add to make each set equal to -1? Explain how you know.

5. What chips would you add to each set to write an expression that is equal to -2? Write an equation for each set.

6. If possible, what chips would you *remove* from each set to make it equal to 1? Draw each set and draw an X through each chip that you would remove.

You can use the chip models to solve addition exercises.

Example

In this example, chip models are used to add -6 and 2.

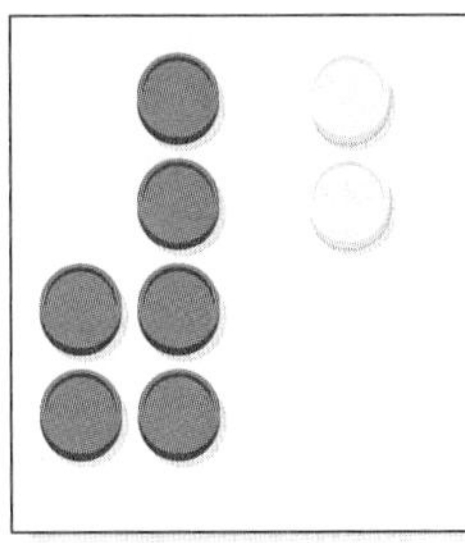

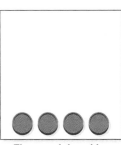

Show original numbers.

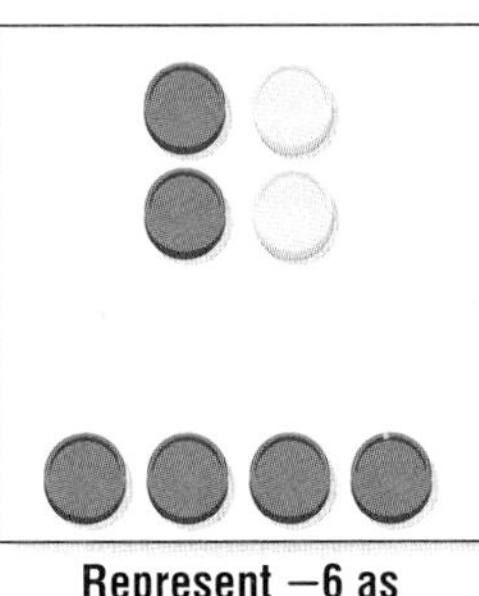

Represent −6 as −4 plus −2. Make a zero pair.

The remaining chips represent the sum.

Step 1.	$-6 + 2$	Write the original numbers.
Step 2.	$[-4 + (-2)] + 2$	Rewrite -6 as -4 plus -2.
Step 3.	$-4 + (-2 + 2)$	Group -2 and 2.
Step 4.	$-4 + 0$	Add -2 and 2 to get 0.
Step 5.	-4	Add -4 and 0.

Real-World Link

With a bank account, a deposit into the account is a positive transaction (credit) while taking money from a bank machine is a withdrawal, or negative transaction (debit).

Develop & Understand: B

7. Use chip models to find each sum. Record each step using an expression.

   a. $7 + (-3)$ b. $-5 + 4$ c. $4 + (-10)$

   d. $-1 + (-8)$ e. $-6 + (-2)$ f. $2 + 9$

8. Write an addition equation for each chip mat.

   a. 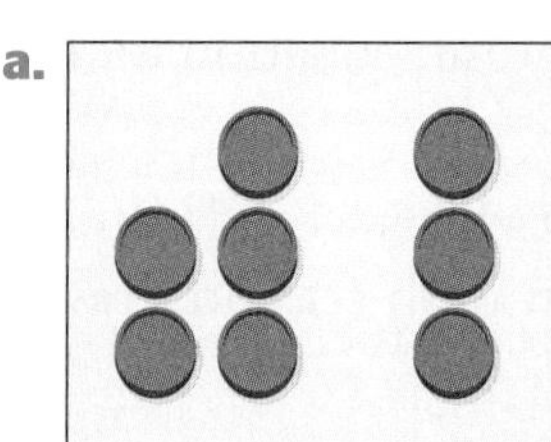 b. 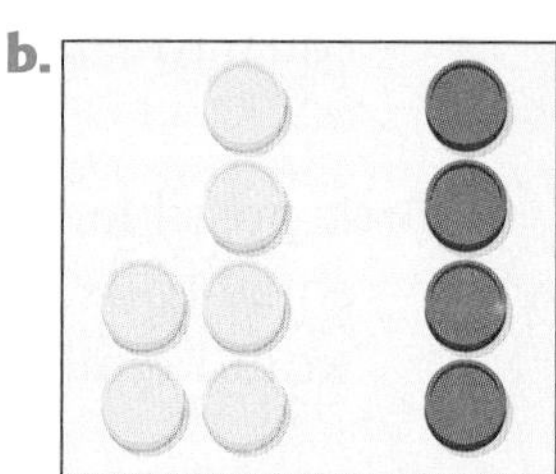 c. 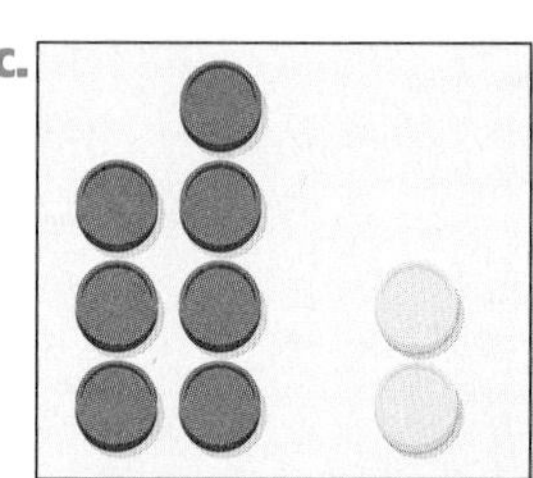

You can also use chip models to solve subtraction exercises. Sometimes it is necessary to add zero pairs.

Example

Use chip models to subtract -2 from 5, or $5 - (-2)$.

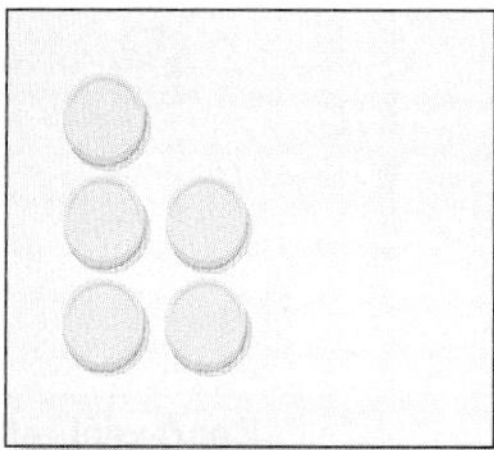

Represent the first number.

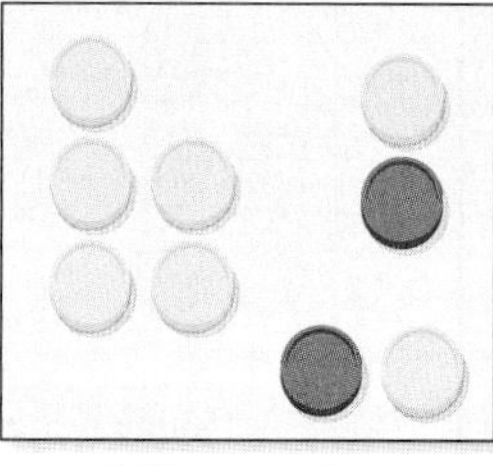

Add zero pairs.

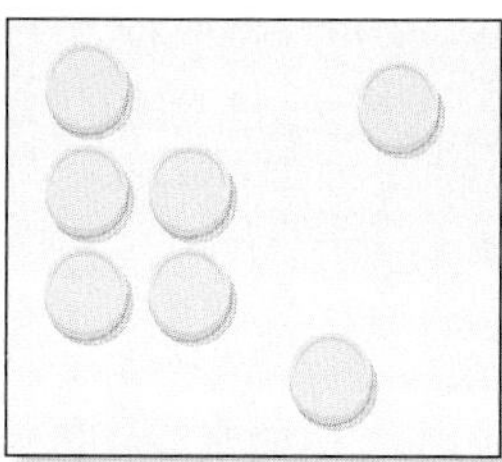

Remove the second number.

Step 1. 5 Write the first number.

Step 2. $5 + [2 + (-2)]$ Add 0 in the form of 2 and -2.

Step 3. $5 + 2 + (-2)$ Subtract -2.

Step 4. 7 Add 5 and 2.

Develop & Understand: C

9. Use chip models to model each expression. Then solve.

   a. $-5 - (-3)$ b. $2 - 8$ c. $-5 - 4$

10. How many zero pairs can you add to any number and not change its value?

Share & Summarize

Explain how you can you use a two-color chip model to subtract $3 - (-5)$.

Inquiry

Investigation 2 A Number Line Model

Materials

- a copy of the plank
- several blank sheets of paper
- a paper arrow

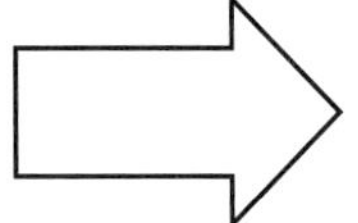

- a direction cube with 3 faces labeled *Ship* and 3 labeled *Shark*
- a walking cube with faces labeled *F1, F2, F3, B1, B2,* and *B3*

Imagine that you are on the crew of a pirate ship commanded by a very generous captain. He has decided to give victims who must walk the plank a chance to survive. He has formulated a game that he can play with his victims, but he has asked your help in testing and revising the game.

The Captain's Game

The captain has set up an eight-foot plank that extends from the ship toward the water. He has drawn lines on the plank at one-foot intervals.

The victim starts by standing in the center of the plank, with four feet in front of him or her and four feet behind him or her. To decide where the victim should move, the captain rolls two cubes, a *direction* cube and a *walking* cube.

Direction Cube If the captain rolls *Ship,* the victim faces the ship. If he rolls *Shark,* the victim faces the sharks in the water.

Walking Cube If the captain rolls *F1, F2,* or *F3,* the victim must walk forward 1, 2, or 3 steps. If he rolls *B1, B2,* or *B3,* the victim must walk backward 1, 2, or 3 steps.

Go on

The captain continues to roll the cubes until the victim walks off the edge of the plank, either into the shark-infested water or onto the safety of the ship's deck.

Try It Out

Play the game with a partner using a drawing of the plank and a paper arrow. Take turns playing the roles of the captain and the victim. For each game, keep track of the number of times the cubes are rolled before the victim falls to the sharks or reaches safety. Then answer the questions.

1. Did the victim in each game make it to the ship, or was he or she forced to jump into the water?
2. How many moves did it take to complete each game?

Make a Prediction

The captain is thinking about changing the length of the plank. If the plank is too short, games will not last long enough to entertain him. If the plank is too long, he will become bored with the game. Your challenge is to determine a plank length that results in an interesting game that is neither too short nor too long.

3. To keep the game interesting, what do you think the ideal number of moves should be?
4. How long do you think the plank should be so that the *average* number of moves is about the right number?

Check Your Prediction

Math Link

To find the average, add up the total number of moves in all trial games. Then divide by the number of trial games.

One way to check your prediction is to collect data. Play at least three games with the plank length that you specified. Record the number of moves for each game. Then find the average number of moves for all the games that you played.

5. Is the average number of moves close to the number you specified in Question 3?
6. If the plank length that you predicted is too long or too short, adjust it. Play the game at least three times with the new length. Continue adjusting the plank length until the average number of moves for a particular length is about what you recommended in Question 3.

You might put the results of your games in a table.

Plank Length	Number of Moves			Average Number of Moves
	Game 1	Game 2	Game 3	

What Did You Learn?

7. In one round of the game, the victim moved two steps in the direction of the ship. What might the captain have rolled when he rolled the two cubes? List as many possibilities as you can.
8. Write a letter to the captain giving him your recommendation. Include the following information in your letter.
   - the number of moves you think is ideal
   - how long the plank should be
   - how you know that this plank length will give your ideal number of moves

Investigation 3 Add and Subtract on the Number Line

Materials

- number line and pointer

In earlier grades, you may have learned to add and subtract positive numbers on the number line. For example, to find $2 + 3$, you start at 2 and move three spaces to the right. The number on which you end, 5, is the sum.

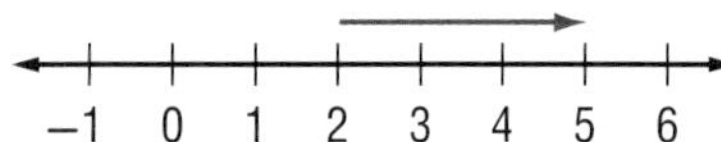

To find $6 - 4$, start at 6 and move four spaces to the left.

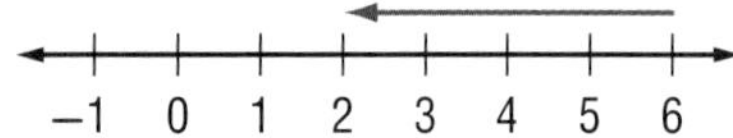

In a similar way, you can "walk the number line" to calculate sums and differences involving negative numbers.

Before you walk the number line, create a large number line that goes from -9 to 9 and a pointer that looks like the one below.

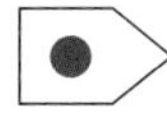

Once you have created your number line and pointer, you are ready to walk the number line.

Here is how to find the solution to $3 + (-2)$ by using your number line.

- Place your pointer, point up, on 3.

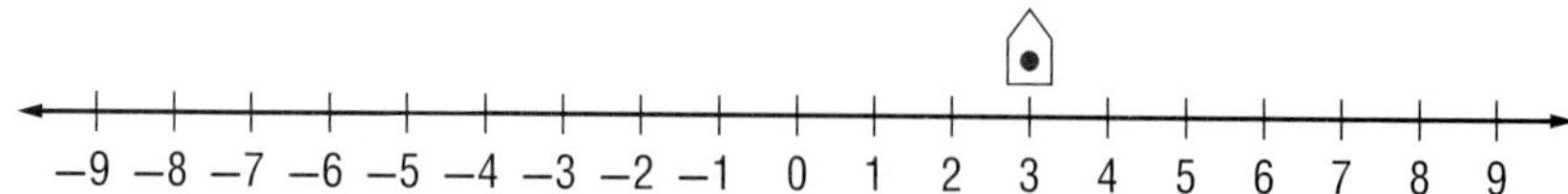

- Look at the operation. Since it is *addition,* face your pointer in the *positive direction,* to the right.

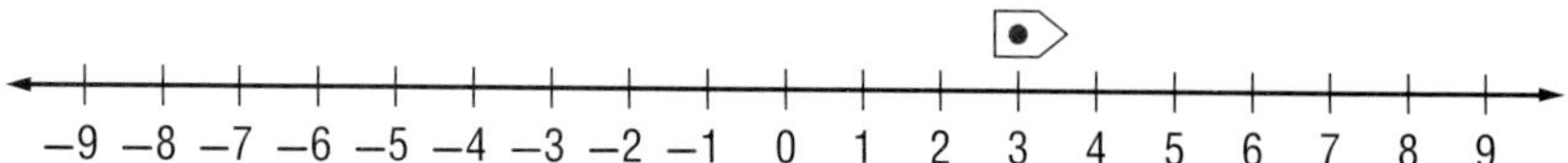

- Look at the number being added, -2. Since it is *negative,* move the pointer *backward,* opposite the direction the pointer is facing, two spaces. The pointer ends on 1, which means the answer is 1. The complete addition equation for this sum is $3 + (-2) = 1$.

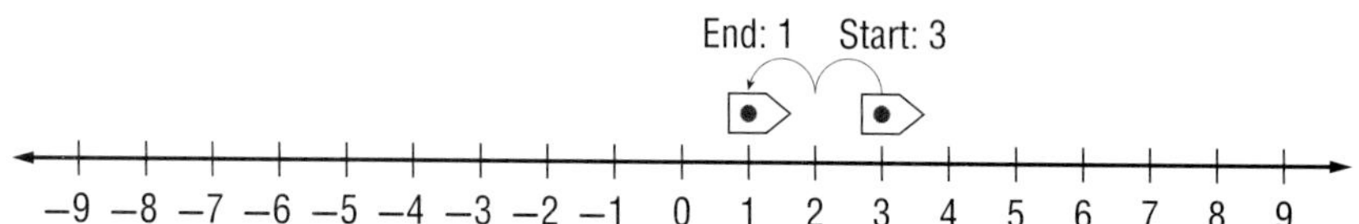

Next, you will consider a subtraction example, $-2 - 6$.

- Place your pointer, point up, on -2.

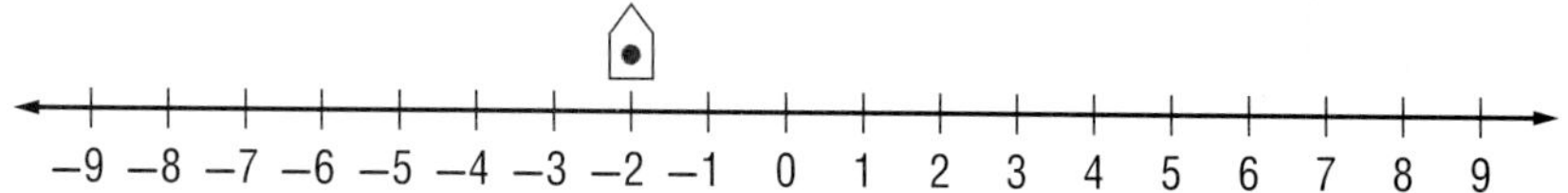

- Look at the operation. Since it is *subtraction,* face your pointer in the *negative direction,* to the left.

- Look at the number being subtracted, 6. Since it is *positive,* move the pointer *forward,* in the direction the pointer is facing, six spaces. The answer is -8. The complete subtraction equation for this difference is $-2 - 6 = -8$.

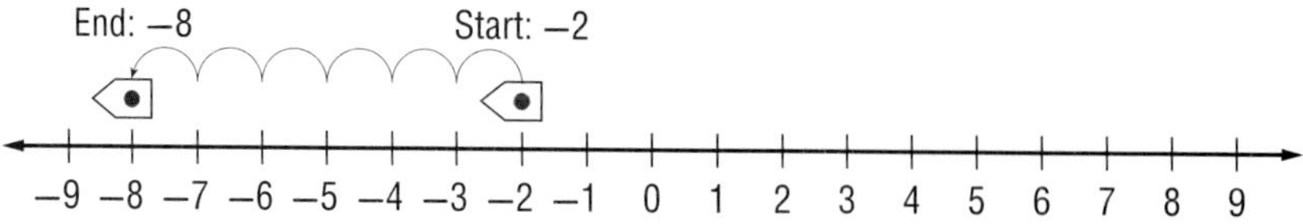

When you walk the number line, the operation determines the direction the pointer faces.

- If the operation is addition, face the pointer to the right in the positive direction.
- If the operation is subtraction, face the pointer to the left in the negative direction.

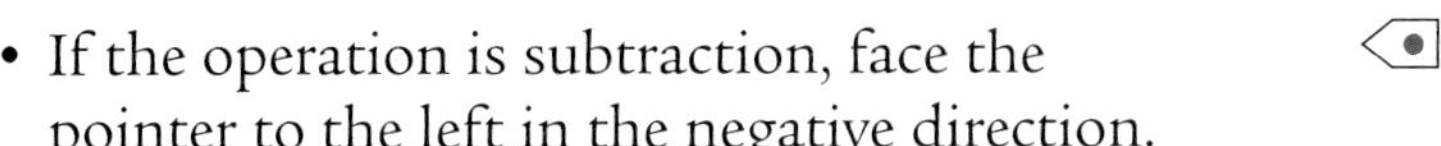

The sign of the number being added or subtracted determines the direction to move the pointer.

- If the number is positive, move the pointer forward, in the direction it is pointing.

or

- If the number is negative, move the pointer backward, opposite the direction it is pointing.

or

Think & Discuss

The number lines below show the steps for finding $5 + (-3)$. Explain each step.

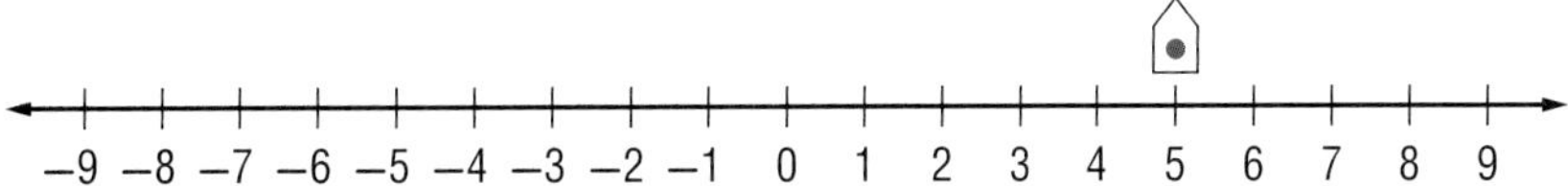

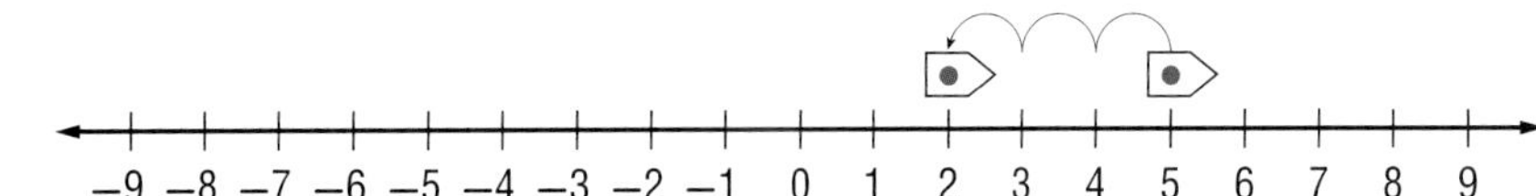

Now you will have an opportunity to compute some sums and differences with the number line.

Develop & Understand: A

For Exercises 1–4, walk the number line to find each sum or difference. Describe your steps in words. Be sure to include the following in your descriptions.

- your starting position on the number line
- the direction the pointer is facing
- whether you move forward or backward and how far you move
- your ending position on the number line, which is the answer
- the complete equation that represents your moves on the number line

1. $5 + (-2)$
2. $-4 + (-3)$
3. $3 - (-2)$
4. $-8 - (-2)$

Estimate each sum or difference. Then, find each sum or difference by walking the number line.

5. $3 + (-5.25)$
6. $5 - 6\frac{1}{3}$
7. $-6 - 2.23$
8. $-\frac{2}{3} - \left(-1\frac{1}{3}\right)$

9. When Bret woke up one morning, the temperature was $-15°$F. By noon, the temperature had risen $7°$F. What was the temperature at noon?

10. Colleen was hiking in Death Valley, parts of which are at elevations below sea level. She began her hike at an elevation of 300 feet and hiked down 450 feet. At what elevation did she end her hike?

You can make drawings to record your moves on the number line. Once you get comfortable walking the number line, you may be able to add and subtract just by making a drawing. Your drawings can show very easily the steps you take to find a sum or a difference.

Example

Make a drawing to record the steps for finding $-3 - (-7)$.

First, draw a pointer over -3. Since you are subtracting, draw the pointer so that it faces left. Make sure the dot on the pointer is directly over -3.

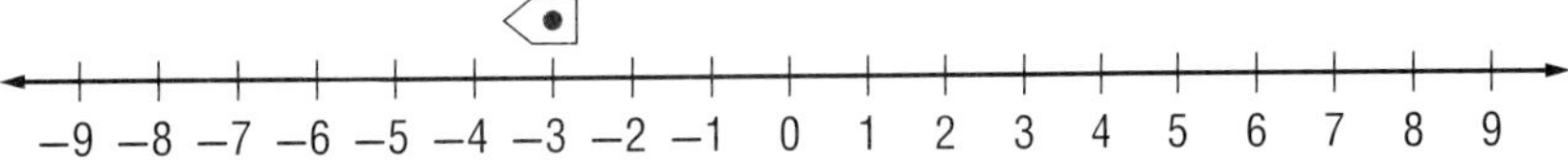

Since the number you are subtracting is -7, move backward seven spaces. Draw an arrow from the pointer to the ending number.

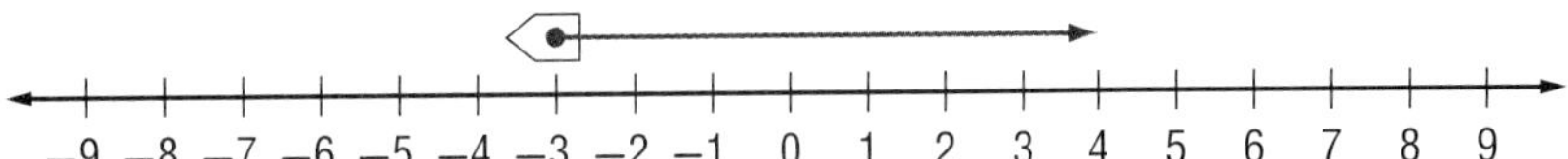

Finish your drawing by adding some labels. Write *Start* and the number where you started above the pointer. Write *End* and the number where you ended above the arrowhead.

Adding labels is especially important when you find sums and differences involving fractions and decimals because it is hard to show the numbers precisely on the number line.

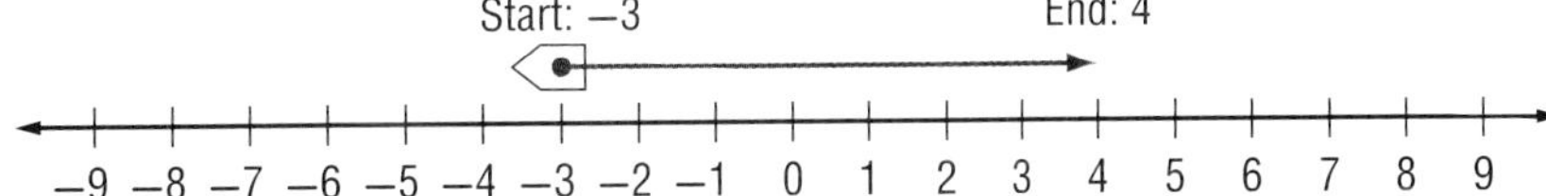

Notice that the final number-line drawing captures all your moves. It shows the number at which you began, the direction of your pointer, whether you moved forward or backward, how far you moved, and your ending number.

Develop & Understand: B

Walk the number line to compute each sum or difference in Exercises 11–16. Record your steps by making a number-line drawing. Write the complete addition or subtraction equation next to each drawing.

11. $-3 + (-2)$

12. $5 - 8$

13. $-5 - (-2)$

14. $-5 + 7$

15. $2 + (-7.2)$

16. $2\frac{2}{3} - \left(-6\frac{1}{3}\right)$

Write an addition or subtraction equation for each drawing.

17.

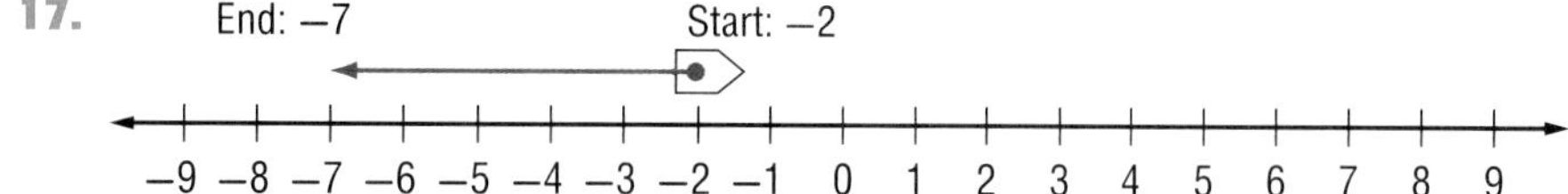

18.

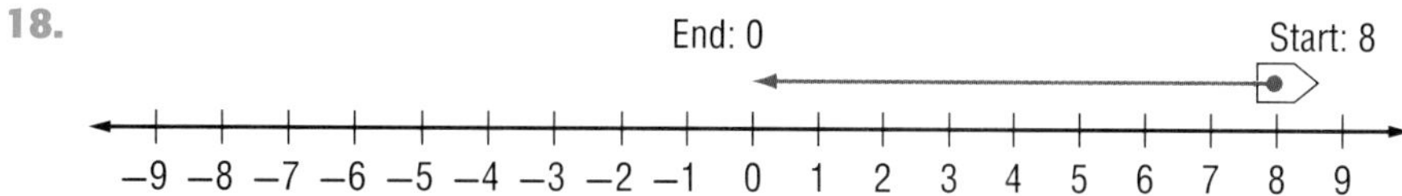

19.

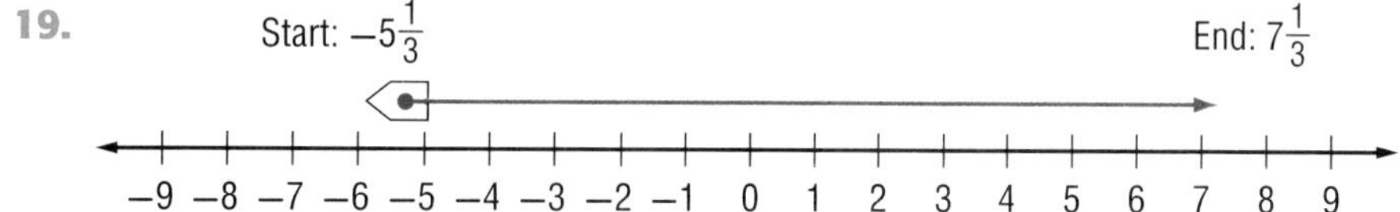

Share & Summarize

Write one addition equation and one subtraction equation. On a separate sheet of paper, make a number-line drawing for each.

Trade drawings with your partner. Write equations that match your partner's drawings. Check with your partner to make sure the equations match the drawings.

Investigation 4 Equivalent Operations

You have practiced writing addition and subtraction equations to match number-line drawings. For each complete number-line drawing, there is only one equation that matches.

What if a number-line drawing shows only the starting and ending values and not the pointer or the arrow?

What number-line drawings start at -3 and end at 2?

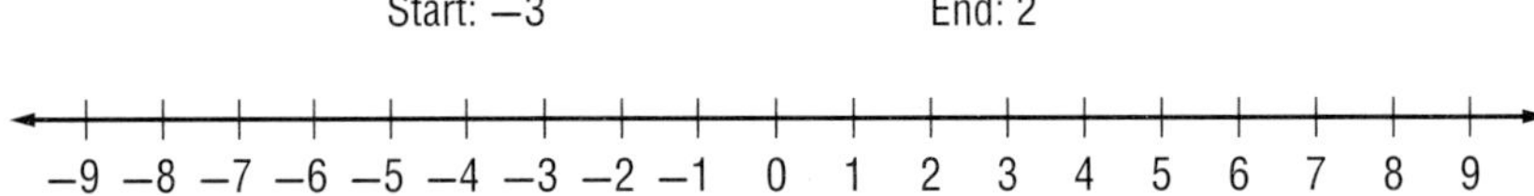

What equation does each number-line drawing represent?

In Exercises 1–6, you will use number-line drawings to explore the relationship between addition and subtraction.

Develop & Understand: A

In Exercises 1–6, make two copies of each number line. Complete one to represent an addition equation and the other to represent a subtraction equation. Write the equations your drawings represent.

1.

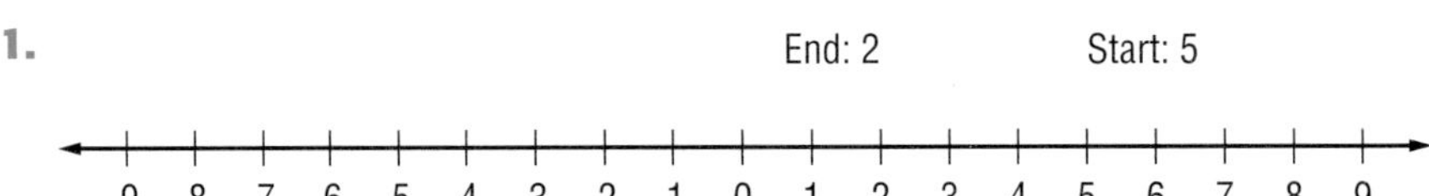

2.

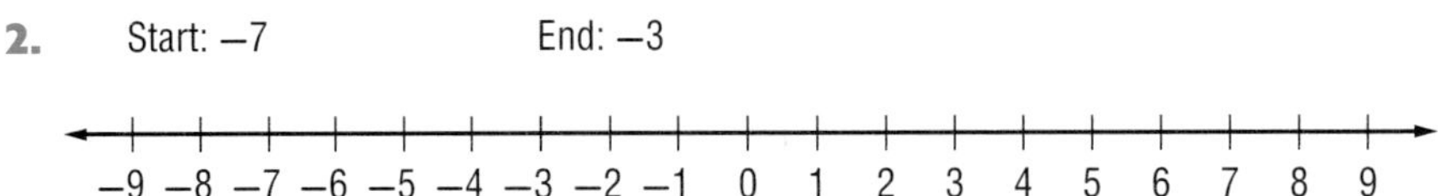

3.

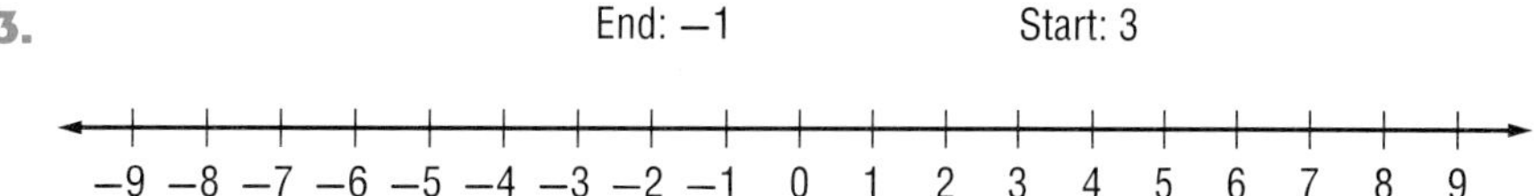

4.

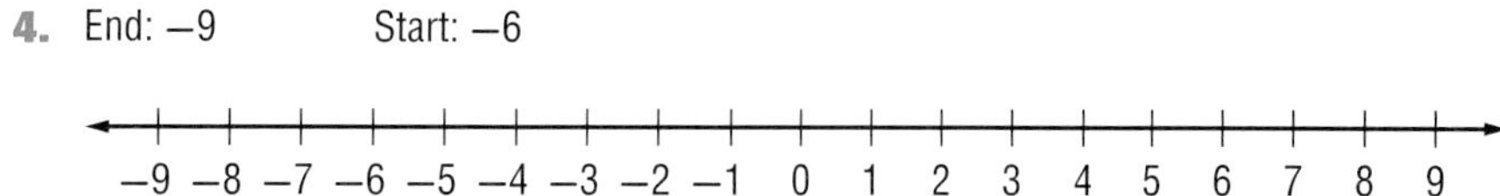

5.

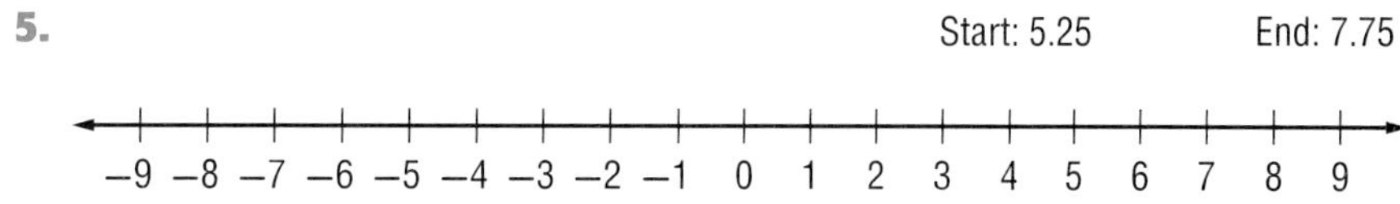

6.

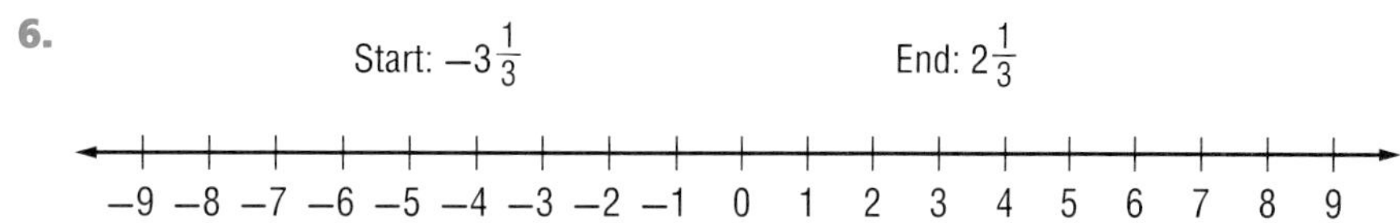

Think & Discuss

Describe the two ways that you can move from a start point to an end point when you walk the number line.

What other move on the number line accomplishes the same thing as facing the negative direction and moving forward?

What other move accomplishes the same thing as facing the positive direction and moving forward?

In Exercises 1–6, you wrote two equations that started and ended at the same place on the number line.

Since you started with the same number and found the same result after adding or subtracting, you were really doing *equivalent* operations.

Develop & Understand: B

Compute each difference. Write your answer in the form of a subtraction equation. Then write an addition equation that is equivalent to the subtraction equation.

7. $2 - 5$

8. $-2 - (-3.2)$

Compute each sum. Write your answer in the form of an addition equation. Then write an equivalent subtraction equation.

9. $-6 + (-5)$

10. $-1 + 7\frac{1}{4}$

In Exercises 11 and 12, write an addition and a subtraction equation that gives the answer to the question.

11. When Sawyer went to sleep, the temperature outside was $-2°F$. The temperature dropped $12°F$ overnight. What was the temperature when Sawyer awoke?

12. **Challenge** Rosa was hiking near the Dead Sea. She began her hike at an elevation of -600 feet and ended the hike at 92 feet. What was the change in her elevation?

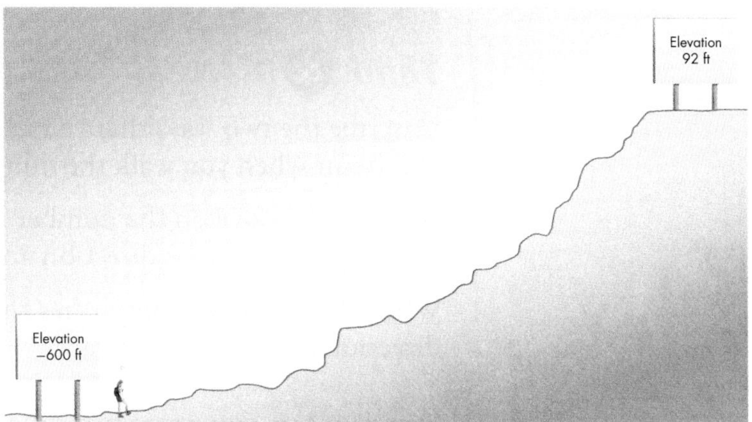

Share & Summarize

Complete each sentence. The first is done for you.

1. Adding a negative is equivalent to subtracting a positive.
2. Subtracting a negative is equivalent to ________________.
3. Adding a positive is equivalent to ________________.
4. Subtracting a positive is equivalent to ________________.
5. Using the number line or anything else you would like, explain why the statements above are true.

Investigation 5 Inequalities and Negative Numbers

Just as with equations, signed numbers can be used with inequalities.

In the next two exercise sets, you will add and subtract with negative numbers as you solve inequalities.

Math Link

An *inequality* is a mathematical statement that one quantity is greater than or less than another. Inequalities use the following symbols.

$>$ greater than

$<$ less than

These symbols indicate that quantities may be equal.

$\geq$ greater than or equal to

$\leq$ less than or equal to

Develop & Understand: A

Tell what number you could put in each blank to make a sum of 0.

1. $3 + \_\_\_\_$
2. $-7.2 + \_\_\_\_$
3. Look at your answers to Exercises 1 and 2.
   a. If you have a positive number, what number could you add to it to get a sum of 0?
   b. If you have a negative number, what number could you add to it to get a sum of 0?

Give one number in each blank to make a sum less than 0.

4. $-6 + \_\_\_\_$
5. $-4.3 + \_\_\_\_$
6. $\frac{1}{2} + \_\_\_\_$
7. $3 + \_\_\_\_$
8. In Exercises 4–7, how did you figure out what number to put in the blank?
9. In Exercise 1, you found a number that gives a sum of 0 when it is added to 3. This is the same as solving the *equation* $3 + x = 0$. In Exercise 7, you found a number that gives a sum less than 0 when it is added to 3. To solve the *inequality* $3 + x < 0$, though, it is not enough to find a single number.
   a. Find three more solutions of the inequality $3 + x < 0$. Draw a number-line picture for each sum.
   b. Describe *all* the x values that are solutions of $3 + x < 0$. It may help to think about walking the number line.

10. Now think about what you might do if you wanted to solve an inequality related to each of Exercises 4–7.

a. What number could you add to a given positive number to get a sum less than 0?

b. What number could you add to a given negative number to get a sum less than 0?

Develop & Understand: B

Tell what number you could put in each blank to make a difference of 0.

11. $\frac{2}{3}$ − _____

12. −3 − _____

13. Look at your answers to Exercises 11 and 12.

a. If you have a positive number, what number could you subtract from it to get a difference of 0?

b. If you have a negative number, what number could you subtract from it to get a difference of 0?

Give one number in each blank to make a difference greater than 0.

14. 7 − _____

15. −2.4 − _____

16. $\frac{3}{4}$ − _____

17. 5 − _____

18. In Exercises 14–17, how did you figure out what number to put in the blank?

19. In Exercise 17, you found a number that, when subtracted from 5, gives a difference greater than 0. That is, you found one solution of the inequality $5 - x > 0$.

a. Find three more solutions of the inequality $5 - x > 0$. Draw a number-line picture for each difference.

b. Describe all the solutions of the inequality $5 - x > 0$. It might help to use a number-line picture.

20. Now think about what you might do if you wanted to solve an inequality related to each of Exercises 14–17.

a. What number could you subtract from a given positive number to get a difference greater than 0?

b. What number could you subtract from a given negative number to get a difference greater than 0?

Consider the following expressions.

Expression 1: $\boxed{n}$ Expression 2: $\boxed{5n + 8}$

For what value of n does Expression 1 equal Expression 2?

For what value(s) of n is Expression 1 greater than Expression 2?

For what value(s) of n is Expression 1 less than Expression 2?

Develop & Understand: C

21. For each pair of expressions, determine the values that make the following statements true.

a. Expression 1 is greater than Expression 2.

b. Expression 1 is equal to Expression 2.

c. Expression 1 is less than Expression 2.

Expression 1	Expression 2
$3a + 6$	15
$2y$	$y + 7$
$3k + 8$	-13
$5m + 1$	$3m + 9$

Share & Summarize

Without doing the calculation, tell whether each sum or difference is positive or negative. Explain how you know.

1. $-25 + 36$

2. $-53 - (-14)$

3. $45 - 87$

4. $123 - (-220)$

5. Choose two of the previous expressions. Find the sum or difference. Explain your steps so that someone who is not in your class could understand them. You may refer to the number-line model or another method that makes sense to you.

Investigation 6 Predict Signs of Sums and Differences

Look over your answers to the Share & Summarize on the previous page. Can you always use your strategy to predict whether a sum will be positive or negative? If you can, you have a way to check that sums and differences have the correct sign.

Develop & Understand: A

1. If possible, give an example in which the sum of a positive number and a negative number is negative. If it is not possible, explain why.
2. If possible, give an example in which the difference between two negative numbers is positive. If it is not possible, explain why.
3. If possible, give an example in which the sum of two negative numbers is positive. If it is not possible, explain why.

Positive and negative numbers can be combined in sums and differences in eight ways.

- positive + positive
- positive + negative
- negative + negative
- negative + positive
- positive − positive
- positive − negative
- negative − negative
- negative − positive

You will now explore each of these combinations.

Develop & Understand: B

Figure out whether the sums or differences in each category are *always negative; always positive;* or *sometimes positive, sometimes negative, and sometimes 0.* Explain how you know you are correct.

4. positive + positive
5. positive + negative
6. negative + negative
7. negative + positive
8. positive − positive
9. negative − positive
10. positive − negative
11. negative − negative

12. Look at your work from Exercises 4–11.

a. Which combinations produce sums or differences that are sometimes negative, sometimes positive, and sometimes 0?

b. Choose one of the combinations from Part a. Without doing the calculation, figure out how you can tell whether a given sum or difference involving that combination will produce a negative number, a positive number, or 0.

Share & Summarize

1. Suppose the sum of two numbers is less than 0.

a. Could both numbers be negative? If so, give an example. If not, explain why not.

b. Could both numbers be positive? If so, give an example. If not, explain why not.

c. Could one number be negative and one positive? If so, give an example. If not, explain why not.

2. Suppose the difference between two numbers is greater than 0.

a. Could both numbers be positive? If so, give an example. If not, explain why not.

b. Could both numbers be negative? If so, give an example. If not, explain why not.

c. Could one number be negative and one positive? If so, give an example. If not, explain why not.

On Your Own Exercises

Lesson 3.1

Practice & Apply

In Exercises 1–5, draw chip models to show how to find each sum or difference.

1. $-2 + 7$ **2.** $3 - 11$ **3.** $8 - (-3)$ **4.** $9 + (-4)$ **5.** $-5 - 6$

6. Use the chip board below for Exercises a–c.

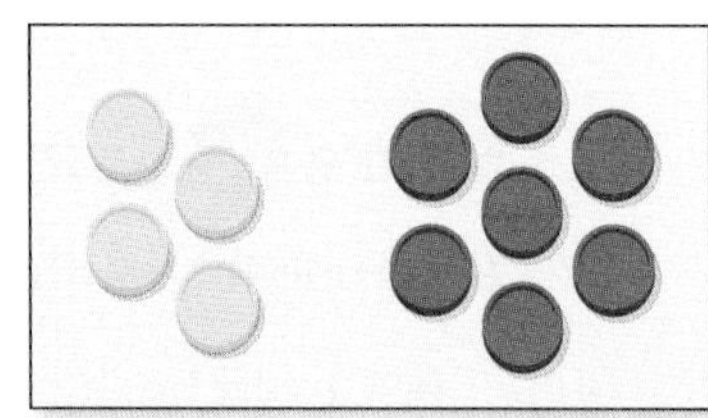

a. What is the value of the board?

b. Explain two ways to change the numbers of chips on the board without changing the value shown on the board.

c. Describe a real-life problem this board could represent.

In Exercises 7–9, walk the number line to find each sum or difference. Describe your steps in words. Include the following in your descriptions.

- where on the number line you start
- the direction the pointer is facing
- whether you move forward or backward and how far you move
- the point on the number line at which you end, which is the answer
- the complete addition or subtraction equation

7. $-8 + 5$ **8.** $8 + (-4)$ **9.** $3 - (-5)$

Walk the number line to compute each sum or difference.

10. $-6 - 3$ **11.** $-2 + (-4)$

12. $2 + (-11)$ **13.** $-9 - (-5)$

14. $4 - 9$ **15.** $-6 + 14$

16. $-5 - (-11)$ **17.** $12 - (-10)$

Estimate each sum or difference. Then, walk the number line to compute each sum or difference.

18. $2\frac{1}{5} - \left(-3\frac{2}{5}\right)$ **19.** $-8.25 + 1.75$

20. $-\frac{13}{2} - \left(-\frac{13}{4}\right)$ **21.** $4.9 - (-3.2)$

Walk the number line to compute each sum or difference. Record your steps by making a number-line drawing. Write the complete addition or subtraction equation next to the drawing.

22. $-2 + (-7)$

23. $6 - (-2)$

24. $5 - 7$

25. $-3 + 11$

26. $2.2 - 1.4$

27. $-5\frac{1}{3} + 2\frac{5}{6}$

In Exercises 28–31, write the addition or subtraction equation represented by the drawing.

28.

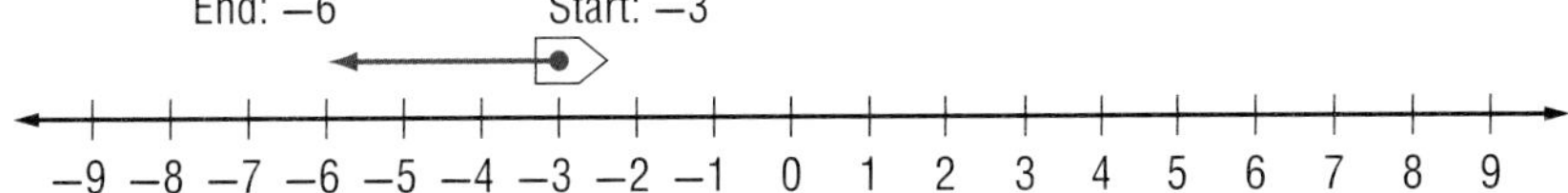

29.

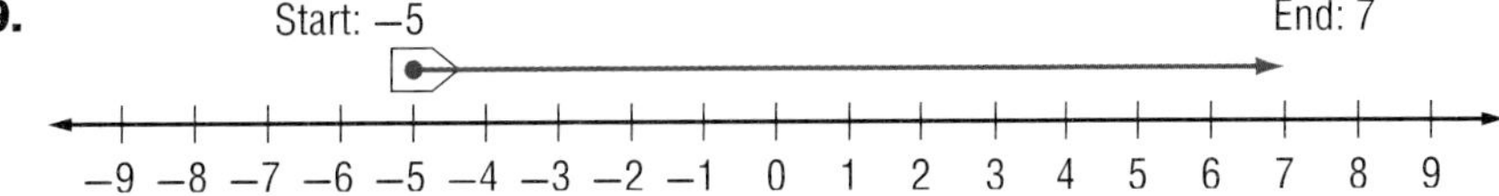

30.

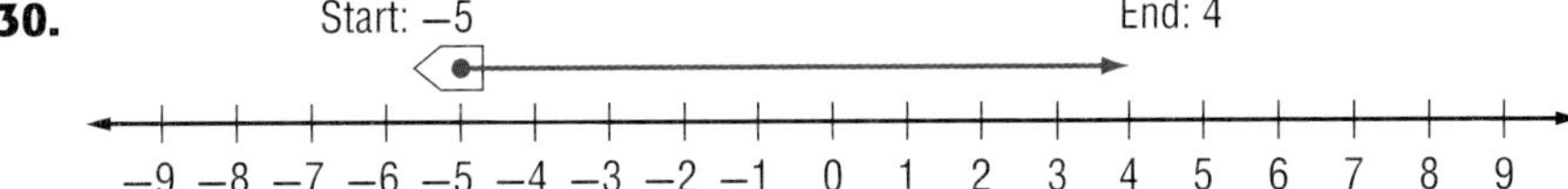

31.

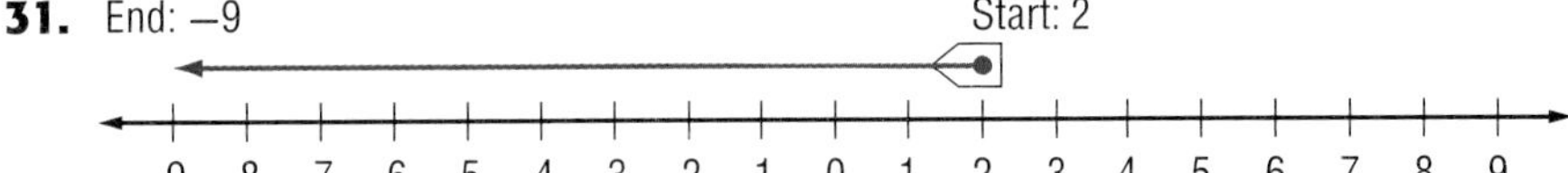

32. The sum of two numbers is -8. What could the numbers be? Give three possibilities.

Write an addition equation and a subtraction equation for each situation. You might want to make a drawing.

33. Start value: -7

End value: 9

34. Start value: $2\frac{1}{7}$

End value: $-3\frac{3}{7}$

Compute each difference. Write your answer in the form of a subtraction equation. Then write an equivalent addition equation.

35. $25.2 - (-3.4)$

36. $-43 - 18$

Compute each sum. Write your answer in the form of an addition equation. Then write an equivalent subtraction equation.

37. $-6 + (-9)$

38. $-21.8 + 17.4$

39. Expression 1 equals $5x$. Expression 2 equals $3x + 8$. Find values for x such that:

- Expression 1 is greater than Expression 2
- Expression 1 equals Expression 2
- Expression 1 is less than Expression 2

40. Algebra In this exercise, solve the inequality $-5 - y < 0$.

a. Find three positive values of y that make the inequality true.

b. Find three negative values of y that make the inequality true.

c. Describe *all* the values of y that make the inequality true.

41. If possible, give an example in which a negative number is subtracted from a positive number and the result is negative. If it is not possible, explain why.

42. Number Sense The sum of two numbers is greater than 0.

a. Could both numbers be negative? If so, give an example. If not, explain why not.

b. Could both numbers be positive? If so, give an example. If not, explain why not.

c. Could one number be negative and one positive? If so, give an example. If not, explain why not.

43. The difference between two numbers is less than 0.

a. Could both numbers be positive? If so, give an example. If not, explain why not.

b. Could both numbers be negative? If so, give an example. If not, explain why not.

c. Could one number be negative and one positive? If so, give an example. If not, explain why not.

Connect & Extend

Complete each computation by walking the number line. Explain each step you take.

44. $3 + (-4) + (-2) - 5$

45. $-2 - 9 + 2 - 4 - (-15)$

46. Statistics The chart shows the high temperature, in degrees Fahrenheit, for the first ten days of the year in a small Alaskan town. What is the average of these temperatures?

Date	1/1	1/2	1/3	1/4	1/5	1/6	1/7	1/8	1/9	1/10
Temp (°F)	3	1	13	5	2	2	−4	−7	−2	−2

47. Lazaro said, "If you add -3 to my sister's age, you will get my age." Lazaro is 12. How old is his sister?

Without calculating each sum or difference, figure out whether it is less than 0, greater than 0, or equal to 0.

48. $-8 - 9$

49. $-3 - (-\pi)$

50. $-78 + 2^6$

51. Copy and complete the flowchart.

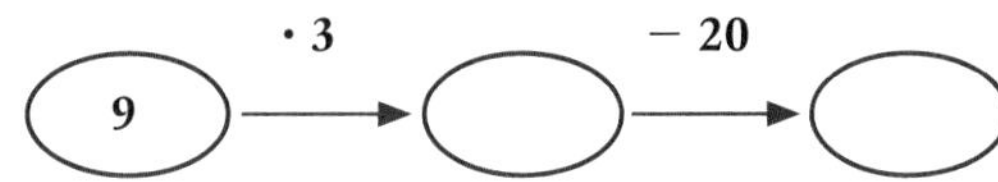

52. *Integers* are counting numbers, their opposites, and zero ... , $-3, -2, -1, 0, 1, 2, 3, \ldots$. The sum of two negative integers is -6.

a. What could the integers be? List all the possibilities.

b. Find two more pairs of negative numbers (not integers) whose sum is -6.

Algebra Use the distributive property to rewrite each expression as simply as you can.

53. $6x - (-2x)$

54. $3z - 5z$

55. $7x + (-2x)$

Math Link

The *absolute value* of a number, its distance from 0, is represented with vertical lines. $|5| = 5$ because 5 is 5 units from 0. $|-7| = 7$ because -7 is 7 units from 0.

56. Prove It! Marta thinks that for all values of x and y, the following equation is true.

$$|x + y| = |x| + |y|$$

If x is -5 and y is -6, the two sides of the equation become

$$|x + y| = |-5 + -6| = |-11| = 11 \text{ and}$$
$$|x| + |y| = |-5| + |-6| = 5 + 6 = 11.$$

Is Marta correct? If so, explain why. If not, give an example of values for x and y such that $|x + y| \neq |x| + |y|$.

57. Choose a number. Write down both the number and its opposite.

a. What happens when you add the two numbers?

b. Will the same thing happen when you add *any* number to its opposite? Explain why or why not.

Algebra Solve each equation by using backtracking or another method.

58. $x + (-9) = 1.5$ **59.** $y - 3 = -0.5$ **60.** $3 - x = 8.4$

61. What values of y make each inequality true?

a. $2 + y < 0$

b. $-10 < 2 + y$

c. $-10 < 2 + y$ and $2 + y < 0$

62. Copy each exercise. Fill in the blank with a number that gives a sum between $-1\frac{1}{2}$ and $-\frac{1}{2}$.

a. $10 +$ _____ **b.** $-1.2 +$ _____

c. $-2 +$ _____ **d.** $-84 +$ _____

e. In Parts a–d, what did you need to do to make the answer between $-1\frac{1}{2}$ and $-\frac{1}{2}$?

Algebra Without solving each equation, determine whether the value of x is less than 0, equal to 0, or greater than 0.

63. $x - 3 = -5$ **64.** $-x + 3 = 5$

65. $4 - x = 9$ **66.** $x - 3 = 12$

67. In this exercise, you will investigate the possible results for the sum of two positive numbers and a negative number.

a. Choose two positive numbers and one negative number. Is their sum positive, negative, or 0?

b. Suppose you repeat Part a with different numbers. Will the sum be always positive; always negative; or sometimes positive, sometimes negative, and sometimes 0? How do you know?

68. In Your Own Words Write a letter to a student a year younger than you. Explain how to subtract a negative number from another negative number. Include examples.

69. Preview In the next investigation, you will explore the multiplication and division of signed numbers.

a. Complete the equation: $(-2) + (-2) + (-2) =$ ____.

b. Write a multiplication expression that represents the sum in Part a.

c. What is the sign of the first factor in the multiplication expression? Of the second factor? Of the product?

Mixed Review

70. Evaluate $\frac{1}{2}k^2$ for each value of k.

a. 0 **b.** $\frac{1}{2}$ **c.** 1.2 **d.** 4

71. Evaluate $10m^2$ for each value of m.

a. 0 **b.** $\frac{1}{2}$ **c.** 1.2 **d.** 4

72. Evaluate $3p^2 + 10$ for each value of p.

a. 0 **b.** $\frac{1}{2}$ **c.** 1.2 **d.** 4

Find each sum. Simplify your answers as much as possible.

73. $\frac{1}{25} + \frac{4}{25}$ **74.** $\frac{1}{10} + \frac{6}{10}$ **75.** $\frac{2}{5} + \frac{1}{5} + 3$

76. $\frac{1}{2} + \frac{1}{3}$ **77.** $\frac{4}{9} + \frac{5}{6}$ **78.** $1\frac{1}{5} + 2\frac{3}{4}$

79. Statistics The table shows the approximate distribution of blood types in the U.S. population. "Positive" and "negative" indicate whether a person's blood has a particular component called the *Rh factor.*

a. What percentage of people in the United States have Rh-negative blood?

b. The projected U.S. population in 2050 is 391,000,000 people. If the distribution of blood types remains fairly constant, how many people in 2050 will have AB-negative blood?

Blood Types in the U.S.

Blood Type	Percent
O positive	38
O negative	7
A positive	34
A negative	6
B positive	9
B negative	2
AB positive	3
AB negative	1

Multiply and Divide with Negative Numbers

In this lesson, you will use what you have learned about addition and subtraction of signed numbers to help you understand how to multiply and divide them.

Inv 1	Multiply a Positive and a Negative	155
Inv 2	Multiply Two Negatives	157
Inv 3	Divide with Negative Numbers	160
Inv 4	Signed Numbers and Data	162

Explore

The table lists the lowest temperatures in degrees Celsius for each day during the month of November at a weather station in Alaska.

The first row lists the different low temperatures. The second row tells the number of days in November with each daily low temperature. For example, the first column indicates that the low temperature was −4°C two days during the month.

November Temperatures in Alaska

Daily Low Temperature (°C)	−4	−3	−2	−1	0	1	2	3	4
Days at This Temperature	2	1	2	4	2	3	6	6	4

Use what you learned in Lesson 3.1 to find the mean, or average, low temperature during the month of November.

How did you compute the mean?

If you calculated the total of the temperatures by using only addition, you probably found the process to be rather tedious. In this lesson, you will learn how to multiply and divide with negative numbers so you can perform calculations like this one quickly and efficiently.

Again, try to do the calculations without a calculator. This will help you understand why the operations work the way that they do.

Investigation 1 Multiply a Positive and a Negative

When you first learned to multiply positive numbers, you thought of multiplication as repeated addition. For example, the product of 3 and 5 can be thought of as three 5s, or $5 + 5 + 5$. It can also be thought of as five 3s, or $3 + 3 + 3 + 3 + 3$.

Develop & Understand: A

Use the way of thinking described above to calculate each product.

1. $2 \cdot (-8)$
2. $-3 \cdot 2$
3. $-5 \cdot 4$
4. $3 \cdot (-7.5)$
5. $-5\frac{1}{3} \cdot 3$
6. $-1 \cdot 6$

Think & Discuss

Look for patterns in your computations in Exercises 1–6.

Can you see a shortcut that you could use to compute the product of a positive and a negative number?

Use your shortcut to find the following products.

$2 \cdot (-45)$ $\qquad$ $-32 \cdot 11$

Develop & Understand: B

Now you will investigate products in which the positive number is a fraction or a decimal.

7. $\frac{1}{2} \cdot (-10)$
8. $\frac{1}{3} \cdot (-9)$
9. $\frac{2}{3} \cdot (-9)$
10. $\frac{4}{5} \cdot (-15)$
11. $-12 \cdot 0.25$
12. $1\frac{1}{2} \cdot (-20)$
13. $-5 \cdot 1.2$
14. $3.2 \cdot (-1.1)$
15. $-\frac{2}{3} \cdot \frac{3}{8}$
16. $\frac{1}{7} \cdot \left(-\frac{14}{3}\right)$

Multiplying with negative numbers can help you explore some interesting situations. Use what you learned in Exercises 1–16 to complete Exercises 17–24.

Real-Word Link

The maximum depth recommended for recreational scuba diving is generally 130 feet. Beyond that depth, divers need special training and equipment to keep themselves safe from the effects of pressure.

Develop & Understand: C

You might want to use a calculator to help you answer Exercises 17 and 18.

17. Cora jumps into the ocean from a boat. She starts at an elevation of 0 feet, and her elevation decreases 50 feet every minute.

a. What is her elevation after one minute?

b. What is her elevation after five minutes?

c. What is her elevation after n minutes?

You know that you can convert Celsius temperatures to Fahrenheit temperatures with the formula $F = \frac{9}{5}C + 32$, where C is the temperature in degrees Celsius and F is the temperature in degrees Fahrenheit.

18. Temperatures on Mercury, the planet closest to our sun, range from $-173°$C to $427°$C. Convert these temperatures to find the range of temperatures on Mercury in degrees Fahrenheit.

19. Create a word-problem exercise that requires calculating $3 \cdot (-8)$.

20. The product of two integers is -14. (Remember: The *integers* are the whole numbers and their opposites, … , $-3, -2, -1, 0, 1, 2, 3,$ … .)

a. What could the integers be? List all the possibilities.

b. Find three more pairs of numbers, not necessarily integers, that have a product of -14.

Use guess-check-and-improve to solve each equation.

21. $3x = -6$

22. $-2y = -12$

23. $4x + 15 = 3$

Share & Summarize

Is the product of a positive number and a negative number:

- always negative?
- always positive?
- sometimes positive, sometimes negative, and sometimes 0?

Investigation 2 Multiply Two Negatives

In Investigation 1, you used addition to figure out how to multiply a negative number by a positive number. You cannot use that strategy with two negative numbers because you cannot add a negative number of times. However, the pattern in products of a positive number and a negative number can help you figure out how to multiply two negative numbers.

Develop & Understand: A

1. Find each product.
   a. $-3 \cdot 4$
   b. $-3 \cdot 3$
   c. $-3 \cdot 2$
   d. $-3 \cdot 1$
   e. $-3 \cdot 0$

2. In Exercise 1, what happens to the product from one expression to the next? Why?

3. Now use your calculator to compute these products.
   a. $-3 \cdot (-1)$
   b. $-3 \cdot (-2)$
   c. $-3 \cdot (-3)$
   d. $-3 \cdot (-4)$
   e. $-3 \cdot (-5)$

4. Did the pattern you observed in Exercise 2 continue?

Think & Discuss

Look for patterns in your computations in Exercises 1–4. What do you think the rule is for finding the product of *any* two negative numbers?

Use your rule to find each product. Check your results with your calculator.

$-5 \cdot (-7)$ $-10 \cdot (-90)$ $-2 \cdot (-4.4)$ $-1.2 \cdot (-7)$

There are four ways to combine positive and negative numbers in products.

- positive • positive
- positive • negative
- negative • positive
- negative • negative

You now know how to multiply each combination and what kind of results to expect.

Develop & Understand: B

If you know the result of a multiplication, can you figure out what the factors might have been? Try these exercises.

5. Suppose the product of two integers is 12. What could the integers be? List all the possibilities.

6. Suppose the product of *three* integers is 12. What could the integers be? List all the possibilities.

Use your understanding of signed numbers and the guess-check-and-improve method to solve each equation.

7. $-4x = 12$

8. $-3x + 5 = 11$

9. $3 - 4x = -17$

When you use exponents to indicate repeated multiplication of a negative number, you need to be careful about notation. Put the negative number *inside* the parentheses, and put the exponent *outside* the parentheses.

Example

Calculate $(-2)^4$ and -2^4.

Notation	Meaning	Calculation
$(-2)^4$	-2 to the fourth power	$(-2) \cdot (-2) \cdot (-2) \cdot (-2) = 16$
-2^4	the opposite of 2^4	$-(2 \cdot 2 \cdot 2 \cdot 2) = -16$

Develop & Understand: C

Evaluate each expression.

10. $(-3)^2$

11. -3^2

12. -4^2

13. $(-4)^2$

14. $(-2)^1$

15. $(-2)^2$

16. $(-2)^3$

17. $(-2)^4$

18. $(-2)^5$

19. $(-2)^6$

20. Look for patterns in your answers to Exercises 14–19.
    a. For what values of n is $(-2)^n$ positive?
    b. For what values of n is $(-2)^n$ negative?

21. Dylan said the solution of $x^2 = 16$ is 4. Felipe stated that there is another solution. Is Felipe correct? If so, find the other solution.

Solve each equation. (Hint: Each equation has two solutions.)

22. $x^2 = 36$

23. $x^2 = \frac{1}{36}$

Share & Summarize

1. Is the product of two negative numbers:
    - always negative?
    - always positive?
    - sometimes positive, sometimes negative, and sometimes 0?
2. Consider the expression $(-3)^m$.
    a. If m is an even number, is $(-3)^m$ positive or negative? How can you tell?
    b. If m is an odd number, is $(-3)^m$ positive or negative? How can you tell?

Investigation 3 Divide with Negative Numbers

You can solve any division exercise by thinking about a corresponding multiplication exercise. Look at how Kate solves $30 \div 5$.

Develop & Understand: A

Use Kate's method to solve each division exercise.

1. $21 \div (-3)$
2. $-64 \div (-2)$
3. $-24 \div 48$
4. $-2 \div (-32)$
5. $-\frac{6}{5}$
6. $\frac{9}{-27}$
7. $-2.16 \div (-54)$
8. $\frac{-3}{-0.3}$

Think & Discuss

Look for patterns in your computations in Exercises 1–8.

- Can you see a shortcut for computing a quotient when exactly one of the numbers is negative?
- Can you see a shortcut for computing the quotient of two negative numbers?

Use your shortcut to find each quotient.

$44 \div (-2.2)$ $\qquad -\frac{7}{3}$ $\qquad -56 \div (-2)$

Develop & Understand: B

Real-World Link

Water exerts pressure on a scuba diver. The deeper the diver, the greater the pressure. Divers are trained to make the pressure in the body's air spaces, the lungs, sinuses, and ears, equal to the outside water pressure.

9. Suppose a scuba diver begins his dive at an elevation of 0 feet. During the dive, his elevation changes at a constant rate of −2 feet per second. How long will it take for him to reach an elevation of −300 feet?

10. The Marianas Trench, south of Guam, contains the deepest known spot in the world. This spot, called Challenger Deep, has an elevation of −36,198 feet.

    Suppose Jamie entered the ocean above this spot. He started at an elevation of 0 feet and moved at a constant rate of −50 feet per minute.

    If Jamie was able to go deep in the ocean without being affected by the pressure and his tank contained enough air, how long would it take him to reach the bottom of Challenger Deep?

11. Write four division expressions with a quotient of −4.

Solve each equation.

12. $\frac{30}{x} = -15$

13. $\frac{x}{-4} = 20$

Share & Summarize

Which of the following statements is true for each situation? Explain how you know.

- The result is always negative.
- The result is always positive.
- The result is sometimes negative, sometimes positive, and sometimes zero.

1. Dividing a positive number by a negative number
2. Dividing a negative number by a positive number
3. Dividing a negative number by a negative number

Investigation  Signed Numbers and Data

You have been learning how to add, subtract, multiply, and divide with integers. In this investigation, you will analyze data by finding range, mean, median, and mode for the data.

Think & Discuss

In golf, "par" is a number assigned to an individual hole or to a course of holes. The par tells the number of strokes, or hits, it should take a golfer to get a golf ball in the hole. A golfer's score is reported as the number of strokes above or below par that it took to finish the course. If par for a course is 72, then a golfer who finishes with 75 strokes is *3 over par.* A golfer who finishes with 69 strokes is *3 under par.*

75 — 3 over par
72 — par
69 — 3 under par

Suppose you relate golf scores to integers.

What number would you use to represent par?

What number would you use to represent *3 under par?*

What number would you use to represent *3 over par?*

Is a negative or positive score better? Why?

Math Link

The *range* of a data set is the difference between the minimum and maximum values.

The *mode* is the value that occurs most often.

The *median* is the middle value when all values are listed from least to greatest.

The *mean* is the distribution of the total of the values in a data set among the members of the data set.

Develop & Understand: A

Sondra plays for her school's golf team. Last spring, she played on the school's home golf course 8 times.

Her scores are shown below.

2 over par	par	3 over par	5 under par
1 over par	2 under par	par	1 over par

1. Rewrite Sondra's scores using positive and negative integers with par being zero.
2. What was Sondra's minimum, or best, score?
3. What was Sondra's maximum, or worst, score?
4. What is the range of Sondra's scores? Show your work.
5. What is the mode of Sondra's scores?
6. What is the median of Sondra's scores? Show your work.
7. What is the mean of Sondra's scores? Show your work.

8. Find the mean score for four other players. Their scores are listed below. Which players score below par on the average?

   a. Lakita 4, −2, 6, −1, −2, 0, −1, 2

   b. Dana −1, 0, 8, −2, 2, 3

   c. Victoria −3, −1, 0, −1, 1, −2

   d. Lynn 5, −3, 2, −5, −2, −1

9. Lakita played another round of golf. What must have been her score to have a mean score of par for all nine rounds?

10. Padma played seven rounds of golf. The mean, median, and mode of all seven scores were par, or 72 strokes. What could have been her scores?

Develop & Understand: B

One January, a science class from a Maine high school recorded the town's overnight low temperatures. The low temperatures are listed in the following *frequency table.*

Temperature (°F)	−8	−6	−5	−3	−2	0	3	8	9	12	14	15	17	20	23
Frequency	1	2	1	3	1	1	1	3	1	3	4	1	3	5	1

Real-World Link

The record low temperature ever recorded in the United States was −79.8°F (rounded off to −80°F) at Prospect Creek Camp in the Endicott Mountains of northern Alaska on January 23, 1971.

11. What is the mode temperature reading?

12. For how many days was the temperature recorded?

13. What is the range of temperatures?

14. Since the data is in a frequency table, what will you need to do to find the median temperature? What is the median temperature?

15. Since the data set is in a frequency table, what do you need to do to find the mean temperature? What is the mean temperature?

16. The minimum value of a set of 6 numbers is −10, and the maximum value is 20. If mean is 5, what are the 4 numbers? Use 4 different values.

Think & Discuss

If you were visiting Maine during January, which statistics about the night lows would be most helpful?

What does each number tell you about the data?

Suppose the minimum temperature dropped to −20°F instead of −8°F. Which measure of center (mean, median, or mode) is affected? By how much?

Real-World Link

In 1980, Earl Campbell, of the Houston Oilers, averaged 5.2 yards per carry.

In football, a running back carries the ball in hopes of gaining yards. He loses yards when he is tackled behind the line where the play began, or the *line of scrimmage.*

During Friday night's football game, Kanye carried the ball 12 times during the second half of the game. In the numbers below, a positive number indicates a gain and a negative number indicates a loss.

$$2 \quad 1 \quad 6 \quad -1 \quad 12 \quad 2 \quad -3 \quad 11 \quad 5 \quad 2 \quad -1 \quad 6$$

17. What was Kanye's worst carry (minimum)?

18. What was Kanye's best carry (maximum)?

19. What is the range of yardage of Kanye's carries? Show your work.

20. What is the mode?

21. What is the median yardage of the 12 carries? Show your work.

22. Add two additional carries so that the median is 3.

23. What is the mean yardage of the 12 carries? Show your work.

24. Add another carry that makes the mean -3 yards.

Share & Summarize

1. Explain the difference between mean, median, and mode.
2. How does including both positive and negative integers in a set of data affect finding the range of the data?
3. How does including both positive and negative integers in a set of data affect finding the mean of the data?
4. Is there ever a time that you would find the mean of a set of data by dividing by a negative number? Why or why not?

Practice & Apply

Compute each product.

1. $-3 \cdot 52$
2. $6.2 \cdot (-5)$
3. $-36 \cdot \frac{1}{2}$
4. $-2 \cdot 4 + 3 \cdot \left(-\frac{1}{3}\right)$
5. The product of two integers is -15.
   a. What could the integers be? List all the possibilities.
   b. Find three more pairs of numbers, not necessarily integers, with a product of -15.
6. Solve $-3x + 4 = -5$.

Compute each product.

7. $-3 \cdot (-32)$
8. $-\frac{1}{2} \cdot (-35)$
9. $-5 \cdot (-0.7)$
10. $-2.5 \cdot (-7)$
11. $-2^5 \cdot (-3)$
12. $-3 \cdot (-5^3)$
13. $-3 \cdot (-5)^3$
14. $-3^2 \cdot (-3)^2$

Without calculating each product, determine whether it is less than 0 or greater than 0.

15. $3 \cdot (-2)$
16. $3 \cdot 2$
17. $-3 \cdot 2$
18. $-3 \cdot (-2)$
19. The product of two integers is 9. What could the integers be? List all the possibilities.
20. Solve $-5y + 13 = 3$.

Find each quotient.

21. $-3 \div 2.5$
22. $-3 \div (-25)$
23. $45 \div (-360)$
24. $\frac{-45}{-90}$
25. One number divided by a *negative* number is 0.467. What could the two numbers be? Think of three possibilities. Write them as division equations.

Solve each equation.

26. $5x = -20$
27. $3y = -\frac{6}{5}$
28. $\frac{w}{9} = -5$

For Exercises 29–33, use the following data list collected from Fahrenheit temperatures in Alaska. Show your work to support each answer.

$$-12, 8, 11, -6, -3, -15, 8, 5, -2, -9, 8$$

29. What is the range of the data?

30. What is the median temperature?

31. What is the mean temperature? What does the mean tell about the temperature for these eleven days?

32. What is the mode temperature?

33. How would the median change if another temperature, 12°, is added to the data? How would the mean change?

34. As a parachutist descends into Death Valley, her elevation decreases 15 feet every second. Five seconds before she lands, she is at an elevation of −127 feet. What is the elevation of her landing place?

Real-World Link

In November 2001, the National Weather Service implemented a new windchill temperature formula using the latest advances in computer modeling. The previous formula was based on the 1945 Siple and Passel Index.

35. Physical Science When it is cold outside and the wind is blowing, it feels colder than it would at the same temperature without any wind. Scientists call this outside temperature the *windchill.*

For example, if the temperature outside is 15°F and the wind speed is 20 mph, the equivalent windchill temperature is about −2°F. That means that even though the thermometer reads 15°F, it *feels* as cold as it would if you were in no wind at −2°F.

Meteorologists use the following formula to calculate windchill. The variable s represents wind speed in mph, and t represents actual temperature in °F.

$$\text{windchill} = 35.74 + 0.6215t - 35.75(s^{0.16}) + 0.4275t(s^{0.16})$$

This formula works for wind speeds between 5 mph and 60 mph.

a. If the wind blows at 20 mph and the temperature is −5°F, what is the windchill?

b. If the wind blows at 30 mph and the temperature is −10°F, what is the windchill?

c. Challenge If the wind blows at 25 mph and the windchill is −51°F, what is the actual temperature?

36. Pedro said, "I'm thinking of a number. When I multiply my number by −2 and subtract 4, I get −16." What is Pedro's number?

Without calculating, predict whether each product is less than 0, equal to 0, or greater than 0.

37. $(-4)^3 \cdot (-2)^2$ **38.** $(-3)^5 \cdot (-6)^3$ **39.** $(-1)^5 \cdot 5^2$

40. $(-2)^2 \cdot (-972)^{90}$ **41.** $-45^2 \cdot (-35)^5$ **42.** $(-2)^8 \cdot (-5)^8$

Challenge If possible, solve each equation. If it is not possible, explain why.

43. $y^3 = -8$ **44.** $x^2 = -9$

Describe the values of *b* that make each inequality true.

45. $3b < 0$ **46.** $3(b - 2) < 0$

Without calculating, determine whether the quotient in Exercises 47–55 is less than −1, between −1 and 0, between 0 and 1, or greater than 1.

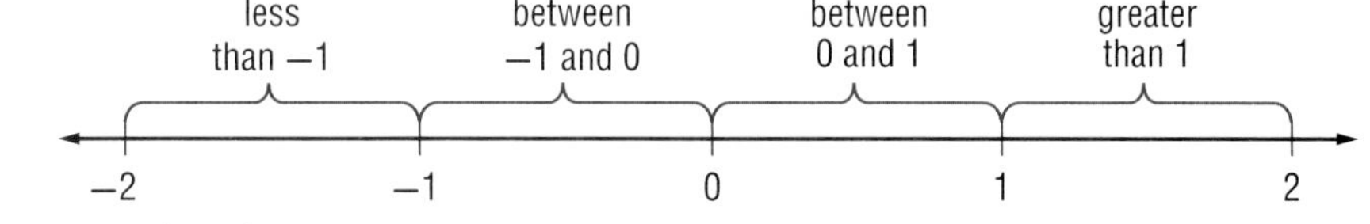

47. $1 \div \left(-\frac{3}{5}\right)$ **48.** $\frac{-555.233}{-5{,}552}$ **49.** $78.3636 \div (-25.33)$

50. The mean of 10 temperatures is −21°F. Assuming that the temperatures are not all the same, what could the temperatures be?

51. The median of 6 temperatures is −21°F. Assuming that the temperatures are not all the same, what could the temperatures be?

Use guess-check-and-improve to solve each equation. Be careful, some equations have two solutions.

52. $\left(\frac{1}{y}\right)^2 = \frac{1}{9}$ **53.** $y^2 = \frac{1}{9}$

Challenge Solve each equation.

54. $\left(\frac{1}{z}\right)^2 = 16$ **55.** $-\frac{5}{x} + 4 = 2$

For Exercises 56–59, use the data below that compares actual rainfall to the normal rainfall for each month. The amount of rainfall is reported as "inches above or below normal rainfall."

Jan	Feb	Mar	Apr	May	June	July	Aug	Sep	Oct	Nov	Dec
+1.0	−0.5	−0.2	+0.7	+1.1	+0.3	−1.6	−2.0	−0.3	+0.5	0.0	−1.2

56. What does the November entry of 0.0 suggest about November's rainfall?

57. Was the biggest change from normal rainfall an amount below normal or an amount above normal? Explain.

58. For the year, was the rainfall above or below normal rainfall? How did you determine this?

59. On average, how did the actual rainfall compare to the normal rainfall per month for this year?

60. In Your Own Words Write a letter to a student a year younger than you. Explain how to divide a negative number by another negative number. Include examples.

Mixed Review

61. Which of the following expressions are equal to each other?

a. $m + m + m + m + m$ **b.** $(5m)^5$

c. $5m$ **d.** m^5

e. $3 \cdot 3 \cdot 3 \cdot 3$ **f.** $4 \cdot 3$

g. 4^3 **h.** 3^4

62. Match each point on the graph to the coordinates given.

a. $(-2, 0)$

b. $(1, -3)$

c. $(-2, 3)$

d. $(-2, -4)$

e. $(2, 1)$

f. $(3, 4)$

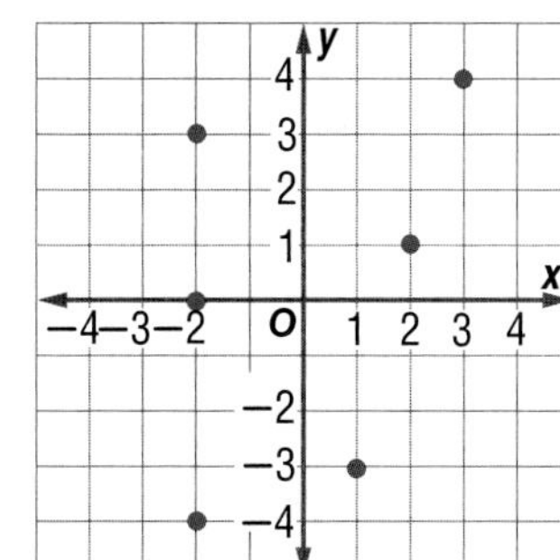

Write each expression in exponential form.

63. $t \cdot t \cdot t$

64. $n \cdot n \cdot n \cdot n$

65. $\pi \cdot r \cdot r$

66. $\pi \cdot r \cdot r \cdot r$

67. $y \cdot y \cdot y \cdot y \cdot y$

68. $z \cdot z \cdot z \cdot z \cdot z \cdot z \cdot z$

CHAPTER 3

Review & Self-Assessment

Vocabulary

absolute value

Chapter Summary

You began this chapter learning about operations with signed numbers. Using the number-line model, you thought about addition and subtraction as facing a particular direction on a number line and moving forward or backward. You discovered some rules for these calculations and then used them to develop rules for multiplication and division. In addition, you used signed numbers when finding the mean, median, and mode of data sets.

Strategies and Applications

The questions in this section will help you review and apply the important ideas and strategies developed in this chapter.

Add and subtract with negative numbers

Using the "walk the number line" model or any other strategy you know, describe how to compute each sum or difference.

1. $2 - (-9)$

2. $-9 + 3$

3. Describe a real-world situation that can be solved in two ways using signed numbers, one involving subtraction and one involving addition. Then write an equation to represent each solution.

Multiply and divide with negative numbers

Describe how to compute each product or quotient.

4. $-3.5 \cdot 4$

5. $-9 \div (-4.5)$

Without computing, determine whether each expression is greater than 0, less than 0, or equal to 0. Explain how you know.

6. $(-33)^5 \cdot (-27)^2$

7. $\dfrac{-5^{58} \cdot (-12)^{23}}{-5^3}$

Demonstrating Skills

Compute each sum, difference, product, or quotient.

8. $3 \cdot (-3)$

9. $-2.3 - 7.9$

10. $5.2 \div (-2.6)$

11. $3.4 + (-5)$

12. $-3.2 - (-0.9)$

13. $-0.2 + 9.1$

14. $-7 \cdot (-8.3)$

15. $(-2)^4 \cdot (-5)$

Amiri kept track of his weight over a year's time. In the chart below, a positive number indicates a weight gain. A negative number indicates a weight loss.

Jan.	Feb.	Mar.	Apr.	May	June	July	Aug.	Sept.	Oct.	Nov.	Dec.
-3	2	0	-5	-4	-2	0	0	-2	-2	1	3

16. What is the range of Amiri's weight loss and gain?

17. What is Amiri's mean weight change?

18. Over the whole year did Amiri gain or lose weight? By how much?

19. What is the mode?

20. What is the median?

21. Models Write an expression to show the sum of 4 and -2. Simplify your expression.

22. Models Create a two-color chip model to show the sum of 3 and -4. Yellow represents positive values. Red represents negative values.

Walk the number line to compute each sum or difference. Record your steps by making a number-line drawing.

23. $-1 + (-4)$

24. $6 - (-3)$

Compute each difference. Then write an addition equation that is equal to the subtraction equation.

25. $17 - 21$

26. $-2 - 12$

Compute each sum. Then write an equivalent subtraction equation.

27. $-15 + 11$

28. $6 + (-7)$

For Exercises 29 and 30, write an addition and a subtraction equation can be used to find the answer to each question.

29. The Tiger football team started with the ball on the 23-yard line and lost 5 yards on its first play. On what yard line did the team start its next play?

30. The temperature yesterday was $-4\,°F$. There is a $15\,°F$ increase in temperature today. What is today's temperature?

Consider the following expressions.

Expression 1: $\boxed{2x}$ Expression 2: $\boxed{x + 7}$

31. For what value(s) of x is Expression 1 equal to Expression 2?

32. For what value(s) of x is Expression 1 greater than Expression 2?

33. For what value(s) of x is Expression 1 greater than Expression 2?

Test-Taking Practice

SHORT RESPONSE

1 Evaluate the following expression. Use the order of operations. Show all of your steps.

$$3 + (-2) + (-4)(2) - (-9) \div (-3)(4 - 5)$$

Show your work.

Answer ____________________

MULTIPLE CHOICE

2 Find the difference.

$$27 - (-5)$$

A −32
B 12
C 22
D 32

3 Evaluate the expression. Use correct order of operations.

$$(5 + 6) - (7 - 3)^2 + 2 \cdot 4$$

F 3
G 11
H 13
J 35

4 Find the value of the expression.

$$\frac{8(-3)(7)}{(-4)(-6)}$$

A $-\frac{6}{5}$
B −7
C $\frac{35}{24}$
D 7

5 Find the quotient.

$$72 \div (-2) \div 4 \div (-3)$$

F −3
G $\frac{8}{3}$
H 3
J 9

Magnitude of Numbers

Contents in Brief

4.1	Scientific Notation	174
4.2	Negative Exponents	194
	Review & Self-Assessment	207

Real-Life Math

The Solar System The center of the solar system is the Sun, and planet distances are given in terms of each planet's distance from the Sun. Astronomical information, such as planets' distances, can be found in reference books and on Internet sites. Planet distances, in miles, are shown below.

Mercury:	36,000,000	Venus:	6.7×10^7
Earth:	93,000,000	Earth:	9.3×10^7
Jupiter:	484,000,000	Mars:	1.42×10^8
Saturn:	887,000,000	Uranus:	1.78×10^9
Neptune:	2,794,000,000		

Think About It Suppose you want to list the planets in order according to their distances from the Sun. How can you use the data to order the distances from least to greatest?

Take the **Chapter Readiness Quiz** at glencoe.com.

Dear Family,

Chapter 2 used repeater machines to model expressions such as 2^4. The class learned that the 2 is called the *base,* and the raised 4 is called the *exponent.*

$$2^4 = 2 \cdot 2 \cdot 2 \cdot 2 = 16$$

In this chapter, your student will study **scientific notation**. This notation is a mathematical shorthand that is often used to represent the magnitude of very large and very small numbers, like the distance between planets and the size of an atom.

Key Concept–Scientific Notation

In scientific notation, a number between 0 and 10 is multiplied by a power of 10. The following repeater machines model changing the scientific notation expressions 1.5×10^3 and 2.1×10^{-2} to standard form.

Input: 1.5 [× 10, 3] Output: 1,500

$$1.5 \times 10^3 = 1{,}500$$

Input: 2.1 [× 10, −2] Output: 0.021

$$2.1 \times 10^{-2} = 0.021$$

In the expression 2.1×10^{-2}, the exponent is negative. The class will explore negative exponents in Lesson 4.2.

Chapter Vocabulary

relative error **scientific notation**

Home Activities

- Discuss real-world situations that involve very large and very small, or greater and lesser numbers.
- Use scientific notation to express numbers found in newspapers, magazines, and print advertisements.

Scientific Notation

How far is it from your house to your school? You may be close enough to walk to school. Or, you may travel a farther distance, perhaps a few miles, by bus. Think about how far the Apollo astronauts traveled to go to the moon and how far the *Mars Pathfinder* traveled to Mars.

In this lesson, you will learn a new way of writing and working with very large numbers.

Inv 1 Powers of 10 175

Inv 2 Work with Scientific Notation 178

Inv 3 Scientific Notation on a Calculator 182

Inv 4 Relative Error 185

Think & Discuss

What is funny about each of these cartoons?

Investigation 1 Powers of 10

Previously, you have worked with 1 million, 1 billion, and 1 trillion. These numbers can be expressed as powers of 10.

$$1 \text{ million} = 1{,}000{,}000 = 10^6$$

$$1 \text{ billion} = 1{,}000{,}000{,}000 = 10^9$$

$$1 \text{ trillion} = 1{,}000{,}000{,}000{,}000 = 10^{12}$$

It is often useful to talk about large numbers in terms of powers of 10. Imagine $\times 10$ repeater machines when working with such powers.

Math Link

Here are some other big numbers.

10^{15}	quadrillion
10^{18}	quintillion
10^{21}	sextillion
10^{24}	septillion
10^{27}	octillion
10^{30}	nonillion

Think & Discuss

Your factory is so successful, you plan a celebration. To make decorations, you send streamers through stretching machines.

Find the length of the streamer that exits each machine. Give your answers in centimeters.

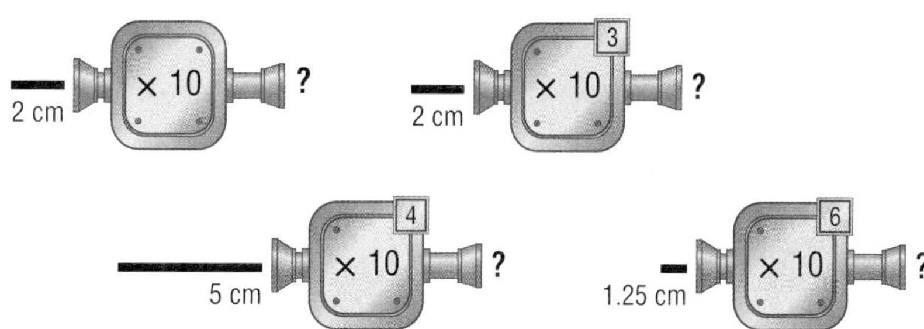

If you know the length of the input streamer and the exponent on the $\times 10$ repeater machine, how can you find the length of the output streamer?

Find the exponent of each repeater machine. (Note: The output streamer is not drawn to scale.)

If you know the length of the input and output streamers, how can you find the exponent of the $\times 10$ repeater machine?

Develop & Understand: A

1. The beverage straw below exited a $\times 10^2$ machine. Without measuring, estimate the length of the input straw in centimeters. It may be helpful if you copy the straw onto your paper. Explain how you found your answer.

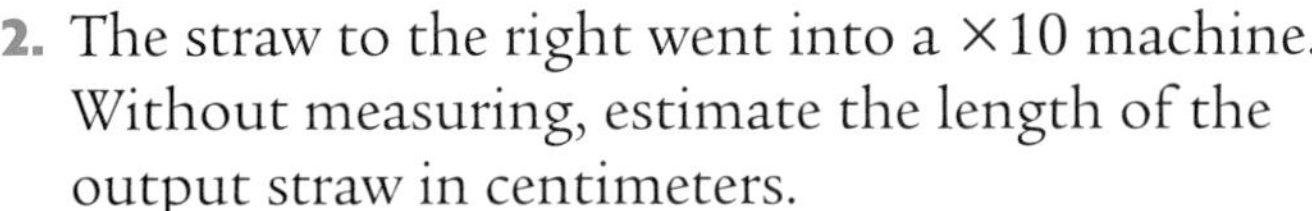

2. The straw to the right went into a $\times 10$ machine. Without measuring, estimate the length of the output straw in centimeters.

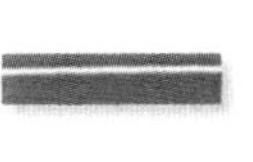

Find the length of each output straw.

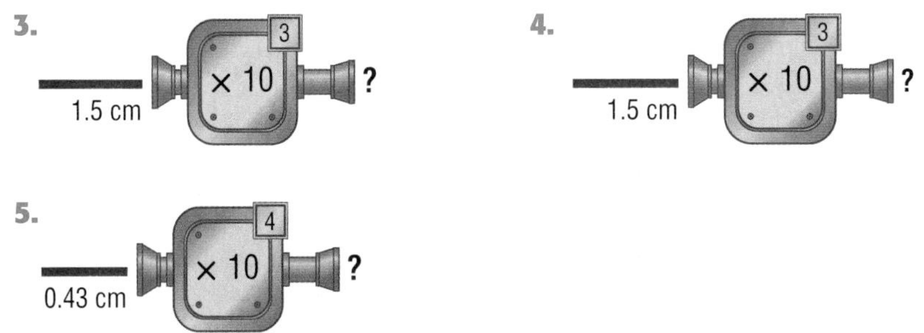

Find each exponent.

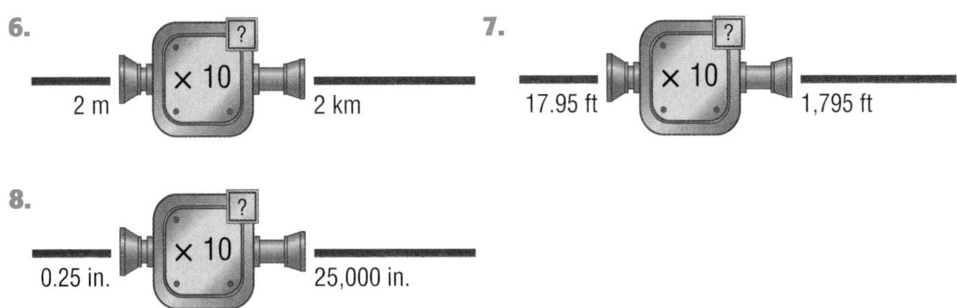

Find the length of each input straw.

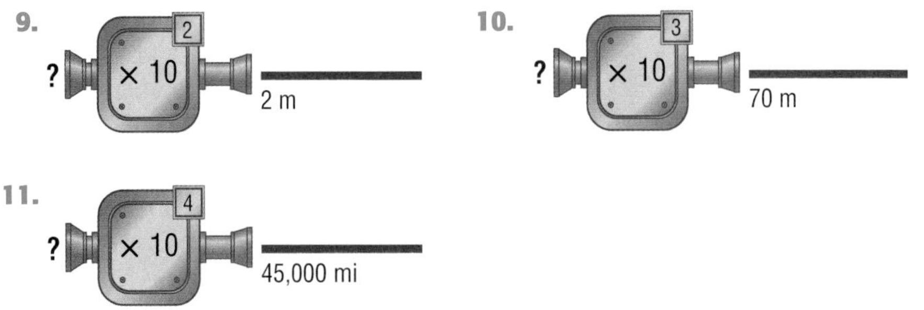

In Exercises 12–31, you will work with expressions involving powers of 10. As you work, it may help to think of repeater machines. For example, you could use the machine below to think about 5×10^3.

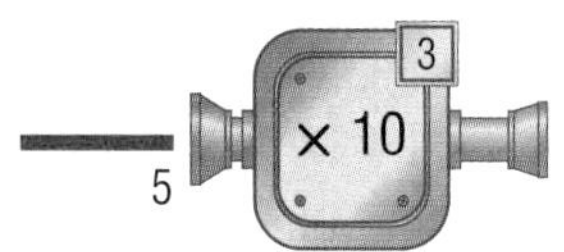

Real-World Link

The human eye blinks an average of 4,200,000, or 4.2×10^6, times a year.

Develop & Understand: B

Find each product.

12. 8×10^5
13. 18.6×10^8
14. $9{,}258 \times 10^4$
15. $1{,}620 \times 10^5$
16. 4×10^{12}
17. 14.9×10^{11}

Find the value of N in each equation.

18. $6 \times 10^N = 600$
19. $8.6 \times 10^N = 86{,}000$
20. $54 \times 10^N = 5{,}400{,}000$
21. $2{,}854 \times 10^N = 285{,}400$
22. $N \times 10^2 = 800$
23. $N \times 10^3 = 47{,}000$
24. $N \times 10^6 = 54{,}800{,}000$
25. $N \times 10^4 = 3{,}958$
26. $85 \times 10^4 = N$
27. $9.8 \times N = 9{,}800{,}000$
28. $0.00427 \times 10^N = 42.7$
29. $N \times 10^{11} = 24{,}985{,}000{,}000$
30. $5 \times 10^N = 5{,}000{,}000{,}000{,}000$
31. $N \times 10^{11} = 3{,}568{,}000{,}000{,}000$

Share & Summarize

1. What repeater machine would stretch a 6-inch straw to 6 million inches?
2. What repeater machine would stretch a 3-inch straw to 3 billion inches?
3. What repeater machine would stretch a 2.3-million-inch straw to 2.3 billion inches?

Investigation 2 Work with Scientific Notation

Vocabulary

greater

lesser

scientific notation

Powers of 10 give us an easy way to express very large and very small, or **greater** and **lesser** numbers, without writing a lot of zeros. This is useful in many fields, such as economics, engineering, computer science, and other areas of science.

For example, astronomers need to describe great distances, like the distances between planets, stars, and galaxies. Chemists often need to describe small measurements, like the sizes of molecules, atoms, and quarks.

In this investigation, you will focus on a method for expressing large numbers. In Lesson 4.2, you will see how this method can be used to express small numbers, as well.

Real-World Link

The Centaurus system includes Alpha Centauri, the star, other than our own Sun, that is closest to Earth.

Example

Astronomical distances are often expressed in lightyears. For example, the Centaurus star system is 4.3 lightyears from Earth. A *lightyear* is the distance light travels in one year. One lightyear is approximately 5,878,000,000,000 miles.

There are many ways to express the number of miles in a lightyear as the product of a number and a power of 10. Here are three.

$$5{,}878 \times 10^9 \qquad 58.78 \times 10^{11} \qquad 5.878 \times 10^{12}$$

Can you explain why each of these three expressions equals 5,878,000,000,000?

The third expression above is written in scientific notation. A number is in **scientific notation** when it is expressed as the product of a number greater than or equal to 1 but less than 10, and a power of 10.

at least 1 but less than 10

a power of 10

$$\mathbf{5.878 \times 10^{12}}$$

The chart lists the outputs in scientific and standard notation of some repeater machines.

Machine	Scientific Notation	Standard Notation	Example of Number
3 × 10 [2]	3×10^2	300	length of a football field in feet
6.5 × 10 [7]	6.5×10^7	65,000,000	years since dinosaurs became extinct
2.528 × 10 [13]	2.528×10^{13}	25,280,000,000,000	distance from Earth to Centaurus star system in miles

Scientific notation can help you compare the sizes of numbers.

Think & Discuss

Here are three numbers written in standard notation.

74,580,000,000,000,000,000

8,395,000,000,000,000,000

242,000,000,000,000,000,000

- Which number is greatest? How do you know?
- Which number is least? How do you know?

Here are three numbers written in scientific notation.

3.723×10^{15} $\quad$ 9.259×10^{25} $\quad$ 4.2×10^{19}

- Which number is greatest? How do you know?
- Which number is least? How do you know?

Math Link

When writing expressions in scientific notation, × is used instead of • to indicate multiplication.

Develop & Understand: A

Tell whether each number is written correctly in scientific notation. For those that are not written correctly, describe what is incorrect.

1. 6.4535×10^{52}
2. 41×10^{3}
3. 0.4×10^{6}
4. 1×10^{1}

Write each number in standard notation.

5. 1.28×10^{6}
6. 9.03×10^{5}
7. 6.02×10^{23}
8. 5.7×10^{8}

Write each number in scientific notation.

9. 850
10. 7 thousand
11. 10,400,000
12. 659,000
13. 83 million
14. 27 billion
15. The table shows the average distance from each planet in the solar system to the Sun.

Planet	Distance from Sun (km) Standard Notation	Distance from Sun (km) Scientific Notation
Earth	149,600,000	1.496×10^{8}
Jupiter	778,300,000	
Mars	227,900,000	
Mercury	57,900,000	
Neptune	4,497,000,000	
Saturn	1,427,000,000	
Uranus	2,870,000,000	
Venus	108,200,000	

a. Complete the table by expressing the distance from each planet to the Sun in scientific notation.

b. Order the planets from closest to the Sun to farthest from the Sun.

Scientific notation is useful for ordering numbers only if the convention is followed properly, that is, the first part of the number is between 1 and 10. Otherwise, the power of 10 may not tell you which number is the greatest.

Math Link

Here are some really big numbers.

10^{45}	quattuordecillion
10^{57}	octadecillion
10^{60}	novemdecillion
10^{63}	vigintillion
10^{100}	googol

The expressions below show the same number written in four ways as the product of a number and a power of 10. However, none of these expressions is in scientific notation. If you compare these numbers by looking at the powers of 10, you might not realize that they are all equal.

0.0357×10^8 $\quad 35.7 \times 10^5$ $\quad 3{,}570 \times 10^3$ $\quad 357{,}000 \times 10^1$

You can change numbers like these into scientific notation without changing them into standard form first.

Example

Write 357×10^4 in scientific notation.

The first step is to change 357 to a number between 1 and 10. To do this, divide by 100, or 10^2. To compensate, you need to multiply by 100, or 10^2.

The last step is to simplify $10^2 \times 10^4$ by applying the product law of exponents.

$$\begin{aligned} 357 \times 10^4 &= \frac{357}{100} \times 100 \times 10^4 \\ &= 3.57 \times 10^2 \times 10^4 \\ &= 3.57 \times 10^6 \end{aligned}$$

Develop & Understand: B

Write each number in scientific notation.

16. 13×10^2

17. 0.932×10^3

18. 461×10^4

19. 59×10^5

20. 98.6×10^9

21. 197×10^6

Share & Summarize

How can you tell whether a number greater than 1 is written in scientific notation?

Investigation 3 Scientific Notation on a Calculator

There is a well-known story about a ruler who placed rice on the squares of a chessboard as a reward for one of his subjects. The first square had two grains of rice, and every square after that had twice as many as the previous square.

The table describes the number of grains of rice on each square.

Square	Number of Grains (as a product)	Number of Grains (as a power of 2)
1	$1 \cdot 2$	2^1
2	$1 \cdot 2 \cdot 2$	2^2
3	$1 \cdot 2 \cdot 2 \cdot 2$	2^3
4	$1 \cdot 2 \cdot 2 \cdot 2 \cdot 2$	2^4
5	$1 \cdot 2 \cdot 2 \cdot 2 \cdot 2 \cdot 2$	2^5
6	$1 \cdot 2 \cdot 2 \cdot 2 \cdot 2 \cdot 2 \cdot 2$	2^6
7	$1 \cdot 2 \cdot 2 \cdot 2 \cdot 2 \cdot 2 \cdot 2 \cdot 2$	2^7

Think & Discuss

There are 2^{40} grains of rice on square 40. Evaluate this number on your calculator. What does the calculator display? What do you think this means?

Numbers too large to fit in a calculator's display are expressed in scientific notation. Different calculators show scientific notation in different ways. When you entered 2^{40}, you may have seen one of these displays.

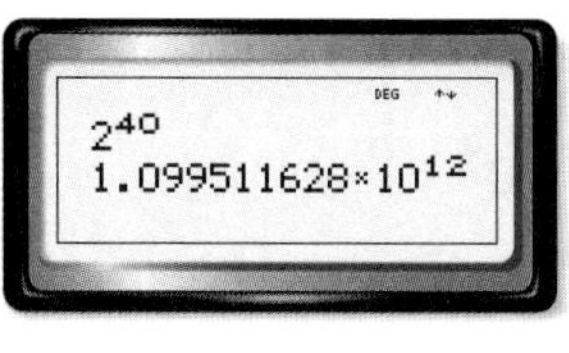

Both of these represent $1.099511628 \times 10^{12}$. When you read the display, you may need to mentally insert the "× 10."

If you continued the table, what power of 2 would represent the number of grains of rice on square 64, the last square on the chessboard?

Use your calculator to find the number of rice grains on square 64. Give your answer in scientific notation.

Develop & Understand: A

Real-World Link

A *day* is the time it takes for one rotation of a planet. A *year* is the time it takes for a planet to make one revolution around the sun. On Jupiter, a day is about 10 Earth hours long, and a year is about 12 Earth years.

Use your calculator to evaluate each expression. Give your answers in scientific notation.

1. 3^{28}
2. 4.05^{21}
3. 7.95^{12}
4. 12^{12}
5. It would be difficult to order the numbers in Exercises 1–4 as they are given. It should be much easier now that you have written them in scientific notation. List the four numbers from least to greatest.
6. Light travels at a speed of about 186,000 miles per second.
   a. How many miles does light travel in a day?
   b. How many miles does light travel in a year?

When the result of a calculation is a very large number, your calculator automatically displays it in scientific notation. You can also enter numbers into your calculator in scientific notation. Different calculators use different keys, but a common one is [EE].

For example, to enter 2.4×10^{12}, press

2.4 [EE] 12 [ENTER].

Develop & Understand: B

Estimate the value of each expression. Then use your calculator to evaluate it. Give your answers in scientific notation.

7. $5.2 \times 10^{15} + 3.5 \times 10^{15}$
8. $(6.5 \times 10^{18}) \cdot (1.8 \times 10^{15})$
9. $(8.443 \times 10^{18}) \div 2$
10. Estimate the value of each expression. Then use your calculator to evaluate it. Give your answers in scientific notation.
    a. $6 \times 10^{12} + 4 \times 10^{2}$
    b. $8.5 \times 10^{20} + 1.43 \times 10^{45}$
    c. $4.92 \times 10^{22} - 9.3 \times 10^{5}$
    d. How do your results in Parts a–c compare to the numbers that were added or subtracted?
    e. Write the numbers in Part a in standard notation. Then add them. How does your result compare to the result you found in Part a? If the answers are different, try to explain why.

Develop & Understand: C

In 1945, a librarian at Harvard University estimated that there were ten million books of "worthwhile" printed material. Suppose you wanted to enter all of this printed material into a computer for storage.

Use your calculator to solve the following exercises. When appropriate, record your answers in scientific notation. Make these assumptions as you work.

- The average book is 500 pages long.
- The average page holds 40 lines of text.
- The average line holds 80 characters.

Real-World Link

In 2007, the U.S. Library of Congress held over 20 million books. Its collections also include audio materials, maps, microforms, and visual materials.

11. How many characters are on one page of an average book?
12. How many characters are in the average book?
13. How many characters were in all the "worthwhile" books in 1945?
14. One *byte* of computer disk space stores about one character of text. A *gigabyte* of computer disk space is 1×10^9 bytes, so a gigabyte can store about 1×10^9 characters. How many gigabytes are needed to store all the "worthwhile" books from 1945?

Share & Summarize

1. Explain, in a way that a fifth grader could understand, why a calculator sometimes displays numbers in scientific notation.
2. Explain why sometimes when you add two numbers in scientific notation on your calculator, the result is just the greater of the two numbers.

Inquiry

Investigation  Relative Error

Vocabulary

relative error

Materials

- scientific or graphing calculator

Most things in life are not perfect. Errors occur all around us everyday. If you open a carton of eggs at the grocery store, you may find a broken one. If you take a test at school, you may miss a question. Mail is sometimes delivered to the wrong address. Errors do occur. The magnitude of the error depends on the *relative size* of the error to the overall situation.

Make a Prediction

If you miss three points on a test, what might you be able to tell about your performance?

1. Suppose you missed three points out of a total ten points. What percent of points did you miss?

2. What percent of points did you miss if the test had 100 total points?

3. If you have 1,000 points on tests throughout the semester and miss three points, what percent of points did you miss?

Missing three points on the 100-point test is a small error relative to the size of the test. Missing three points on the ten-point test is a large error relative to the size of the test. To analyze the "seriousness" of an error, you have to think about it relative to the size of the object or population being measured.

4. In general, what can you say about the relative size of an error as compared to the overall population?

Try It Out

Before the presidential election of 2004, forecasters predicted that 1.18×10^8 people would vote. When the election was over, 62,040,610 votes had been cast for President Bush, and 59,028,111 votes had been cast for Senator Kerry.

5. What was the total number of people who voted in the election?

6. What was the difference, in absolute value, between the prediction and the actual voter turnout?

7. Divide the difference from Question 6 by the total number of actual voters from Question 5. This number is the **relative error**. Express this value to four decimal places, in scientific notation, and as a percent error to the nearest percent.

Math Link

The absolute value of a number is its distance from 0 on the number line. The absolute value of both 2 and −2 is 2.

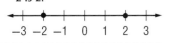

−3 −2 −1 0 1 2 3

Go on 

Real-World Link

Presidential elections are held every four years.

8. Suppose the relative error is 1. What does this tell you about the percent of error?

9. Suppose a state election predicted a voter turnout of six million, but only 2,931,279 actually voted.

a. What is the difference, in absolute value, between the prediction and actual voter turnout?

b. Divide the difference from Part a by the number of people who actually voted. Express this relative error to four decimal places, in scientific notation, and as a percent error to the nearest percent.

10. In Question 7, you found the relative error for predicting voter turnout in the presidential election was 2.53×10^{-2} compared to a relative error in the state election of 1.05×10^{0}. In which scenario was the relative error smaller. That is, which estimate for voter turnout was better? Explain.

What might appear to be a really large error, such as an estimation that is off by more than three million, may be *relatively* small when compared to the population being estimated. However, there are times when an error appears to be quite small but becomes a large *actual* error when magnified over a long period of time or in several situations.

Try It Again

Each town in Baxter County reported its population in scientific notation to the county seat. This sparked a debate about how much of an error would be created between the actual population and the reported population in the county.

11. Complete the table. Use your calculator to do the calculations.

Town	Actual Population	Population in Scientific Notation	Actual Error in Report |actual population − population in scientific notation|	Relative Error in Report [actual error/ actual population]
Springfield	1,414,999	1.41×10^6	4,999	0.0035
Cedarville	806,499			
Hillsboro	545,489			
Plainview	17,848			
Charleston	48,543			
Lewis City	6,024			

12. Is the relative error for any town greater than 1%?

13. Because of rounding, the scientific notation population numbers for all of these towns are less than the actual populations. Can you predict the possible maximum error, above or below, that is possible for a town's population if the population is expressed in scientific notation?

14. Is it possible for a town's actual population to be less than the population expressed in scientific notation? Explain.

15. Find the actual population in the entire county by adding the entries in the "actual population" column. Find the actual error in the report for the entire county by adding the entries in the "actual error in report" column. The error is the amount by which the county underestimated its population.

In your opinion, is it a small error or a large error? What if the county gets $1,000 per person in state tax money? How much would the county stand to lose by underestimating its population in this way?

What Did You Learn?

16. Suppose you hear that a person on your street will receive an annual raise of $5,000 today. What else would you need to know to decide how substantial this raise is? Use the term *relative* in your answer.

On Your Own Exercises

Lesson 4.1

Practice & Apply

Math Link

1 cm = $\frac{1}{100}$ m

1 mm = $\frac{1}{1{,}000}$ m

1 km = 1,000 m

Supply each missing value. For Exercises 3 and 4, give the answer in kilometers.

1.

2.

5
× 10
19 ft
?

3.

6
× 10
683 cm
?

4.

9
× 10
3.95 mm
?

Find each product.

5. 4×10^3 **6.** 62×10^4 **7.** 15.8×10^2

Find each exponent.

8.

?
× 10
6 in.
600 in.

9.

?
× 10
15 m
150 km

Find the value of *N* in each equation.

10. $8 \times 10^N = 80$

11. $53 \times 10^N = 5{,}300$

12. $9.9 \times 10^N = 99{,}000$

13. $6.13 \times 10^N = 6{,}130$

Find the length of each input.

14.

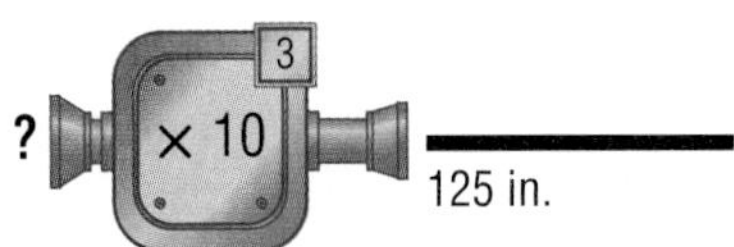

15.

5

? × 10

280,000 mi

16.

4

? × 10

75.45 km

Find the value of *N* in each equation.

17. $N \times 10^2 = 900$

18. $523 \times N = 52{,}300$

19. $0.0614 \times N = 6.14$

20. $N \times 10^6 = 39{,}650{,}000$

Write each number in standard notation.

21. 8×10^1

22. 5×10^3

23. 7.5×10^2

24. 3.54×10^8

25. 3.4×10^5

26. 6.19×10^7

Write each number in scientific notation.

27. 300

28. ten thousand

29. 158,000

30. 8,350,000,000

31. 183 billion

32. 421,938,000,000,000

33. four trillion six billion

34. 192,850,000,000

Real-World Link
Venus is the brightest planet in the solar system. It can often be seen in the morning and at night. Venus is sometimes called the Morning Star or the Evening Star.

35. Astronomy The table shows the mass of the planets in the solar system, the Sun, and Earth's moon.

Celestial Body	Mass (kg) Standard Notation	Mass (kg) Scientific Notation
Sun	1,990,000,000,000,000,000,000,000,000,000	1.99×10^{30}
Mercury	330,000,000,000,000,000,000,000	
Venus	4,870,000,000,000,000,000,000,000	
Earth	5,970,000,000,000,000,000,000,000	
Mars	642,000,000,000,000,000,000,000	
Jupiter	1,900,000,000,000,000,000,000,000,000	
Saturn	568,000,000,000,000,000,000,000,000	
Uranus	86,800,000,000,000,000,000,000,000	
Neptune	102,000,000,000,000,000,000,000,000	
Moon	73,500,000,000,000,000,000,000	

a. Write the mass of each planet and the moon in scientific notation.

b. Order the planets and the moon by mass, from least to greatest.

c. Which planet has about the same mass as Earth?

36. Refer to the table of masses in Exercise 35. Find how many times greater the mass of Earth is than the mass of each body.

a. Mercury **b.** Venus **c.** The moon

37. Refer to the table of masses in Exercise 35. Find how many times greater the mass of each body is than the mass of Earth.

a. The Sun **b.** Jupiter **c.** Saturn

d. Uranus **e.** Neptune

Estimate the value of each expression. Then use your calculator to evaluate it. Give your answers in scientific notation.

38. $7.35 \times 10^{22} - 1.33 \times 10^{21}$

39. $6.42 \times 10^{45} + 4.55 \times 10^{45}$

40. $6.02 \times 10^{23} \times 15$

Connect & Extend

15.7 cm

41. A dollar bill is approximately 15.7 cm long.

a. How long would one thousand dollar bills laid end-to-end be?

b. How long would one million dollar bills laid end-to-end be?

c. The distance from Earth to the Sun is about 1.5×10^8 km. How many dollar bills, taped together end-to-end, would it take to reach the Sun?

Find the length of each output.

42.

Math Link

12 inches = 1 foot

5,280 feet = 1 mile

43.

21 ft

3
× 10
× 4
?

44.

16 mm

2
× 10
÷ 2
?

45.

276 km

2
× 10
2
÷ 10
?

Find the length of each output.

46.

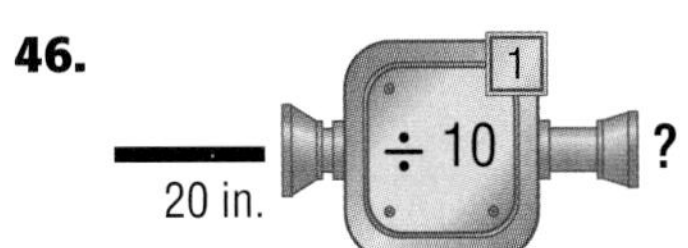

47.

48.

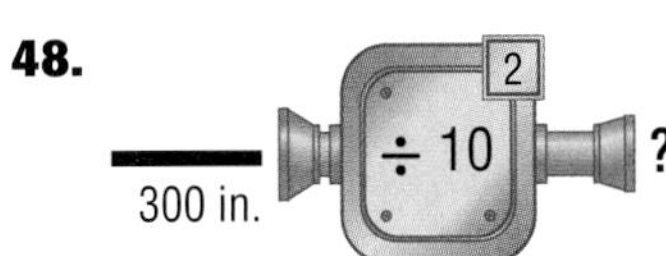

49.

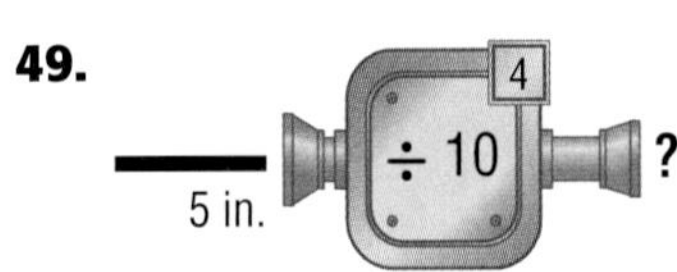

Find the value of N in each equation.

50. $N \times \frac{1}{10^2} = 8$ **51.** $N \times \frac{1}{10^4} = 12$ **52.** $N \times \frac{1}{10^6} = 4.927$

53. $17 \div 10^1 = N$ **54.** $128.4 \div 10^3 = N$ **55.** $714 \div 10^8 = N$

56. List these six numbers from least to greatest.

a quarter of a billion 2.5×10^7 two thousand million

10^9 half a million 10^5

57. Geometry Carson arranged 10^3 one-centimeter cubes into a large cube.

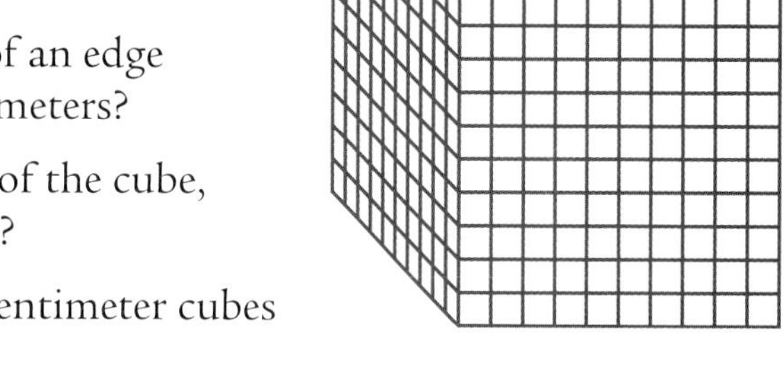

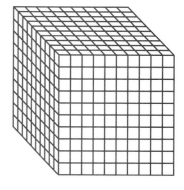

a. What is the length of an edge of the cube, in centimeters?

b. What is the volume of the cube, in cubic centimeters?

Sophie arranged 10^6 one-centimeter cubes into a large cube.

c. What is the edge length of the new cube, in centimeters? What is the volume of the new cube, in cubic centimeters?

d. What is the volume of the new cube, in cubic meters? What is the volume of the first large cube, in cubic meters?

e. What is the area of a face of the new cube, in square centimeters?

58. Geometry Irena stacked 10^6 two-centimeter cubes into a tower. Part of the tower is shown here. What would be the height of the tower in meters?

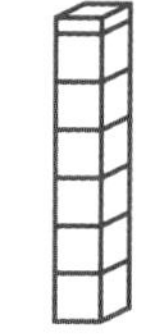

59. Preview Fill in the exponents below. (Hint: Fill in the exponents you know first. Then fill in the others by continuing the pattern.)

$$5{,}000 = 5 \times 10^?$$
$$500 = 5 \times 10^?$$
$$50 = 5 \times 10^?$$
$$5 = 5 \times 10^?$$
$$0.5 = 5 \times 10^?$$
$$0.05 = 5 \times 10^?$$

60. Economics In October 2007, the U.S. national debt was $9.1 trillion, or $\$9.1 \times 10^{12}$.

a. If the government could pay back $1 million each year and the debt did not increase, how long would it take to pay off the entire debt?

b. A typical railway freight car measures 3 meters by 4 meters by 12 meters. A dollar bill is approximately 15.7 cm long and 6.6 cm wide. A stack of 100 dollar bills is one centimeter high.

How many freight cars would it take to carry the 2007 U.S. federal deficit in $1 bills? In $100 bills? (Hint: a $100 bill is the same size as a $1 bill).

c. In 2007, the estimated population of the United States was 301,000,000. Suppose that every person in the country gave the government $1 each minute. How long would it take to pay off the deficit?

d. Invent another example to illustrate the size of the federal deficit.

61. In Your Own Words Explain why scientific notation is a useful way to write greater and lesser numbers.

Mixed Review

62. Find 33% of 140.

63. Find 14% of 31.

64. What percent of 250 is 15?

65. What percent of 12 is 0.12?

Write each expression in the form c^m.

66. $5^3 \cdot 5^2$

67. $2^7 \cdot 2^2$

68. $a^2 \cdot a^3$

69. Use the expression given to complete the table. Then write another expression that gives the same values. Check your expression with the values in the table.

m	1	2	3	4	5	100
$2(m-1)$						

Find each product.

70. $-4 \cdot 12$

71. $8 \cdot (-2.5)$

72. $-2 \cdot 4 \cdot (-8)$

73. $3 \cdot (-\frac{1}{6})$

Find each quotient.

74. $-50 \div 4$

75. $-4 \div 48$

76. $-8 \div (-64)$

77. $144 \div (-10)$

Negative Exponents

You have seen that negative numbers can be used to show amounts that are less than 0, such as low temperatures, debts, and elevations below sea level. Negative numbers can also be used as exponents.

Inv 1 Model Negative Exponents 194

Inv 2 Evaluate Expressions with Negative Exponents 198

Inv 3 Laws of Exponents and Scientific Notation 200

You have learned that positive exponents mean repeated multiplication. For example, 3^4 means $3 \cdot 3 \cdot 3 \cdot 3$. But what does 3^{-4} mean?

In this lesson, you will see that you can extend what you know about positive exponents to help you understand and work with expressions involving negative exponents.

Think & Discuss

In Chapter 2, you used stretching and shrinking machines as a way to think about exponents.

Take a look at this connection.

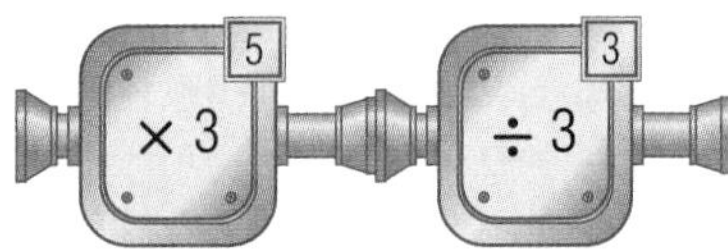

If you put a one-inch stick of modeling clay through this connection, how long will it be when it exits? Describe at least two ways that you could figure this out.

Investigation 1 Model Negative Exponents

You will soon use shrinking machines to help you think about negative exponents. First, though, you will review connections of machines with positive exponents.

Develop & Understand: A

If possible, find a single repeater machine that will do the same work as the given connection.

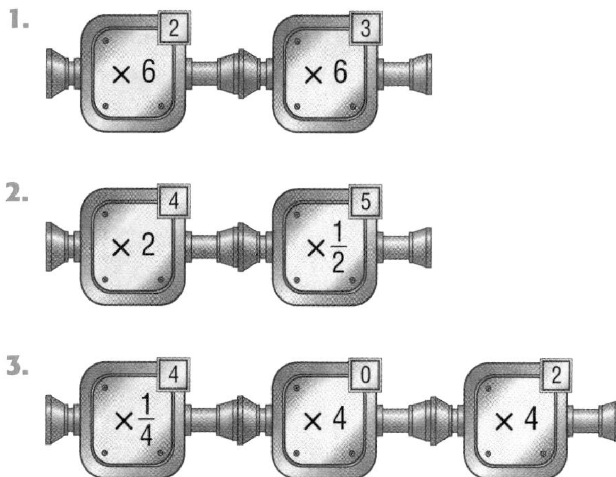

Jordan is a new employee at the resizing factory. He is assigned to all the machines with base 3. A clown troupe has asked him to stretch some ribbons for its balloons. To become familiar with the machines, Jordan sends one-centimeter pieces of ribbon through them and records the lengths of the outputs.

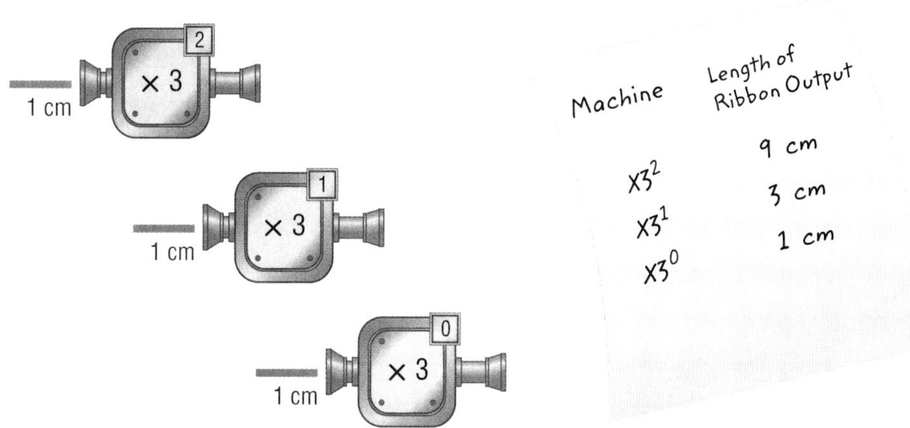

Think & Discuss

Jordan noticed a pattern in the outputs. Each time the number of repeats is reduced by 1, the resulting length is $\frac{1}{3}$ the previous length. Why?

Some of Jordan's base 3 machines have negative exponents. When he puts one-centimeter ribbons through a $\times 3^{-1}$ machine, a $\times 3^{-2}$ machine, and a $\times 3^{-3}$ machine, the pattern in the output length continues.

Develop & Understand: B

4. Continue Jordan's pattern in the chart below.

What Went In	Machine	What Came Out
1-cm ribbon	$\times 3^{2}$	9-cm ribbon
1-cm ribbon	$\times 3^{1}$	3-cm ribbon
1-cm ribbon	$\times 3^{0}$	1-cm ribbon
1-cm ribbon	$\times 3^{-1}$	
1-cm ribbon	$\times 3^{-2}$	
1-cm ribbon	$\times 3^{-3}$	

5. After conducting several experiments, Jordan concluded that these three machines do the same thing.

a. Suppose you put an 18-inch length of rope into a $\times 3^{-1}$ machine. Describe two ways that you could figure out how long the resulting piece would be.

b. Describe two other repeater machines, one using multiplication and one using division, that do the same thing as a $\times 3^{-2}$ machine.

c. Describe two other repeater machines, one using multiplication and one using division, that do the same thing as a $\times 3^{-3}$ machine.

6. Jordan repeated the pattern shown in Exercise 4 with two other bases.

   a. With base 2, Jordan found that each time the exponent was reduced by 1, the output length was half the previous output length. Complete the chart on the left below to show what happened to one-centimeter ribbons.

   b. A similar pattern appeared when Jordan put one-centimeter ribbons through base 4 machines. Complete the chart on the right below.

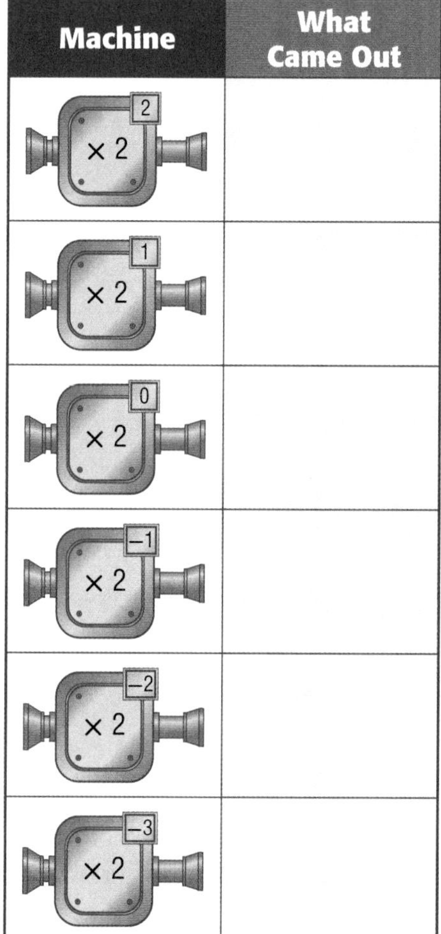

Machine	What Came Out
× 2 (2)	
× 2 (1)	
× 2 (0)	
× 2 (−1)	
× 2 (−2)	
× 2 (−3)	

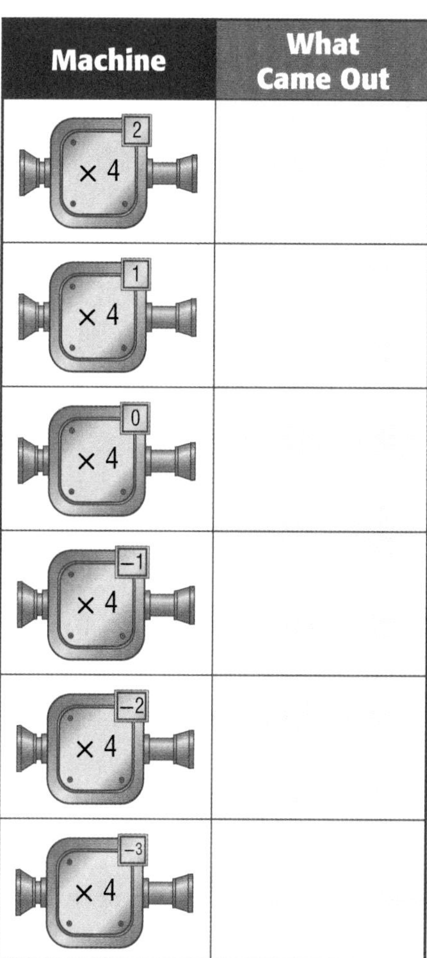

Machine	What Came Out
× 4 (2)	
× 4 (1)	
× 4 (0)	
× 4 (−1)	
× 4 (−2)	
× 4 (−3)	

Machines with other bases also follow the pattern. If you put a 72-inch pole through each machine, how long would each output be?

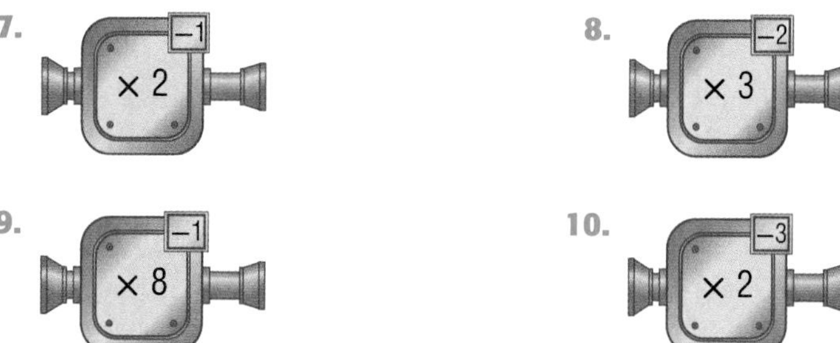

Share & Summarize

In general, when you put something in a multiplication machine with a negative exponent, like $\times 2^{-4}$, how can you find the output length?

Investigation 2 Evaluate Expressions with Negative Exponents

Thinking about shrinking machines can help you evaluate expressions with negative exponents.

Example

Evaluate 4^{-3}.

Think about a $\times 4^{-3}$ machine. Putting something through a $\times 4^{-3}$ machine is equivalent to putting it through a $\times\left(\frac{1}{4}\right)^3$ machine, so $\times 4^{-3} = \left(\frac{1}{4}\right)^3$, or $\frac{1}{64}$.

Evaluate $2^{-3} \cdot 2^2$.

Think about this connection.

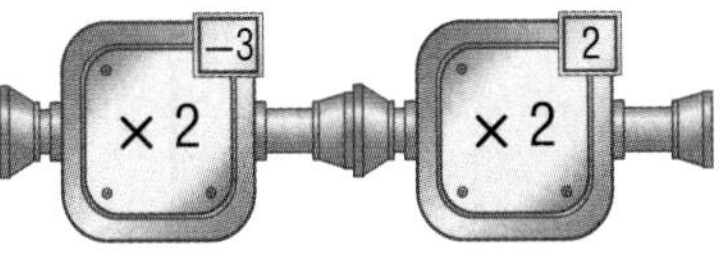

The $\times 2^{-3}$ machine halves the length of an input three times.

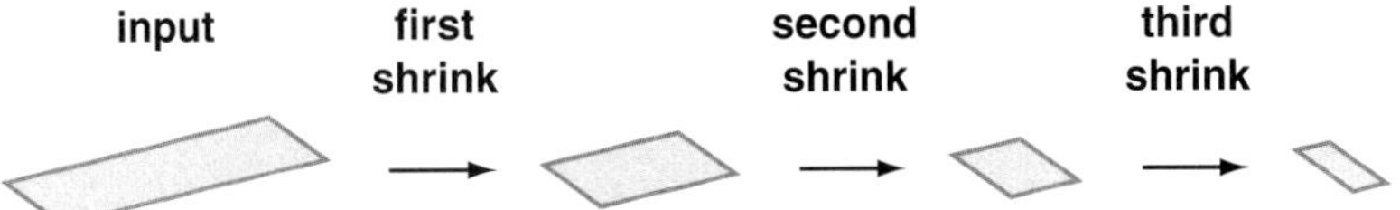

The $\times 2^2$ machine doubles the length twice.

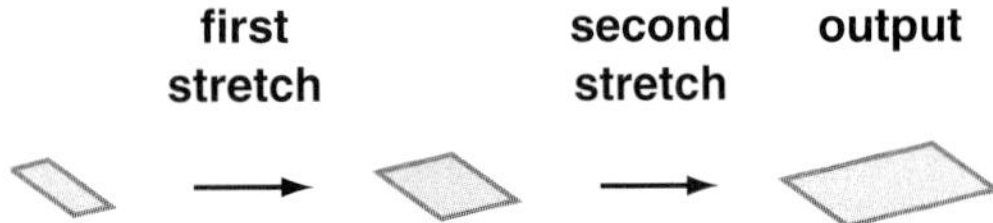

The output is half the length of the original input. The connection is the same as a $\times\frac{1}{2}$ or a $\times 2^{-1}$ machine. So, $2^{-3} \cdot 2^2 = \frac{1}{2}$, or 2^{-1}.

Now you will practice evaluating expressions with negative exponents. Pay attention to how you evaluate the expressions. Be ready to share your approach with your class.

Develop & Understand: A

Evaluate each expression.

1. 3^{-4}
2. 2^{-1}
3. $3^{-3} \cdot 3^{5}$
4. $7^{5} \cdot 7^{-7}$
5. $3^{9} \cdot 3^{-6}$
6. $-49 \cdot 7^{-2}$
7. $256 \cdot 4^{-3}$
8. $32 \cdot 1^{-5}$
9. $-60 \cdot 6^{-1}$
10. $2^{-2} \cdot 3^{-1}$
11. $2^{-1} \cdot 3^{-1} \cdot 5^{-2}$
12. $2^{-2} \cdot 4^{2} \cdot 2^{-1}$
13. Which is greater, 5^{-1} or 5^{-2}? Why?
14. Describe a connection of two repeater machines, one with a negative exponent, that does the same work as a $\times 5^{0}$ machine.
15. Is it possible to shrink something to a length of 0 by using shrinking machines? Why or why not?

Develop & Understand: B

You sometimes need to take extra care to be sure you understand a mathematical expression. For example, in expressions such as -3^{-2} and $(-3)^{-2}$, the two negative symbols give two different kinds of information.

16. Why is $-3^{-2} = -\frac{1}{9}$?
17. Why is $(-3)^{-2} = \frac{1}{9}$?
18. What does -4^{-2} equal?
19. Write an expression that has a negative exponent and equals $-\frac{1}{27}$.

Evaluate each expression.

20. -5^{-2}
21. $(-3)^{3}$
22. -1^{-5}
23. $(-1)^{-5}$
24. $(-5)^{-3}$
25. -2^{4}

Share & Summarize

1. Give two values of n for which $(-2)^{n}$ is positive.
2. Give two values of n for which $(-2)^{n}$ is negative.
3. Give two values of n for which $(-2)^{n}$ is not an integer.
4. Write an expression with a negative exponent that equals $-\frac{1}{64}$.

Investigation 3 Laws of Exponents and Scientific Notation

In Chapter 2, you learned about the laws of exponents, which describe how to simplify expressions with exponents.

Product Laws	Quotient Laws	Power of a Power Law
$a^b \cdot a^c = a^{b+c}$	$a^b \div a^c = a^{b-c}$	$(a^b)^c = a^{b \cdot c}$
$a^c \cdot b^c = (a \cdot b)^c$	$a^c \div b^c = (a \div b)^c$	

Now you will apply these laws to expressions with negative exponents.

Develop & Understand: A

The product laws work for negative exponents as well as for positive exponents. In Exercises 1–6, use the product laws to rewrite each expression using a single base. Use a calculator to check your answers.

1. $50^{-6} \cdot 50^{12}$
2. $2^5 \cdot 2^{-6} \cdot 2^3$
3. $0.5^{-10} \cdot 0.5^{-10}$
4. $3^{-1} \cdot 9$
5. $-3^{-3} \cdot 2^{-3} \cdot 4^{-3}$
6. $10^{-23} \cdot 10^{23}$

The quotient laws and the power of a power law also work for negative exponents. In Exercises 7–12, use the appropriate law to rewrite each expression using a single base. Use a calculator to check your answers.

7. $7^3 \div 7^5$
8. $12^{-4} \div 4^{-4}$
9. $10^{-5} \div 10^3$
10. $(3^{-3})^2$
11. $2.3^2 \div 2.3^{-6}$
12. $(2^{-5})^{-5}$

Rewrite each expression using a single base or exponent.

13. $m^{23} \cdot m^{-17}$
14. $7^{-x} \cdot 7^x$
15. $a^{-3} \div b^{-3}$

Rewrite each expression using only positive exponents.

16. $\left(\frac{2}{3}\right)^{-1}$
17. $\left(\frac{1}{2}\right)^{-3}$
18. 10^{-7}

Simplify each expression as much as you can.

19. $-3n \cdot -3n$
20. $5n \div 2n^{-2}$
21. $4y^2 \cdot 3y^{-2}$

In Lesson 4.1, you used scientific notation to represent very large numbers. You can also use scientific notation to represent very small numbers. For small numbers, scientific notation involves negative powers of 10.

Math Link

In scientific notation, a number is expressed as a power of 10 and as the product of a number greater than or equal to 1 but less than 10. For example, 5,400 is written 5.4×10^3.

Develop & Understand: B

A 5.3-cm drinking straw is sent through each machine. Give the length of the straw that emerges from each machine. Write your answers as decimals.

22. 23.

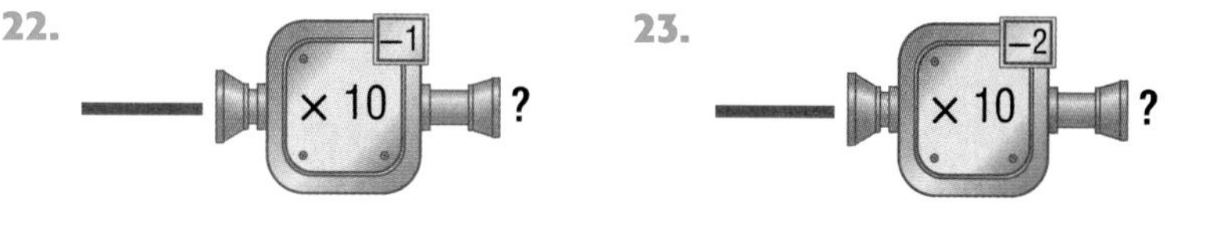

24. 25.

26. Write your answers for Exercises 22–25 in scientific notation.

27. List your answers for Exercises 26 from greatest to least.

Find each missing value.

28. $3 \cdot 10^{?} = 0.03$

29. $4.5 \cdot 10^{?} = 0.00045$

30. $? \cdot 10^{-3} = 0.0065$

31. $1 \cdot ? = 0.00000001$

32. There are 86,400 seconds in a day. How many days long is a second? Express your answer in scientific notation.

Share & Summarize

1. Without calculating the product, how can you tell whether $2^{-4} \cdot 2^3$ is greater than 0 or less than 0?
2. Which is smaller, 1×10^{-6} or 1×10^{-7}? How do you know?

On Your Own Exercises

Lesson 4.2

Practice & Apply

Find the length of the output if an 144-inch input is sent through each machine.

1.

2. 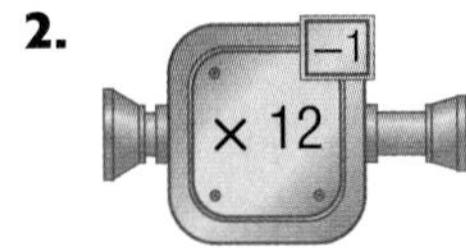

3. 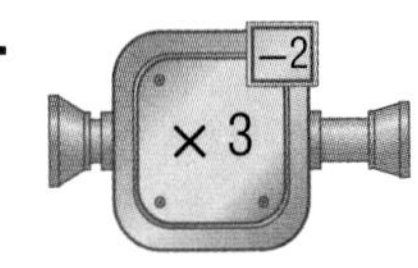

If possible, find a single repeater machine that will do the same work as the given connection.

4. 

5. 

6. 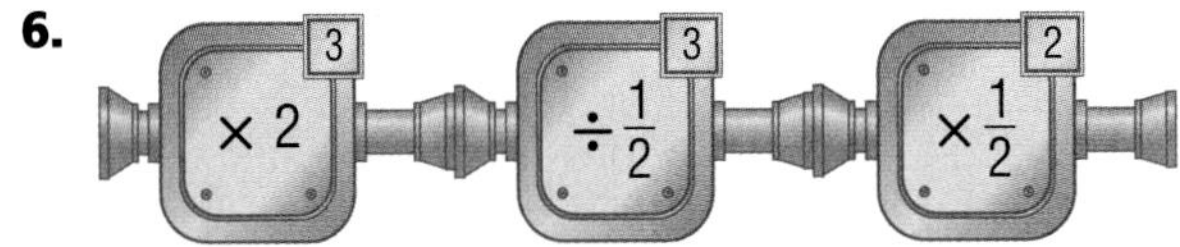

7. 

Describe a machine with a negative exponent that does the same work as the given machine.

8. 

9. 

10. 

11. 

12. 

13. 

14. Without calculating, figure out whether 4^{-3} or 4^{-4} is greater. Explain how you decided.

Evaluate each expression.

15. $-343 \cdot 7^{-2}$

16. $-1{,}375 \cdot 5^{-3}$

17. $128 \cdot 2^{-5}$

18. $-72 \cdot 9^{-1}$

19. $48 \cdot 6^{-1}$

20. $243 \cdot 3^{-4}$

21. $-9 \cdot 1^{-3}$

22. $4{,}096 \cdot 8^{-2}$

23. $-11 \cdot 11^{-1}$

24. How could you rewrite Exercise 22 in division? In multiplication using a positive exponent?

Decide whether each statement is true. If the statement is false, explain why.

25. $2^{-2} \cdot 2^{-1} = 2^{-3}$

26. $5^{-3} \cdot 5^{-1} \cdot 5^{2} = 5^{-2}$

27. $3^{-1} \cdot 4^{-3} = 12^{-4}$

28. $3^{-2} \cdot 3^{-3} = 9^{-5}$

29. $x^{-5} \div x^{7} = x^{-12}$

30. $k^{-2} \div k^{-5} = k^{3}$

31. $w^{-2} \div w^{-5} = w^{-7}$

32. $b^{-8} \div g^{-4} = bg^{-2}$

Without evaluating, tell whether each expression is greater or less than 1.

33. $3^{-4} \cdot 3^{5}$

34. $2^{-1} \cdot 2^{-3}$

35. $1^{-3} \div 2^{-3}$

36. $10^{-5} \div 10^{3}$

Evaluate each expression.

37. -4^{-2}

38. $(-5)^{-1}$

39. -2^{3}

40. $(-10)^{4}$

Rewrite each expression using a single base and a single exponent.

41. $\frac{1}{2^{-2}} \cdot 4^{-2}$

42. $100^{-20} \div -25^{-20}$

43. $-3^{100} \cdot -\frac{5}{6^{100}} \cdot 10^{100}$

44. $\left(\frac{1}{3}\right)^{-a} \cdot \left(\frac{1}{5}\right)^{-a}$

45. $-1{,}000^{0} \div 3^{0}$

46. $15^{-5} \cdot \left(\frac{1}{5}\right)^{-5}$

Find the value of *n* in each equation.

47. $7.24 \times 10^{n} = 0.00724$

48. $5 \times 10^{n} = 0.5$

49. $n \times 10^{-4} = 0.000104$

50. $9.2 \cdot n = 0.0092$

51. A mile is 1.609×10^{5} centimeters. How many miles is a centimeter? Express your answer in scientific notation.

52. There is 7.8125×10^{-3} gallon in an ounce. How many ounces are in a gallon? Express your answer in scientific notation.

Challenge Simplify each expression.

53. $-6a^{-2} \div 4a^{-4}$

54. $-\frac{1}{2}x^{-3} \cdot x^{4}$

55. $z^{2} \cdot 2z^{-3} \div -6z^{-2}$

56. $9b^{4} \cdot 9b^{-8}$

Connect & Extend

Find the length of the output if a meterstick is sent through each connection.

57. 

58. 

59. 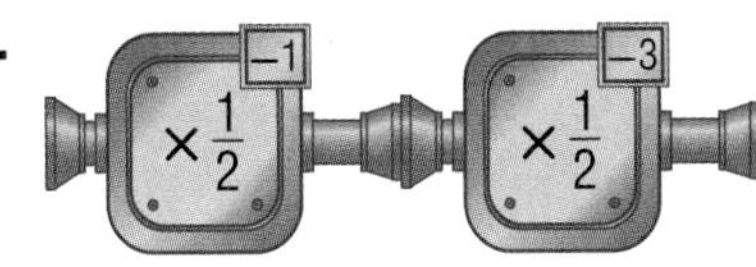

60. 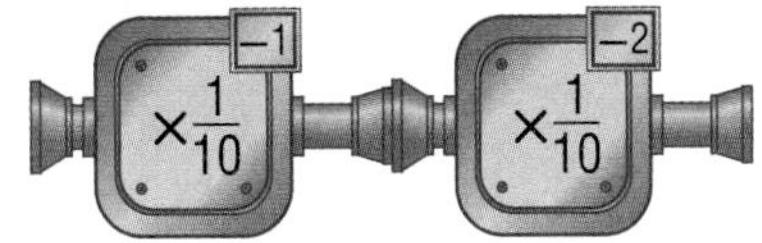

61. A 1-inch length of modeling clay went into each machine. Complete the charts.

Machine	What Came Out
2, $\times\frac{1}{2}$	
1, $\times\frac{1}{2}$	
0, $\times\frac{1}{2}$	

Machine	What Came Out
−1, $\times\frac{1}{2}$	
−2, $\times\frac{1}{2}$	
−3, $\times\frac{1}{2}$	

62. Challenge Describe a machine that has a negative exponent and that does the same work as the machine shown below.

63. Challenge A strand of wire is sent through this connection. It emerges 2 inches long. What was the starting length?

Without calculating, figure out whether each expression is positive or negative. Explain how you decided.

64. $(-4)^{-1}$

65. $(-4)^{-2}$

66. $(-4)^{-3}$

67. $(-4)^{-4}$

68. $(-2)^{-3}$

69. $(-2)^{-2}$

70. $(-5)^{-1}$

71. $(-10)^{-6}$

72. Prove It! It is true that $\left(\frac{1}{2}\right)^{-1} = (2^{-1})^{-1} = 2^1 = 2$. In Parts a–c, you will explain each step in this calculation.

a. Why does $\left(\frac{1}{2}\right)^{-1} = (2^{-1})^{-1}$?

b. Why does $(2^{-1})^{-1} = 2^1$?

c. Why does $2^1 = 2$?

73. Physics Protons, neutrons, and electrons are *stable* particles. They can exist for a very long time. But many other particles exist for only very short periods of time. A *muon,* for example, has an average life of 0.000002197 second.

a. Write this length of time in scientific notation.

b. A particular muon existed for half the average life of this type of particle. How long did it exist?

c. Another muon existed for three times the average life of this type of particle. How long did it exist?

74. In Your Own Words Write a letter to someone in your family who does not know what negative exponents are. In your letter, describe why 3^{-1} is equal to $\frac{1}{3}$.

Mixed Review

Order each group of numbers from least to greatest.

75. $\frac{6}{5}, 0.2, \frac{3}{5}, 0.4, \frac{5}{3}$

76. 0.75, 0.18, 0.25, 0.025, 0.7

Find each product. Express your answers in scientific notation.

77. $(5 \times 10^{14}) \times (11 \times 10^{13})$

78. $(8 \times 10^{9}) \times (6 \times 10^{6})$

Fill in the blanks to make true statements.

79. $0.5 + \_\_\_\_ = 1$

80. $0.5 + 0.25 + \_\_\_\_ = 1$

81. $\frac{1}{8} + \_\_\_\_ = 1$

82. $\frac{8}{5} - \_\_\_\_ = 1$

83. $\frac{9}{12} + \frac{3}{12} - \_\_\_\_ = 1$

84. $\frac{3}{5} + \frac{7}{10} - \_\_\_\_ = 1$

Find the product or quotient.

85. $12(-25)$

86. $-352 \div 11$

87. $\frac{2}{3} \cdot \left(-\frac{4}{5}\right)$

88. $\frac{2}{3} \div \left(-\frac{4}{5}\right)$

89. Hobbies Clara wants to display her collection of 30 first-class stamps in groups of equal size. What group sizes are possible?

Find each product.

90. $4 \cdot \frac{1}{10^2}$

91. $61 \cdot \frac{1}{10^2}$

92. $842 \cdot 10^{-3}$

93. 750×10^{-4}

CHAPTER 4

Review & Self-Assessment

Chapter Summary

In this chapter, you learned about powers of 10. In addition, you found that sometimes it is useful to express large numbers in scientific notation.

You discovered what negative numbers mean as exponents. You learned that the laws of exponents apply when the exponents are negative. You also practiced writing very small numbers in scientific notation.

Vocabulary

greater
lesser
relative error
scientific notation

Strategies and Applications

Expressing large numbers in scientific notation

1. Explain how to write a large number in scientific notation. Use an example to demonstrate your method.

2. Explain how to write $1{,}234 \times 10^9$ in scientific notation.

Evaluating expressions involving negative exponents

3. What are two ways to find the value of $\left(\frac{1}{2}\right)^{-1} \cdot \left(\frac{1}{2}\right)^{-1}$?

4. Use the facts that $2^2 = 4$, $2^1 = 2$, and $2^0 = 1$ to explain why $2^{-1} = \frac{1}{2}$.

5. How can you use the laws of exponents to tell whether $3^4 \cdot 3^{-6}$ is greater than 1 or less than 1?

Demonstrating Skills

Write each number in scientific notation.

6. 123,000

7. 57,700,000

8. 45×10^6

9. 321×10^3

10. 684×10^5

11. 92×10^7

12. Earth is approximately 93,000,000 miles from the Sun. Express this number in scientific notation.

13. Mars is approximately 1.42×10^8 miles from the Sun. Express this number in standard notation.

14. Imagine putting a 5-foot fishing rod into this connection of shrinking machines.

a. Write this situation using multiplication.

b. Write this situation using negative exponents.

c. What is the length of the exiting fishing rod?

Find each missing value.

15. $8.3 \times 10^{?} = 0.0083$

16. $3.7 \times ? = 0.00037$

17. $1.9 \times 10^{?} = 0.19$

18. $3.2 \times ? = 0.00000032$

Write each number in standard notation.

19. 1.4×10^{-2}

20. 6.21×10^{-5}

21. 7.52×10^{-3}

22. 8.119×10^{-6}

Simplify each expression as much as possible.

23. $256 \cdot 2^{-5}$

24. $3y^{2} \cdot y^{-3}$

25. $4^{-1} \cdot 4^{-3}$

26. -1^{-3}

27. $(-5)^{-2}$

28. -4^{3}

29. -5^{-2}

30. -4^{-3}

31. A typical flu virus has a diameter of 2×10^{-7} meter. Write the diameter in standard notation.

32. Consider the following numbers.

1.03×10^{4} $\quad 1.1 \times 10^{-3}$ $\quad 1.03 \times 10^{-4}$

0.13×10^{-2} $\quad 1.1 \times 10^{3}$ $\quad 10.03 \times 10^{-1}$

a. Order the numbers from least to greatest.

b. Which numbers are written in scientific notation?

Test-Taking Practice

SHORT RESPONSE

1 Simplify the expression shown below. Write your answer in standard form.

$$\frac{(5.1 \times 10^3)(2 \times 10^4)}{3 \times 10^9}$$

Show your work.

Answer ____________________

MULTIPLE CHOICE

2 What is 8.2×10^4 written in standard form?

A 0.00082
B 8,200
C 0.82
D 82,000

3 Write 45.78×10^{-4} in scientific notation.

F 4.578×10^{-3}
G 4.578×10^{-4}
H 4.578×10^{-5}
J 4.6×10^{-3}

4 What is 0.00068 written in scientific notation?

A 6.8×10^{-4}
B 6.8×10^{-5}
C 68×10^{-5}
D 0.68×10^{-3}

5 Simplify the expression $(6^4 \div 6^{-3}) \times 6^{-5}$.

F 6^{-4}
G 6^{-2}
H 6^2
J 6^6

6 Which expression is equivalent to 7×3^{-2}?

A $\frac{1}{9^2}$
B $\frac{7}{9}$
C 10^{-2}
D 21^{-1}

7 Simplify the expression $10^{-2} \times 1^{-2} \times 10$.

F 0.01
G 0.1
H 10
J 100

CHAPTER 5

Geometry in Three Dimensions

Contents in Brief

5.1	Surface Area and Volume	212
5.2	Nets and Solids	228
5.3	Mass and Weight	240
	Review & Self-Assessment	256

Real-Life Math

Patterns and Plans Some of the most powerful examples of using two-dimensional drawings to represent three-dimensional objects can be found in different types of designing. Architects must be very skilled at drawing three-dimensional objects so there is no confusion about what the drawings represent.

Before a house is built, an architect creates drawings of what the house will look like. The drawings include elevations that show how the house will look from two sides.

In addition, the plans for a house include blueprints, which show how the interior will be divided into rooms. Blueprints show details such as where doors are and which way they open as well as where to place items like sinks, the stove, and the bathtub. Construction workers and contractors refer to blueprints when they work on projects.

Think About It Can you think of other careers or companies that use two-dimensional drawings to represent three-dimensional objects?

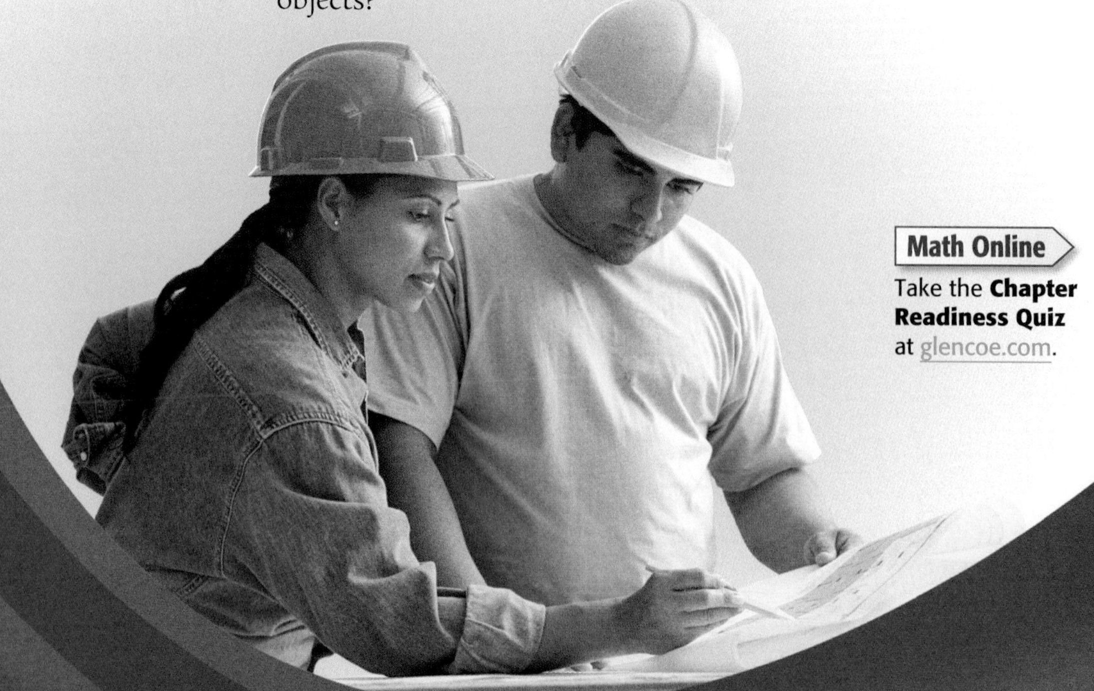

Math Online
Take the **Chapter Readiness Quiz** at glencoe.com.

Dear Family,

In Chapter 5, the class will begin studying three-dimensional geometry, focusing on prisms and cylinders. Students will learn how to use two-dimensional nets to represent three-dimensional solids. They will also learn the relationship between measures of mass and weight.

Key Concept–Nets

A **net** is a flat figure that can be folded to form a closed, three-dimensional object. This type of object is called a geometric **solid**.

Nets made of squares and rectangles can be used to represent cubes and **prisms**. Nets that include circles can be used to represent **cylinders**.

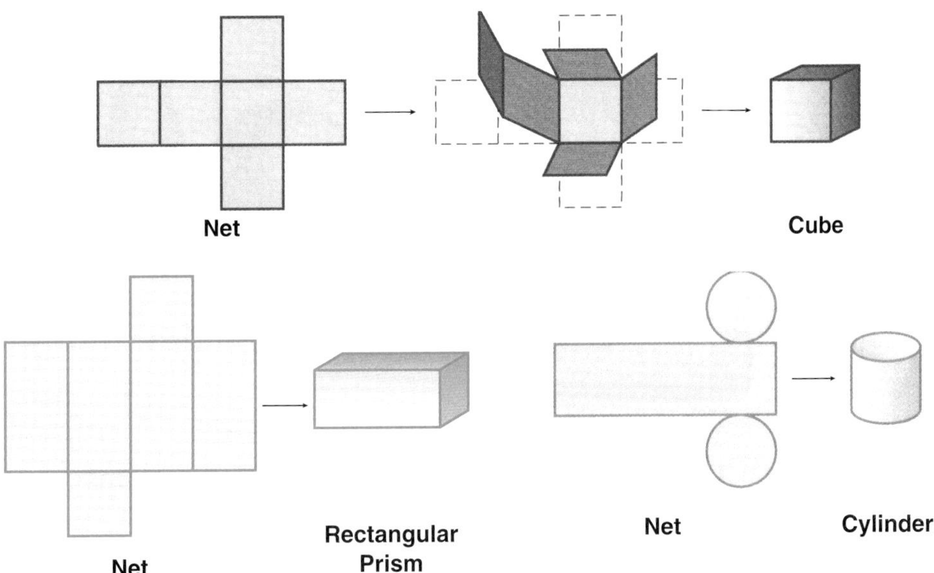

Students will use nets and models to find the surface area and volume of geometric solids. Finally, your student will explore weight and mass.

Chapter Vocabulary

mass	**volume**
surface area	**weight**

Home Activities

- Look at common household objects, such as tissue boxes and canned goods, from different perspectives.
- Make nets from food containers by deconstructing objects, such as empty cereal boxes and movie theater popcorn tubs.
- Estimate the mass and weight of objects, like pencils, books, dinner plates, chairs, and tables. Discuss the appropriate unit of measure for each, such as milligrams, grams, kilograms, ounces, or pounds.

Surface Area and Volume

Inv 1	Measure Prisms	213
Inv 2	Volume of Prisms and Cylinders	216
Inv 3	Package Design	220

Vocabulary

prism

These figures are *prisms.*

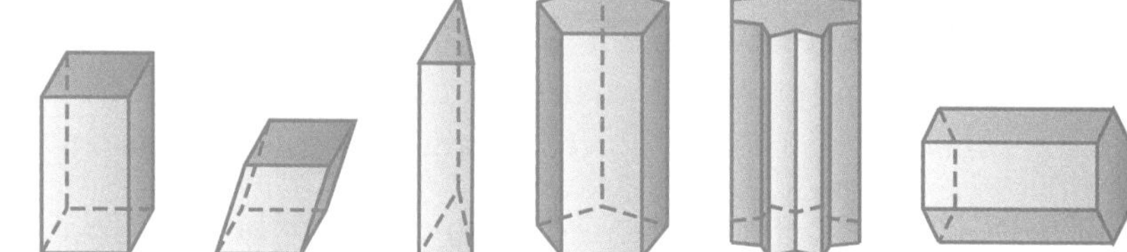

These figures are not prisms.

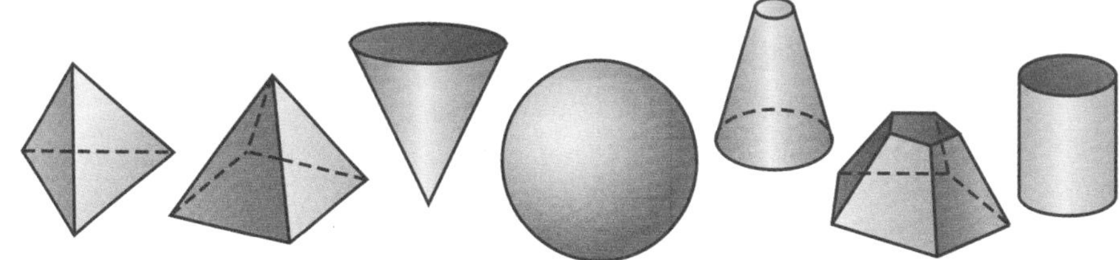

Think & Discuss

What do all the prisms have in common?

How are the nonprisms different from the prisms?

All **prisms** have two identical, parallel faces. These two faces are always polygons. A prism's other faces are always parallelograms.

A prism is sometimes referred to by the shape of the two identical faces on its ends. For example, a *triangular prism* has triangular faces on its ends, and a *rectangular prism* has rectangular faces on its ends.

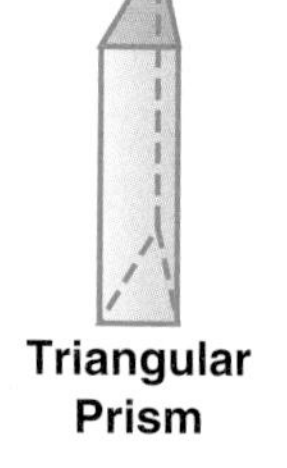

Triangular Prism

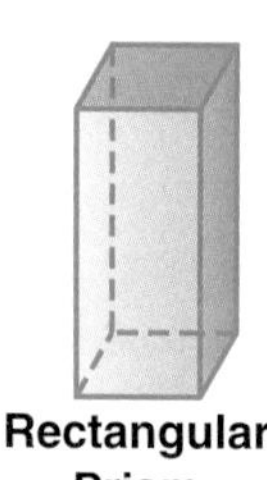

Rectangular Prism

Investigation 1 Measure Prisms

Vocabulary

surface area

volume

You encounter prisms in many everyday situations. Your bedroom is a prism. A cereal box is a prism. The box you receive in the mail with a new book is also a prism. Each prism has different characteristics.

In this investigation, you will explore two specific characteristics of prisms: **surface area** and **volume**.

Think & Discuss

Look around your classroom and choose a prism.

What are the dimensions of your prism? You can estimate the dimensions if it is difficult to measure them.

What does it mean to measure the volume of your prism? Estimate its volume and include units in your answer.

What does it mean to measure the surface area of your prism? Estimate its surface area and include units in your answer.

Develop & Understand: A

1. Consider a box with a bottom that is 16 centimeters by 6 centimeters and a height of 8 centimeters.

   8 cm

   16 cm

   6 cm

   a. How many centimeter cubes would be needed to cover the bottom of the box?

   b. How many layers like the one covering the bottom could be stacked in the box?

   c. How many total cubes would be needed to fill the box?

   d. What calculation can you use to show the number of cubes needed? How is this related to the dimensions of the box?

2. Mrs. White needs to purchase a briefcase. She found two briefcases that she can afford, but she is not sure which will give her room to pack the most belongings. Determine the volume of each briefcase to find out which will hold more.

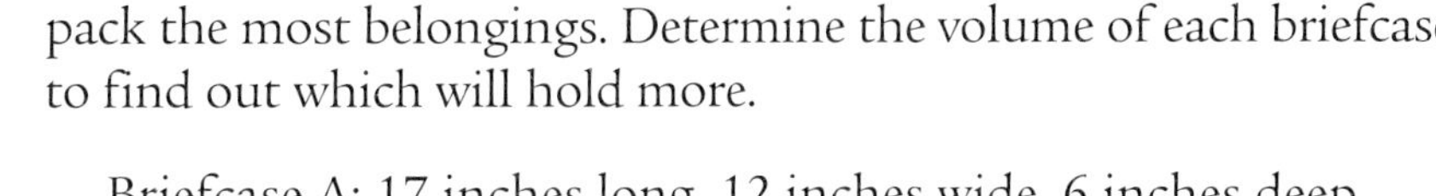

Briefcase A: 17 inches long, 12 inches wide, 6 inches deep

Briefcase B: 18 inches long, 10 inches wide, 6.5 inches deep

3. Nancy plans to store flour in a plastic container in her pantry. Three rectangular prism containers each hold the same amount of flour. Nancy wants to make an environmentally friendly decision by choosing the container that is made from the smallest amount of plastic. Find the surface area of each container to determine which is produced using the least material.

   Container 1: 6 inches by 3 inches by 8 inches

   Container 2: 4 inches by 6 inches by 6 inches

   Container 3: 4 inches by 2 inches by 18 inches

Develop & Understand: B

4. Compare the volumes and surface areas of boxes A, B, and C.

   Box A:

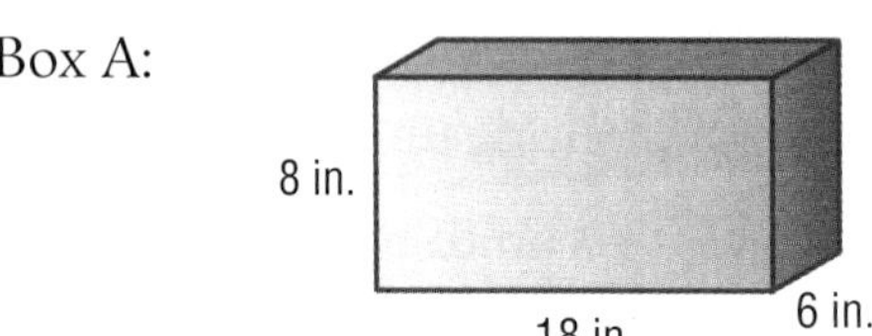

   Box B:

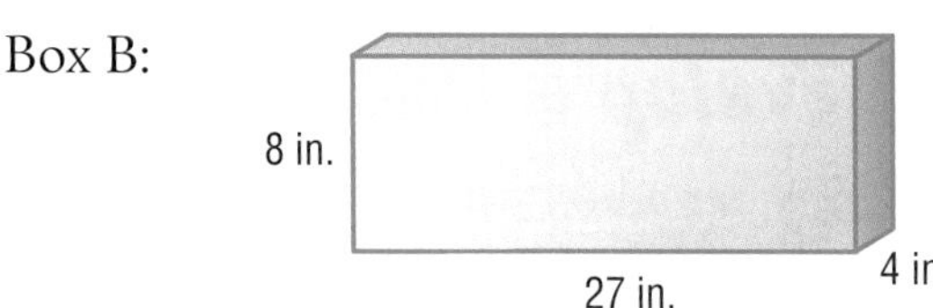

   Box C:

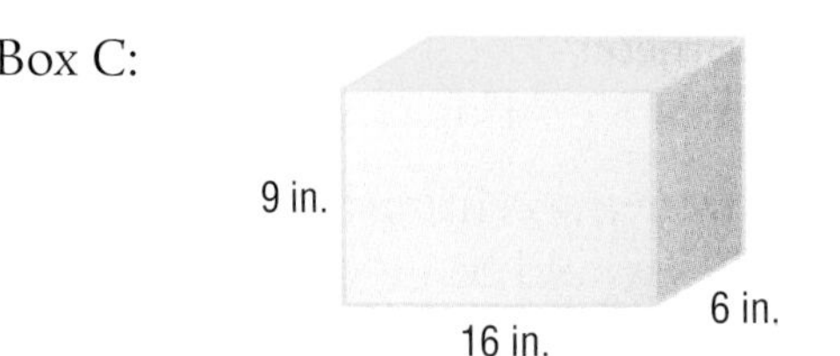

5. Determine the prime factorization of 72.
   a. Use this information to sketch and label two different rectangular prisms with a volume of 72 cubic centimeters.
   b. Which of your prisms has the greatest surface area?

6. Determine the prime factorization of 600.
   a. Use this information to sketch and label two different rectangular prisms with a volume of 600 cubic centimeters.
   b. Which of your prisms has the least surface area?

Real-World Link

One feature of a vehicle given in cubic units is the *cargo room.* This is the space available in the trunk of the vehicle. If an employer provides a car to an employee, there may be a minimum amount of cargo room required to carry materials.

Think & Discuss

Consider your answers to Exercises 5 and 6. Work with a partner to state a formula to find the volume of a rectangular prism.

Can you think of another way to write the formula?

Develop & Understand: C

Use your calculator for the following exercises.

7. Find the volume of a rectangular prism with dimensions 6.5 cm by 4 cm by 10.2 cm.

8. A rectangular prism has a volume of 48 cubic feet and a surface area of 88 square feet. What are its dimensions?

9. A rectangular prism has a volume of 8 cubic yards. Assume the dimensions are whole numbers. What dimensions yield a prism with the greatest surface area? The least surface area?

You have looked at the surface area and volume of many different rectangular prisms. Now you will explore what happens when you change a dimension of the prism by a unit.

10. Complete the following table about a rectangular prism.

Length (inches)	Width (inches)	Height (inches)	Surface Area (in^2)	Volume (in^3)
4	5	6	148 in^2	120 in^3
5	5	6		
6	5	6		
7	5	6		

a. Is there a pattern to the increase in surface area? In volume?

b. Explain why a pattern does or does not exist.

c. What happens if you keep the length as 4 inches, keep the height as 6 inches, and increase the width of the prism by 1-inch increments?

Share & Summarize

1. How do you calculate the volume of a rectangular prism?
2. How do you calculate the surface area of a rectangular prism?
3. If you know the volume of a rectangular prism, can you always determine its surface area? Why or why not?
4. How can prime factorization help you design a rectangular prism with a given volume?

Investigation 2 Volume of Prisms and Cylinders

Vocabulary

base

cylinder

Will your method for finding the volume of a rectangular prism work for other prisms? The exercises that follow will help you draw a conclusion.

Remember that a prism has two identical, parallel faces that can be any type of polygon. These faces are called the **bases** of the prism.

Develop & Understand: A

Math Link

Your block has an edge length of 1 unit and a volume of 1 cubic unit. Each face has an area of 1 square unit.

1. This triangle is half of the face of a square block that has an edge length of one unit.

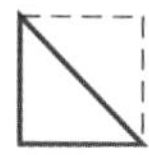

   a. What is the area of this triangle?

   b. If you cut one of your blocks in half as shown here, you could build a structure 1 unit high with the triangle as its base. What would be the volume of this structure?

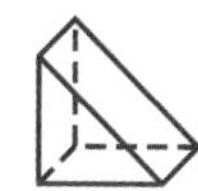

   c. If you built a structure 10 units high that had the triangle as its base, what would be its volume?

2. The dashed lines on the parallelogram below show the relationship between the parallelogram and a face of one of your blocks.

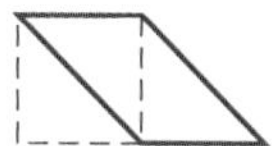

   a. What is the area of the parallelogram?

   b. If you had blocks like the one shown here, you could build a structure 2 units high that had the parallelogram as its base. What would be its volume?

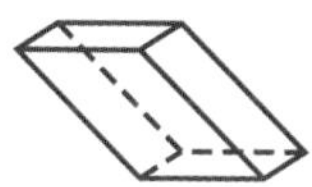

   c. If you could build a structure h units high with this parallelogram as its base, what would be its volume?

A **cylinder** is a three-dimensional object with two parallel, identical circular bases. The following are cylinders.

 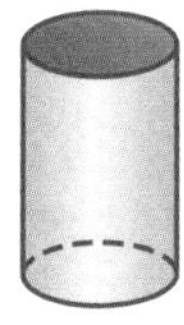 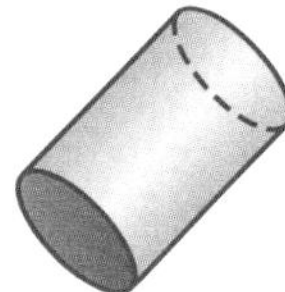 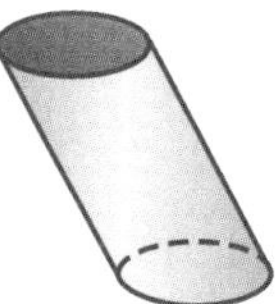

Math Link

The area of a circle is πr^2, where r is the radius and π is about 3.14.

3. The segments in the figure below show the relationship between the circle and a face of one of your blocks.

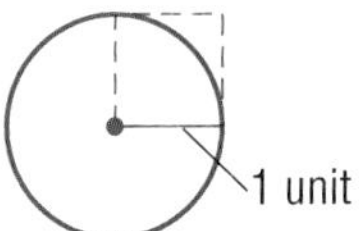

   a. What is the area of the circle?

   b. If you had blocks like this one, you could build a structure 3 units high with the circle as its base. What would be its volume?

   c. If you could build a structure 1,000 units high with this base, what would be its volume?

4. Felix drew a rectangular floor plan for a playroom.

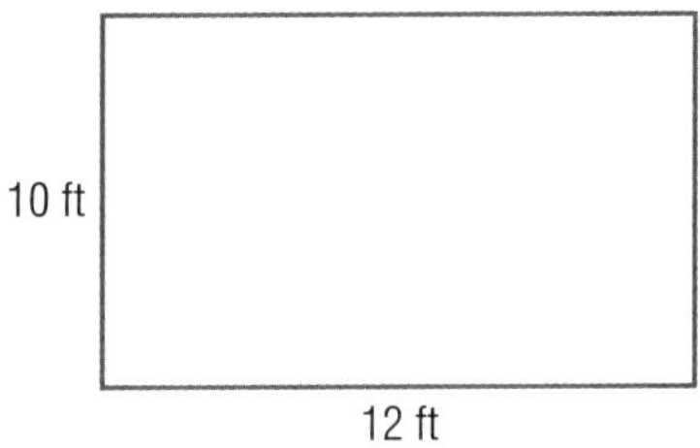

   a. What is the area of the floor?

   b. If the playroom has 8-foot ceilings, what is its volume?

   c. If the playroom has 10-foot ceilings, what is its volume?

5. Here is a design for a large oil tank.

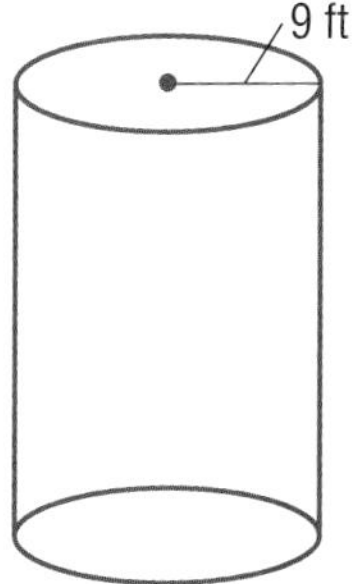

   a. What is the area of the base of the tank?

   b. If the tank is 50 feet high, how much water will it hold (in cubic feet)?

   c. If the tank is h feet high, how much water will it hold?

Think & Discuss

- Describe a single strategy that you think will work for finding the volume of *any* prism. Test your strategy with the prisms in Investigations 1 and 2.
- Will the same strategy work for the volume of any *cylinder?* If not, can you modify the strategy so that it *will* work?

Develop & Understand: B

In the previous exercises, you have been finding the volumes of *right prisms,* where the sides are perpendicular to the base. Some prisms, called *oblique prisms,* are slanted.

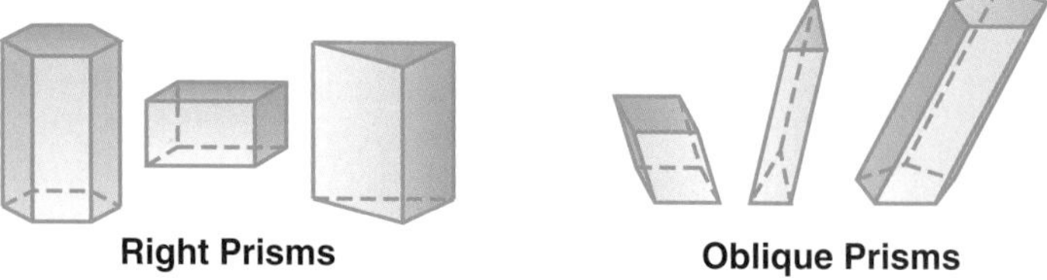

6. This right prism has been sliced into very thin pieces, like a deck of cards. These "cards" have then been pushed into an oblique prism, but the cards themselves have not changed.

   a. If each card is 5 centimeters wide and 8 centimeters long and the stack is 15 centimeters high, what is the volume of the first "deck"?

   b. Is the volume of the second deck *the same as, greater than,* or *less than* the volume of the first deck? Explain your answer.

   c. Is the base of the second deck *the same as, greater than,* or *less than* the base of the first deck? Explain.

   d. The diagram at the right shows how to measure the height of an oblique prism. Is the height of the second deck *the same as, greater than,* or *less than* the height of the first deck? Explain.

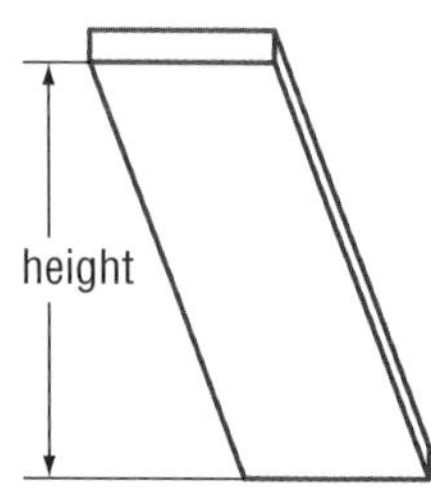

7. Now think about a deck of circular cards.

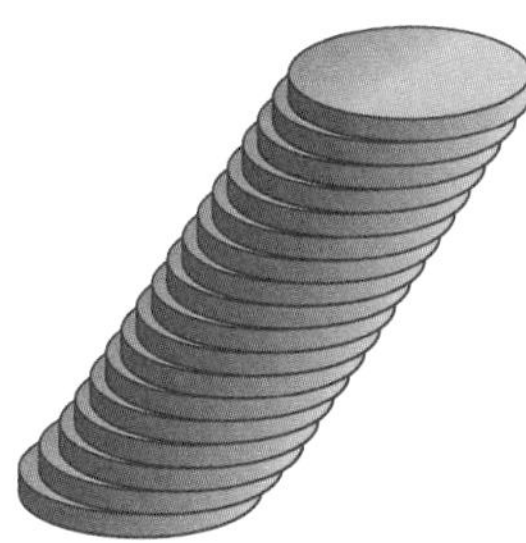

a. If the cards each have radius 4 cm and the stack is 9 cm high, what is the volume of the first deck?

b. Is the volume of the second deck *the same as, greater than,* or *less than* the volume of the first deck? Explain.

c. Is the base of the second deck *the same as, greater than,* or *less than* the base of the first deck? Explain.

d. Is the height of the second deck *the same as, greater than,* or *less than* the height of the first deck? Explain.

8. Will the formula *area of base* × *height* give you the volume of an oblique prism? Explain.

Share & Summarize

1. Write a formula for the volume of a prism based on the area of the base A and the height h. Explain why your formula works.

2. Write a formula for the volume of a rectangular prism, using l, w, and h. Write another formula for the volume of a cylinder, using r and h.

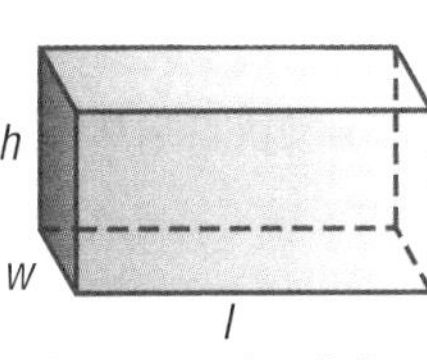

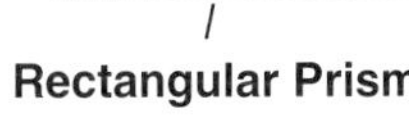

Rectangular Prism

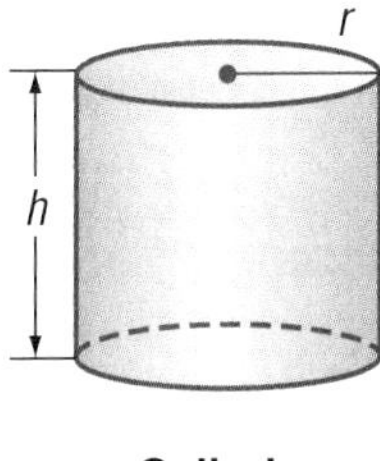

Cylinder

Inquiry

Investigation 3 Package Design

Materials

- materials for making models of beverage containers

Bursting Bubbles beverage company wants to attract attention to its products at an upcoming convention. The company normally packages its carbonated water in 350-milliliter cylindrical cans that are 15 centimeters high.

Bursting Bubbles wants to show creativity at the convention by packaging its carbonated water in new containers of different shapes. The company has commissioned you, a mathematician, to investigate some other sizes and shapes and to make a recommendation.

Company representatives tell you that the new containers can be any height at all, but they must have a volume of 350 milliliters. It is now up to you to create some attention-getting containers.

The company gave the required volume in milliliters (mL). You probably already know that there are 1,000 mL in 1 liter (L). To do this investigation, you also need to know that 1 milliliter has the same volume as 1 cubic centimeter.

Make a Prediction

You decide to design some cylindrical containers first. As the height of the can changes, the base area must also change to keep the volume fixed at 350 milliliters.

1. You could design a really tall container, even as tall as 1 meter (100 cm). To keep the volume 350 milliliters, would the radius of the base circle have to increase or decrease as the height of the can increased?
2. Suppose you designed a very short can, even as short as 2 centimeters. To keep the volume 350 milliliters, would the radius of the base circle have to increase or decrease as the height decreased?

Try It Out

Math Link

The number π, pronounced "pie" and spelled pi, is about 3.14. Your calculator probably has a π key to make these calculations easier.

3. Bursting Bubble's standard cans are 15 centimeters high. What is the radius of the circular base of these cans?

To find the radius, you can use a strategy of systematic trial and error. The table shows the volume for a base radius of 3 centimeters and of 2.5 centimeters. Use a calculator to find the radius that would give a volume of 350 milliliters. Keep searching until you get within 1 milliliter of 350 milliliters.

Base Radius (cm)	Base Area (cm^2) ($A = \pi \times r \times r$)	Volume (mL, want 350) (base area × height)
3	28.27	424.05 mL (too large)
2.5	19.635	294.525 mL (too small)

4. Choose five heights from the table below and make your own table. Complete your table with radius values that give a volume within 1 milliliter of 350 milliliters.

Height (cm)	Radius of Base Circle (cm)	Volume (cm^3)
1		
2		
3		
5		
10		
20		

Height (cm)	Radius of Base Circle (cm)	Volume (cm^3)
30		
50		
75		
90		
100		
150		

5. Were your predictions from Questions 1 and 2 correct?

Try It Again

The committee wants you to consider other shapes, not just cylinders, that might get attention at the convention. You can think about cones, spheres, other prisms, anything you want.

Here are two shapes, with formulas for their volumes. Use this information to answer the next set of questions.

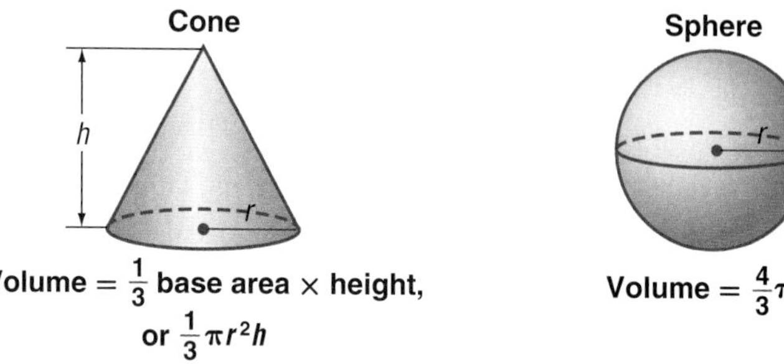

6. Try the cone shape first. Choose at least three heights for the cone. For each height, find the radius needed to give a cone a volume of 350 mL. Get within 1 mL of 350 mL.
7. Using the formula for the volume of a sphere, find as many spheres as you can with a volume of 350 mL.
8. Find at least two other shapes for containers with a volume of 350 mL.

What Did You Learn?

Choose one design to present to the company.

9. Describe your design to the company, including its dimensions.
10. Explain why your design might be successful at the convention.
11. Make a model of your design to show the company.

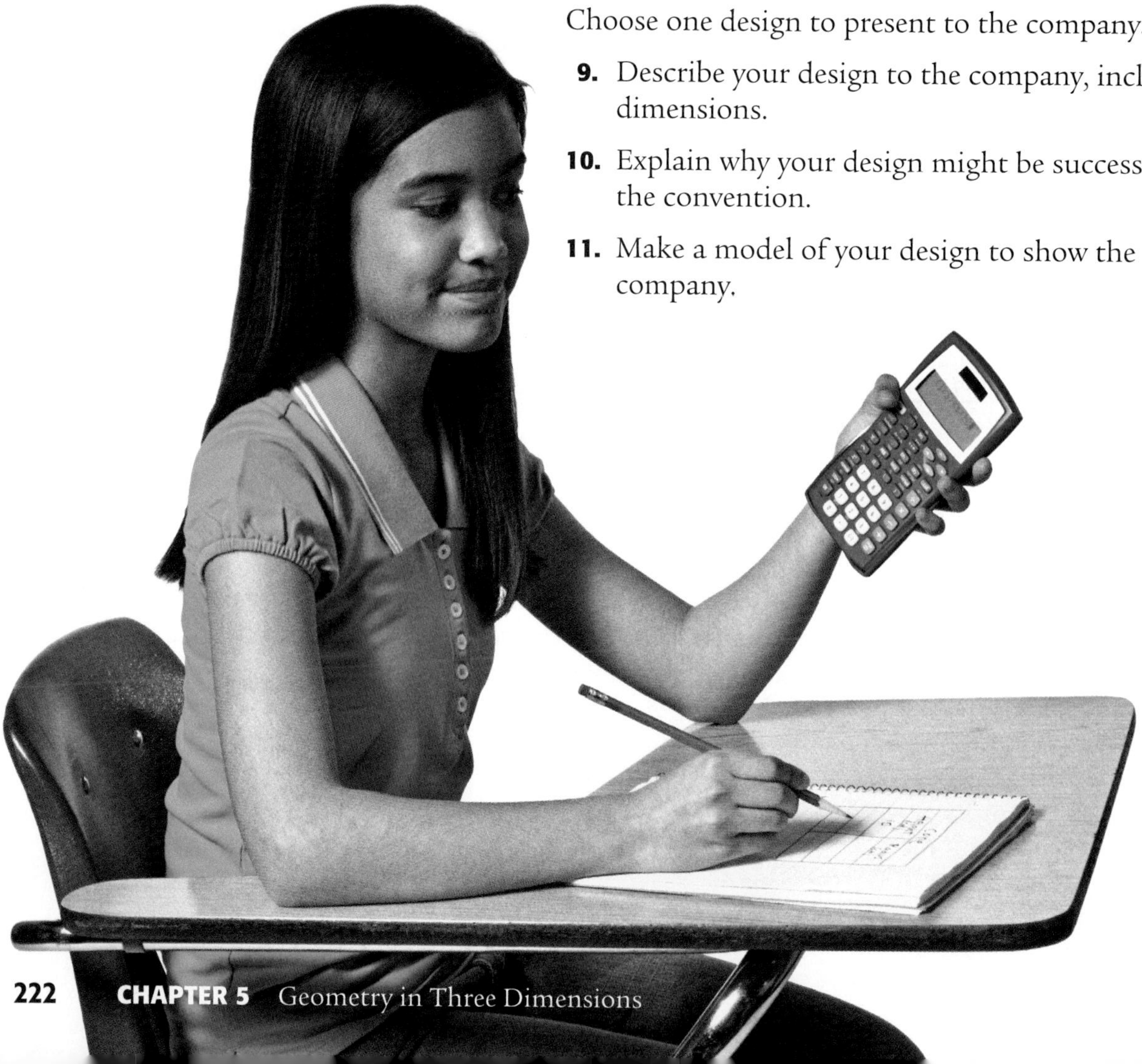

In Exercises 1–4, show your work.

1. What volume of sand will be needed to fill a sandbox that is 3 feet by 4 feet by 1.5 feet?

2. How much material will Sara use to make a cardboard jewelry box that is 6 cm by 6 cm by 3 cm?

3. How much space does a bed take up if it is 30 inches high, 75 inches long, and 50 inches wide?

4. How much space will be covered by paint if you paint the walls of a room that is 30 feet long, 25 feet wide, and 8 feet high?

5. A refrigerator advertises a capacity of 24 cubic feet. Which is more likely to be the dimensions of the refrigerator? Explain.
   a. 12 feet by 2 feet by 1 foot
   b. 6 feet by 2 feet by 2 feet

6. Here is a parallelogram.

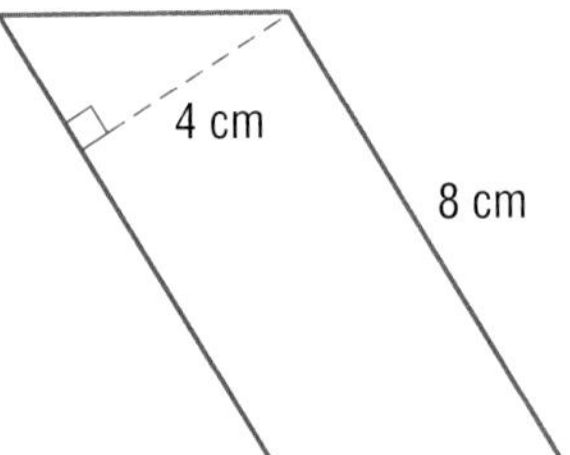

   a. What is the area of the parallelogram?
   b. If you build a prism 1 cm high using this parallelogram as a base, what will be its volume?
   c. Draw two other bases for containers that would have the same volume for a 1 cm height.

Math Link

The formula for the area of a parallelogram is bh, where b is the base and h is the height. The formula for the area of a triangle is $\frac{1}{2}bh$, where b is the base and h is the height.

7. Six equilateral triangles are joined to form a hexagon.

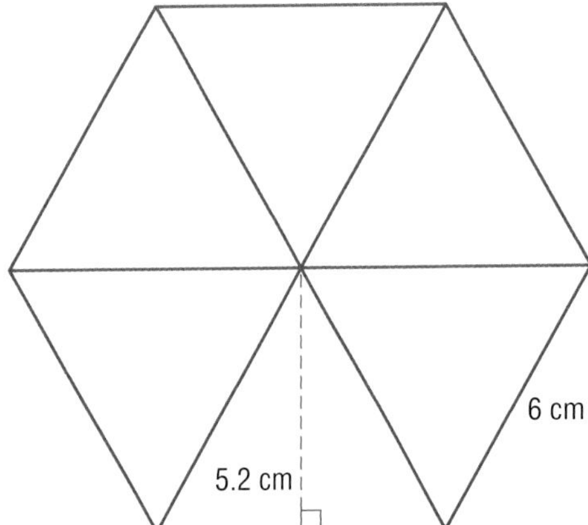

   a. What is the area of the hexagon? Explain how you found your answer.
   b. If you built a prism on this base with a height of h cm, what would be its volume?

8. The base of a cylinder has radius r meters. Write your answers in terms of π.

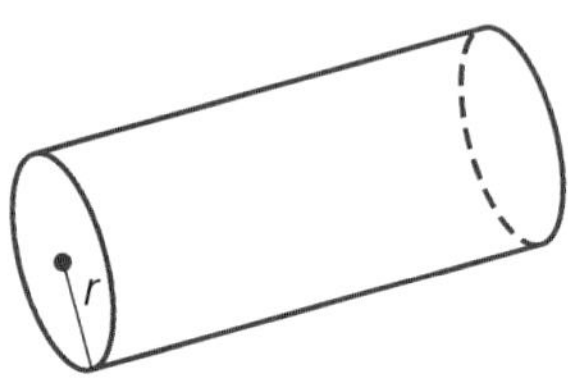

   a. What is the area of the cylinder's base?
   b. If the height of the cylinder is 1 m, what is its volume (in m^3)?
   c. If the height is 10 m, what is the volume?
   d. What is the cylinder's volume if its height is h meters?

9. You can find, or at least make a good estimate of, the volume of a room in your home.
   a. Draw the base (floor) of the room. Show the measurements in feet or meters. Indicate if you measured exactly. If you estimated, describe how you made your estimates.
   b. What is the height of the room? Do you know exactly or did you approximate it?
   c. What is the volume of the room?

10. Give the dimensions of four different containers that each have a volume of 360 cubic centimeters. They should not all be rectangular prisms.

Connect & Extend

11. Let l represent length, w represent width, and h represent height. You know that a formula for finding the volume of a rectangular prism is $V = l \cdot w \cdot h$. You have also learned that this formula can be expressed as $V = A \cdot h$, where A represents the area of the rectangular base of the prism.
    a. As you probably remember, the formula for finding the area of a triangle is $A = \frac{1}{2} b \cdot h$. The formula for finding the volume of a triangular prism is $V = \frac{1}{2} b \cdot h \cdot l$, where l is the height of the triangular prism. How else could the formula be written? Why?
    b. If you have a prism with a base shaped like a parallelogram, what would be the formula for the volume? Write the formula two different ways.

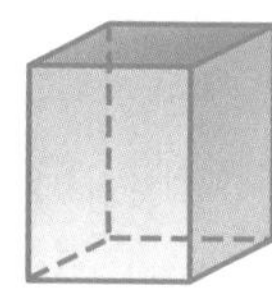

12. Tom is working at a manufacturing facility and has been asked to prepare a presentation regarding the packaging used to ship products to customers. The current shipping box holds five DVDs. On average, the manufacturer sends two or more boxes to each customer.

To reduce the charges for shipping, Tom thinks that the size of the box should be doubled to hold more DVDs. Currently, the shipping box is 6 inches by 10 inches by 8 inches. Tom thinks that the length of each side should be doubled in order to double the volume to ship ten DVDs instead of five.

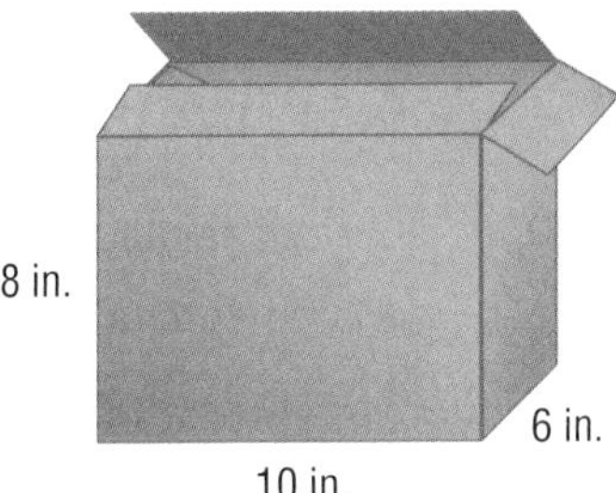

Help Tom evaluate the situation. Create a table like the following to record your results. What if the manufacturer doubles the box size again?

Length (inches)	Width (inches)	Height (inches)	Surface Area (in^2)	Volume (in^3)
6	10	8		
12	20	16		
24	40	32		

a. What happens to the surface area and volume as the box dimensions are doubled?

b. Suppose Tom only doubles one dimension instead of all dimensions. What happens to the surface area and volume? Try doubling different sides.

Length (inches)	Width (inches)	Height (inches)	Surface Area (in^2)	Volume (in^3)
12	10	8		
24	10	8		
48	10	8		

13. Many cereal boxes have dimensions of about 6 cm by 20 cm for the base and are about 27 cm high.

    a. What volume of cereal could this shape box hold?

    b. Give the dimensions (in cm) of four other rectangular boxes that have this same volume.

    c. Give the dimensions of two other containers that have this same volume. They do not have to be rectangular boxes.

14. Here are three cylinders that you probably have in your house. Choose one and find its volume.

    - a penny (It is hard to measure the height of a single penny, but it is not hard to measure the height of 10 pennies and then divide that measurement by ten.)
    - a soup can
    - a strand of spaghetti

Real-World Link

Young children do not understand that glasses of different shapes can hold the same amount of water. If given a choice of several glasses, they might always want the tallest glass or the widest glass because they think it holds the most.

15. Which of these containers holds the most water?

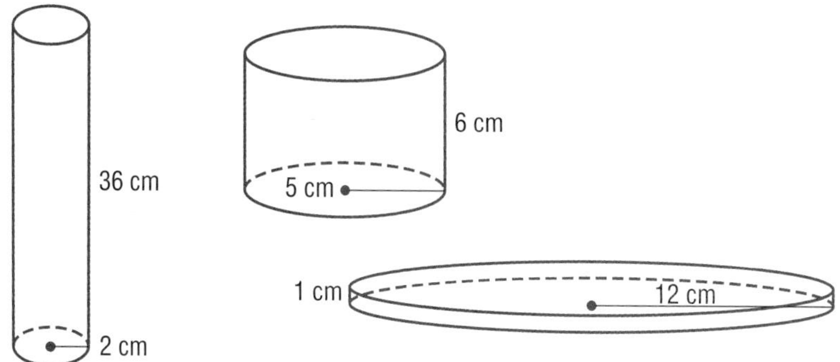

16. Draw two glasses that look different but hold the same amount of water. Show their dimensions.

17. **In Your Own Words** Give two examples of structures with the same volume and different surface areas. Suppose you are making one of these structures for art class, and you want to paint them. Which one will need less paint? Explain.

Mixed Review

Find the value of each expression.

18. $4 - (-3)$

19. $10 - (-4)$

20. $13 - (-10)$

21. $-7 - (-8)$

Rewrite in the form a^b.

22. $6^9 \cdot 6$

23. $4^2 \cdot 4 \cdot 4^7$

24. $10^{11} \cdot 10^5$

25. $3^5 \cdot 3^5$

Rewrite the expression as simply as possible. Use only positive exponents in your answers.

26. $t \cdot t \cdot t^{-2}$

27. $x^7 \cdot x^{-2} \cdot x^{-8}$

28. $(a^{-3}b^2)^{-5}$

29. $\left(\left(\frac{1}{n}\right)^{-4}\right)^{-3}$

Write each number in standard notation.

30. 6×10^4

31. 4.8×10^{-3}

Evaluate without using your calculator. Write your answer in scientific notation.

32. $3.2 \times 10^{-6} + 0.2 \times 10^{-7}$

33. $1.4 \times 10^{-3} + 0.9 \times 10^{-4}$

34. Social Studies In 2000, about 105,400,000 votes were cast in the United States presidential election.

a. Write this number of votes using scientific notation.

b. George W. Bush received about 48% of these votes. Approximately how many votes did he receive?

Nets and Solids

A **net** is a flat figure that can be folded to form a closed, three-dimensional object. This type of object is called a geometric **solid**.

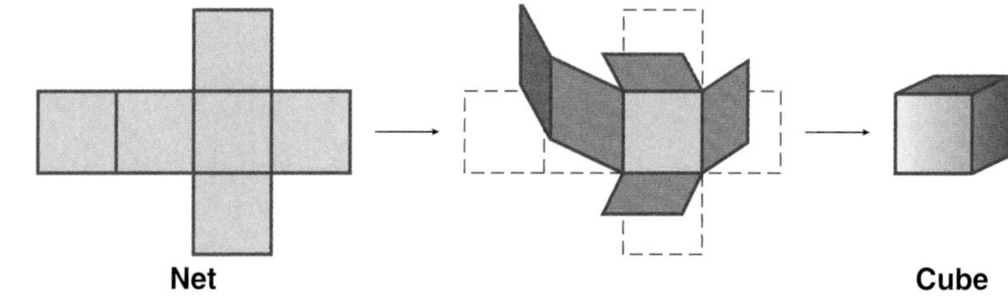

Inv 1 Use a Net 229

Inv 2 Use Nets to Investigate Solids 230

Inv 3 Is Today's Beverage Can the Best Shape? 233

Vocabulary

net

solid

Materials

- scissors
- graph paper

Explore

- Which of these figures are also nets for a cube? That is, which will fold into a cube? Cut out a copy of each figure. Try to fold it into a cube.

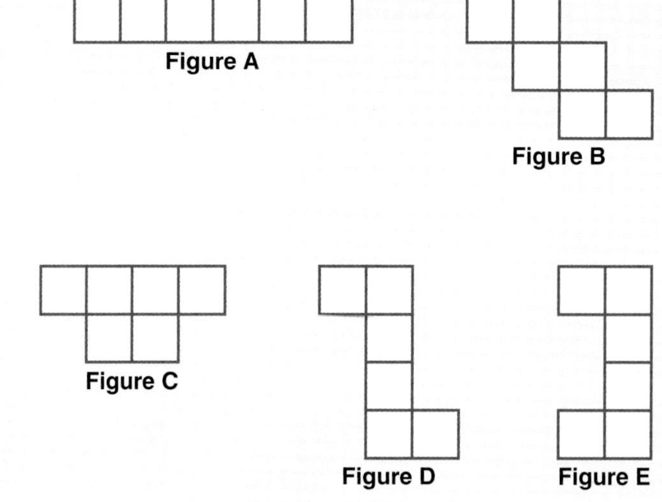

- Find and draw three other nets for a cube.

With your class, compile all the different nets for a cube that were found. How many nets are there? Do you think your class found all of them?

Investigation 1 Use a Net

Materials

- scissors

For a net to form a three-dimensional object, certain lengths have to match. For example, in the nets for cubes, the side lengths of all the squares must be the same. You will now investigate whether other nets will form closed solids. Pay close attention to the measurements that need to match.

Develop & Understand: A

Decide whether each figure is a net, that is, whether it will fold into a solid. One way to decide is to cut out a copy of the figure and try to fold it. If the figure is a net, describe the shape it creates. If it is not a net, explain why not.

1.

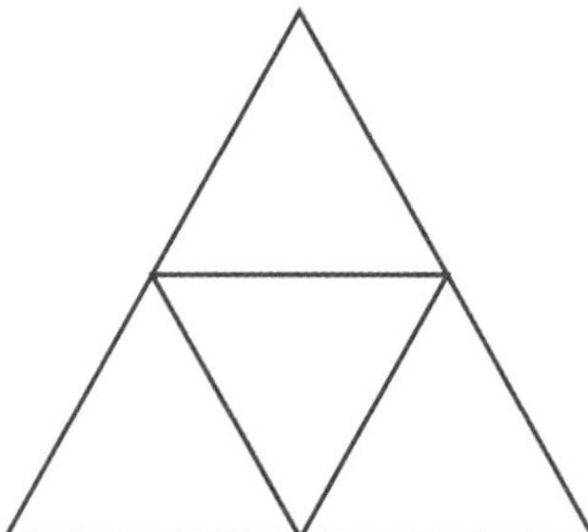

2.

3.

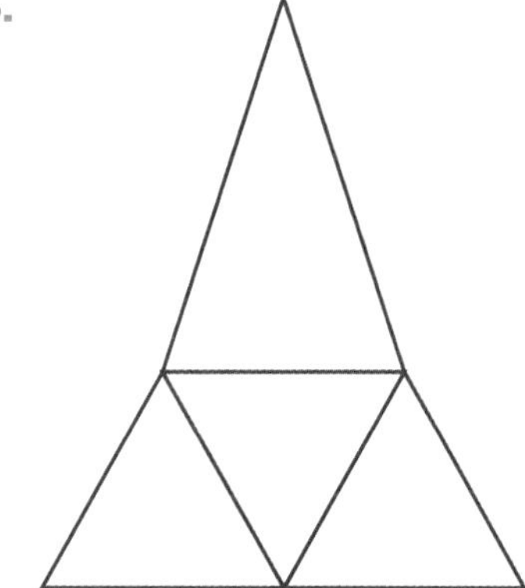

4.

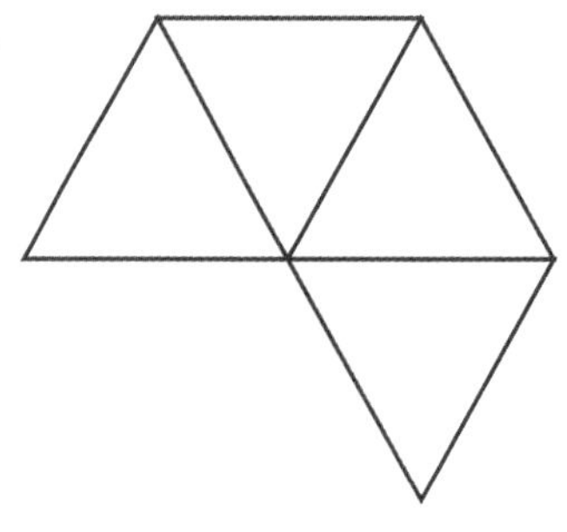

5.

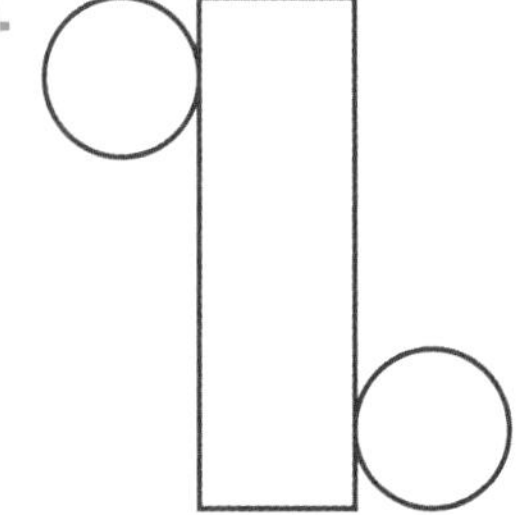

6.

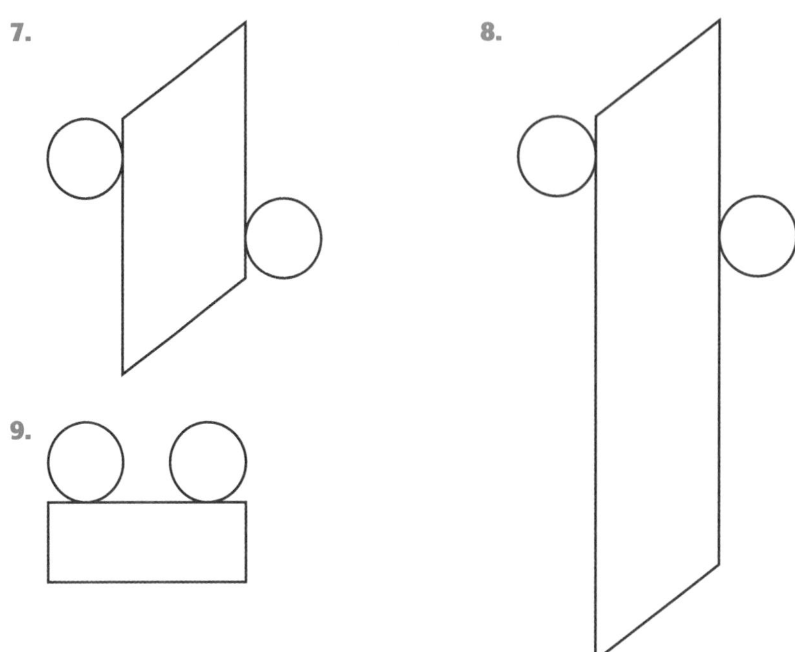

Share & Summarize

1. For a net to fold into a cylinder, the circumference of the circles must be the same as what other length?
2. Draw a net for a figure different from those in Exercises 1–9. Explain how you know your net will fold to form a solid.

Investigation 2 Use Nets to Investigate Solids

Materials

- scissors
- metric ruler

You have calculated surface area by counting the square units on the outside of a structure. If the faces of a solid are not squares, like the figures below, you can find the solid's surface area by adding the areas of all its faces.

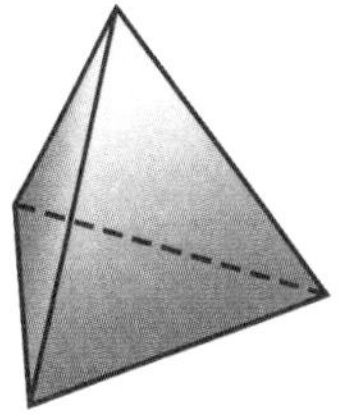

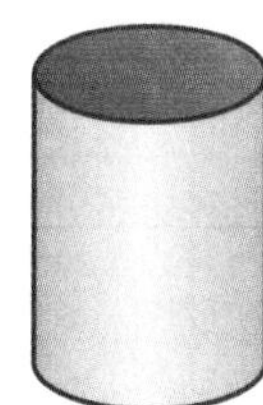

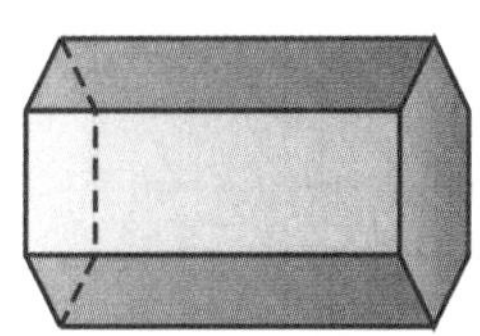

Example

This net folds to form a cylinder.

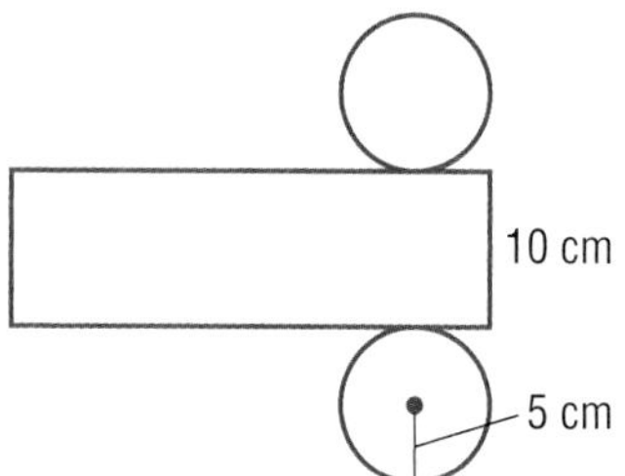

Because the net folds into a cylinder, the circles must be the same size. The area of each circle is πr^2 cm^2, or 25π cm^2. This is about 78.5 cm^2.

The length of the rectangle must be equal to the circumference of each circle, which is 10π cm, or about 31.4 cm. So, the area of the rectangle is about 314 cm^2.

To find the cylinder's surface area, add the three areas.

$$78.5 \text{ cm}^2 + 78.5 \text{ cm}^2 + 314 \text{ cm}^2 = 471 \text{ cm}^2$$

Develop & Understand: A

Each net shown will fold to form a closed solid. Use the net's measurements to find the *surface area* and *volume* of the solid. If you cannot find the exact volume, approximate it. It may be helpful to cut out and fold a copy of the net.

Math Link

The formula for the circumference of a circle is $2\pi r$, where r is the radius; or πd, where d is the diameter.

The formula for the area of a circle is πr^2.

1.

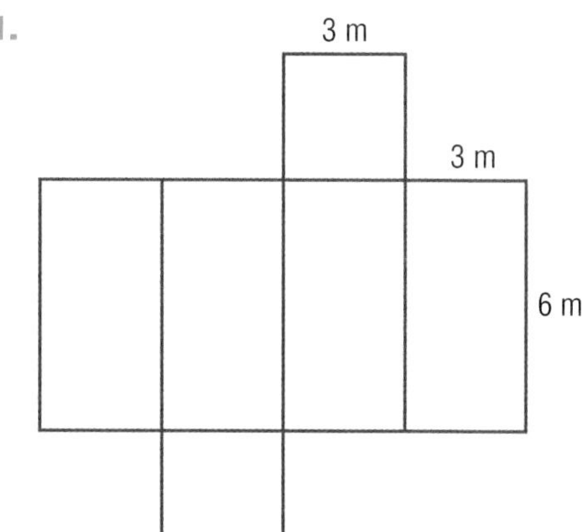

2.

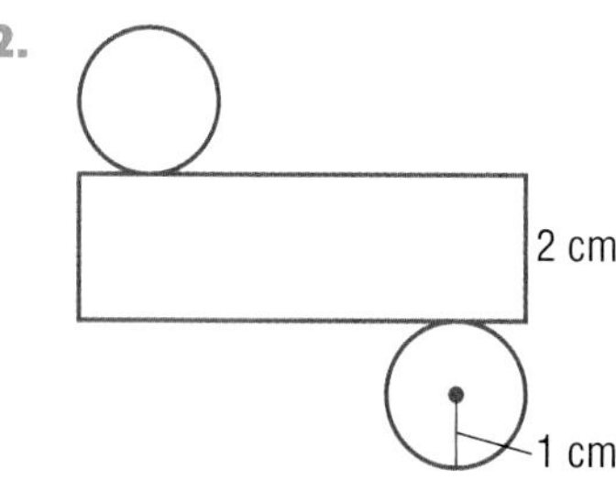

3.

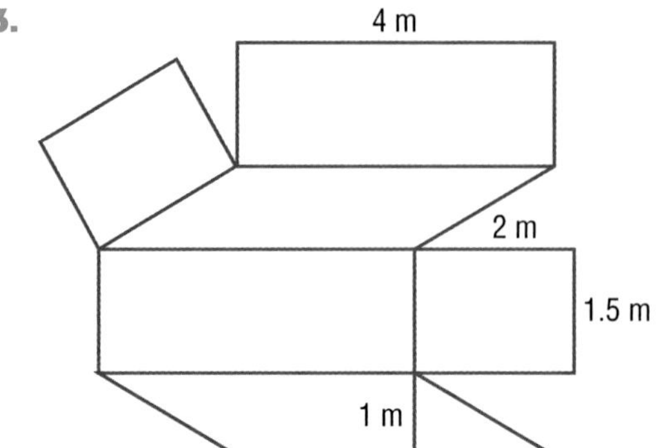

Math Link

The formula for the area A of a triangle is $A = \frac{1}{2}bh$, where b is the base and h is the height.

4.

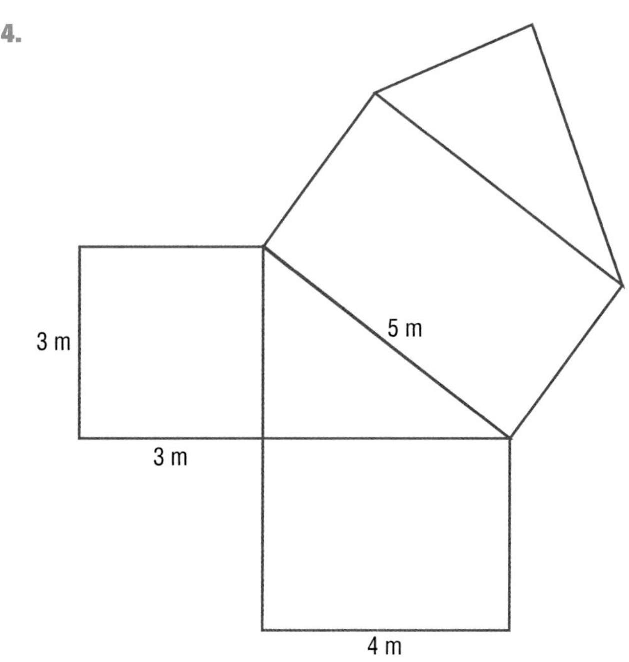

5.

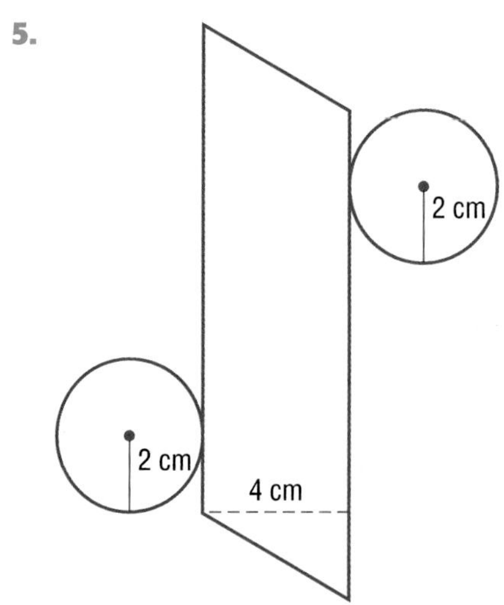

Share & Summarize

1. Explain how to find the surface area of a solid from the net for that solid.
2. Here is a net for a rectangular solid. Take whatever measurements you think are necessary to find the solid's volume and surface area. Explain what measurements you took and how you used them.

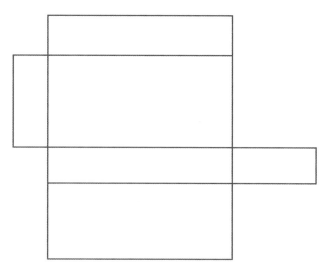

Investigation  Is Today's Beverage Can the Best Shape?

Materials

- 12-oz beverage can
- metric ruler

The Bursting Bubbles company is trying to reduce its manufacturing costs in order to increase its profits. The president of the company wonders if there is a way to use less material to make a beverage can.

In this investigation, you will help Bursting Bubbles design a can that has the minimum surface area for the volume of beverage it contains.

Develop & Understand: A

To get started, inspect a 12-ounce beverage can. Get an idea of its dimensions, volume, and surface area.

1. Find the area in square centimeters of the base of the can. If the can you are using is indented at the bottom, assume that it is not.
2. Measure the can's height in centimeters.

The volume printed on the can is given in milliliters (abbreviated mL). A milliliter has the same volume as a cubic centimeter, so 1 mL = 1 cm^3.

3. Does multiplying the area of the base by the height give the volume that is printed on the can? If not, why might the measures differ?

4. What is the surface area of the can in square centimeters? How did you find it? (Hint: Imagine what a net for the can would look like.)

Develop & Understand: B

Now consider how the can design might be changed. For these exercises, use the volume that you *calculated* in Exercises 1–4, rather than the volume printed on the can.

Real-World Link

As water bubbles up through the earth from natural springs, it collects minerals and carbon dioxide. In the 1700s, chemists perfected a method of adding carbonation and minerals to plain water and bottling the result.

5. Design and describe five other cans that have approximately the same volume as your can. Sketch a net for each can. Record the radius of the base and the height. Include cans that are both shorter and taller than a regular beverage can. (Hint: First choose the height of the can and then find the radius.)

6. Calculate the surface area of each can you designed. Record your group's data on the board for the class to see.

7. Compare the surface areas of the cans that you and your classmates found. Which can has the greatest surface area? Which has the least surface area?

Bursting Bubbles wants to make a can using the least amount of aluminum possible. The company cannot do it by evaluating every possible can, like the five you found, and choosing the best. There are too many, an infinite number, and Bursting Bubbles could never be sure it had found the one using the *least* material.

One strategy that mathematicians use is to gather data, as you did in Exercise 5, and look for patterns in the data. For an exercise like this one, they would also try to show that a particular solution is the best in a way that does not involve testing every case.

8. Use what you know about volume and any patterns that you found in Exercises 5 and 6 to recommend a can with the least possible surface area to the president of Bursting Bubbles. Describe your can completely. Explain why you believe it has the least surface area.

Share & Summarize

Does the shape of standard beverage cans use the minimum surface area for the volume they contain? If not, what might be some reasons companies use the shape they do?

Practice & Apply

Decide whether each figure is a net, that is, whether it will fold into a closed solid. One way to decide is to cut out a copy of the figure and try to fold it. If the figure is a net, describe the shape it creates. If it is not a net, explain why not.

1.

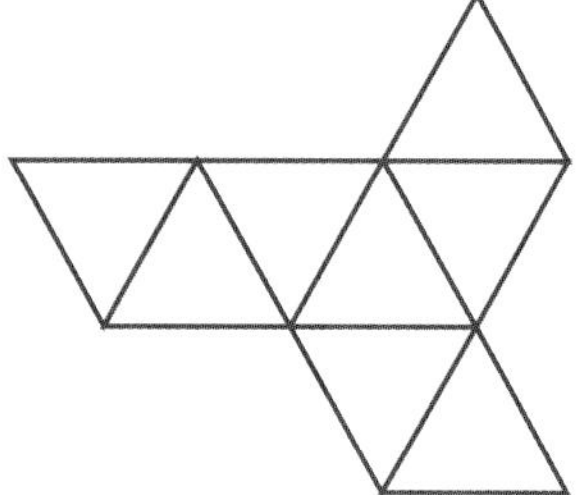

2.

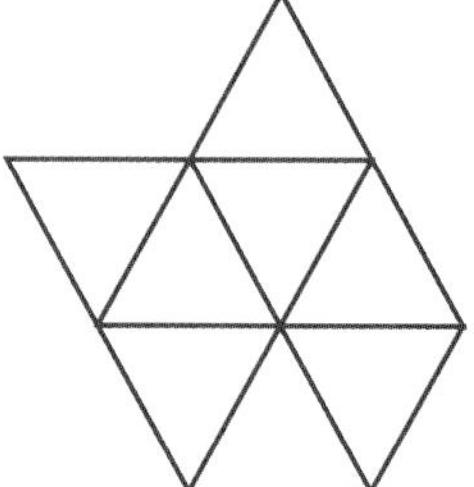

3.

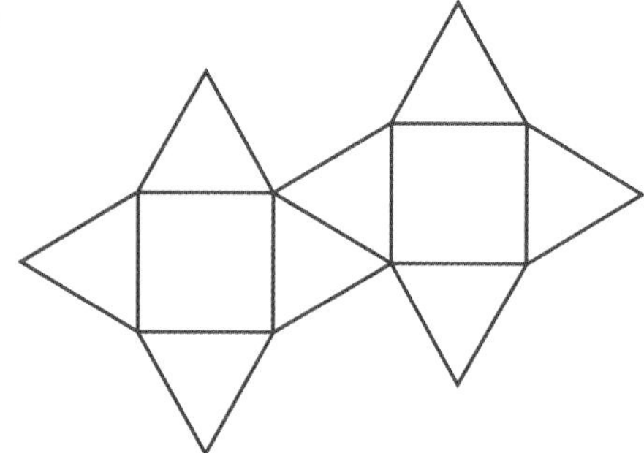

4. A *square pyramid* has a square base and triangular faces that meet at a vertex. Draw at least three nets for a square pyramid.

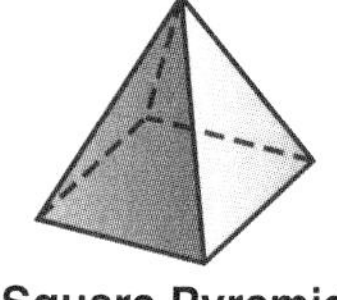

Square Pyramid

In Exercises 5–8, find the surface area of the solid that can be created from the net. Show how you found the surface area.

5. Each triangle is equilateral with sides 5 centimeters and height 4.3 centimeters.

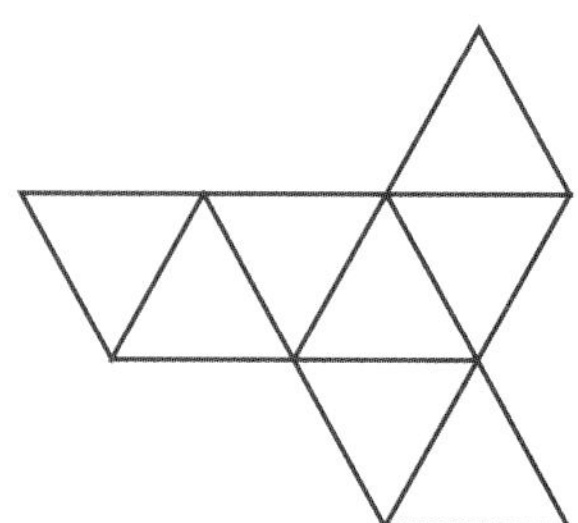

6. The radius of each circle is 2 centimeters. The height of the parallelogram is 2 centimeters.

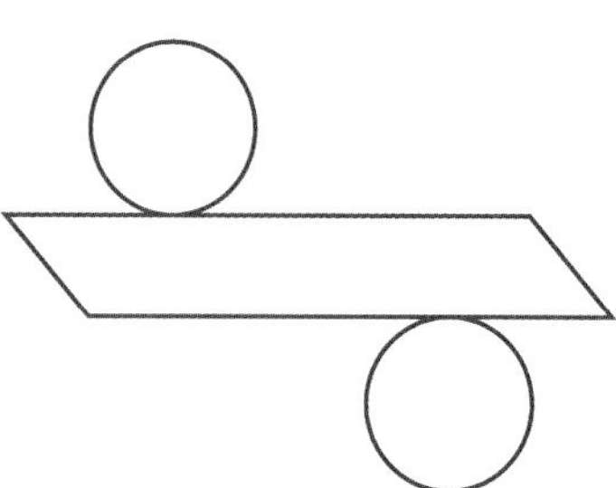

7. The pentagons have side lengths of 3 centimeters. If you divide each pentagon into five isosceles triangles, the height of each triangle is 2 centimeters. The other shapes are squares.

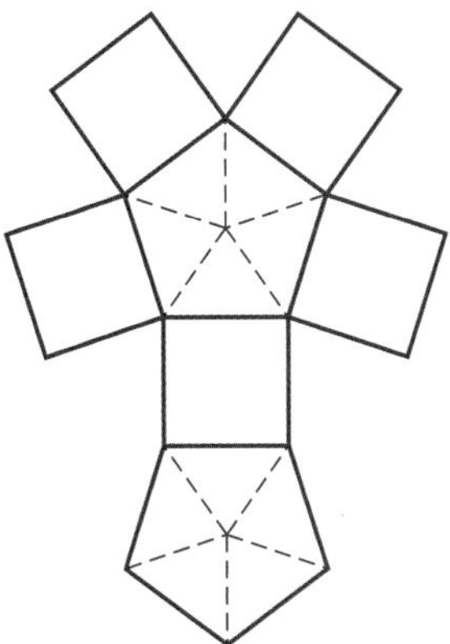

8. Each triangle is equilateral with sides 8 centimeters and height 7 centimeters. The other shapes are squares.

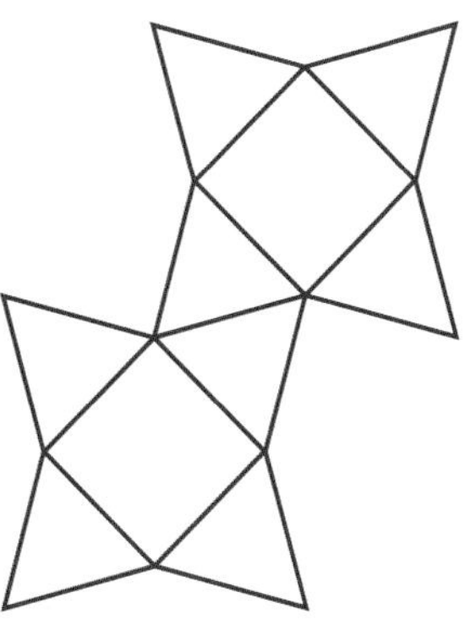

9. A box of tissues has length 24 centimeters, width 12 centimeters, and height 10.5 centimeters.

   a. What is the volume of the box?

   b. What is the surface area of the box?

10. Alana has a juice pitcher that is shaped like a cylinder. It is 30 centimeters high and has a base with radius 6 centimeters. It has a flat lid.

    a. How much juice will her pitcher hold?

    b. What is the surface area of her pitcher and its lid?

11. **Challenge** This net will fold into a *triangular pyramid.* Find the surface area of the pyramid. Estimate its volume. You may want to fold a copy of the net into the pyramid.

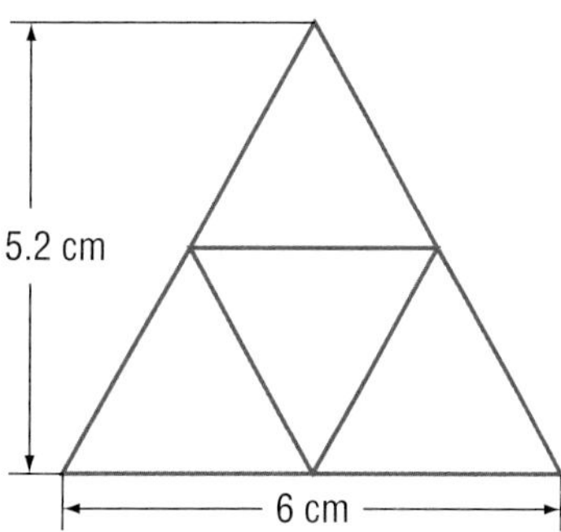

12. Choose a container in your house. You might choose a rectangular box, a can other than a beverage can, or some other object.

    a. Draw a net for the object.

    b. Find the surface area and volume of the object by measuring its dimensions.

    c. Draw nets for three other objects with the same volume as your container. Find the surface area of each. The new objects do not have to be real containers from your home.

    d. Is there a prism or cylinder with less surface area but the same volume as your container? If so, draw a net for it.

Connect & Extend

13. Imagine removing a label from a soup can.

    a. What is the shape of the label?

    b. If the radius of the soup can is r and the height is h, what are the dimensions of the label?

Create a net for each block structure.

14.

15.

16.

17. Challenge A cone has a circular base and a vertex some height away from that base. Make a net for a cone.

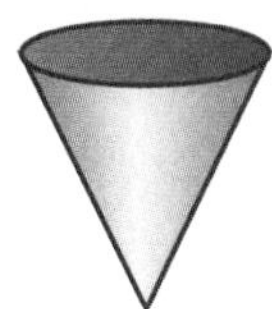

18. Challenge A *tetrahedron* is a solid with four triangular faces. You have made tetrahedrons from nets in this lesson. Nets for a tetrahedron contain only triangles. Can you make a net for another figure using only triangles? (Hint: You might try taping paper triangles together to form a new solid.)

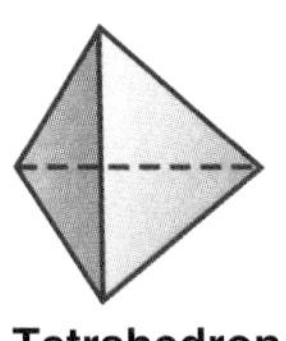
Tetrahedron

19. Kinu's Frozen Yogurt store packs its product into cylindrical tubs that are 20 cm in diameter and 30 cm tall.

a. How much frozen yogurt does a tub hold?

b. Kinu's sells frozen yogurt in cylindrical scoops. If each scoop has base radius 2 cm and height 6 cm, what is the volume of a single scoop?

c. How many scoops of frozen yogurt are in a tub? Show how you found your answer.

d. What is the surface area of a tub?

e. What is the total surface area of the frozen yogurt from one tub *after* it has all been scooped?

f. Which has more surface area, the frozen yogurt in the tub or the scooped frozen yogurt? Which would melt faster?

20. In Your Own Words Explain why nets might be useful in sewing, manufacturing, or some other profession.

21. A circular pond is 100 feet in diameter. In the middle of winter, the ice on the pond is 6 inches thick.

   a. Think of the ice on the pond as a cylinder. What is the volume of ice?

   b. What is the surface area of the floating cylinder of ice?

   c. What would be the edge length of a cube of ice that had the same volume?

   d. What would be the surface area of a cube of ice that had the same volume?

   e. Which has greater surface area, the cylinder of ice or the cube of ice? How much more?

Preview Match the following items with the most appropriate unit of weight.

22. an elephant

23. a cup of yogurt

24. a human being

A. ounces

B. pounds

C. tons

Mixed Review

Find each product or quotient.

25. $6(-12)$

26. $-\frac{1}{5}\left(-\frac{2}{3}\right)$

27. $15 \div (-2)$

28. $-\frac{4}{9} \div \left(-\frac{2}{3}\right)$

29. Copy and complete the chart.

Fraction	$\frac{1}{2}$	$\frac{3}{5}$		
Decimal	0.5		0.65	
Percent	50%			21%

Rewrite each number in standard notation.

30. 8×10^2

31. 1.42×10^6

32. 3.12×10^{-4}

33. 6.13×10^{-2}

34. Suppose you use a photocopying machine to reduce a photograph.

   a. Your photograph is 10 cm wide, and you reduce it to 85% of its original size. What is the width of the reduced image?

   b. You use the same setting to reduce the copy. What is the width of the image after the second reduction?

LESSON 5.3 Mass and Weight

Inv 1	Measure Mass	241
Inv 2	Estimate Mass	244
Inv 3	Measure Weight	247

Vocabulary

mass

weight

You are familiar with weight, or how heavy things are. So, what is mass, and what does it have to do with weight?

Mass is a measurement of the amount of matter in an object. **Weight** is the pull of gravity on that amount of matter. Gravity pulls everything down toward the center of Earth. The more matter that is in an object, the more gravity pulls it down, and the heavier it is. When you pick up a rock and a piece of paper, the rock feels heavier because there is more matter in the rock.

What about the weight of objects on the moon, where the pull of gravity is less? What about the weight of objects in outer space, where there is so little gravity that everything floats?

The same rock feels much heavier on Earth than it does on the moon, but the amount of matter has not changed. In other words, the weight of the rock is different on the moon than it was on Earth, but its mass is the same regardless of where it is.

On Earth, mass and weight are the same if the object is at sea level. Above sea level, the object's weight will be less than its mass. The difference is quite small, however, even on the highest mountains. So, as long as you are talking about objects that are on Earth, you can treat weight and mass as being equivalent.

Real-World Link

Although we typically talk about weight in everyday speech, the distinction between mass and weight is important in sciences such as physics, chemistry, and astronomy.

Think & Discuss

Suppose you weigh 120 pounds on Earth. How much do you think you would weigh on the moon?

In fact, the gravitational pull of the moon is only $\frac{1}{6}$ that of Earth. How close was your Think & Discuss estimate? Approximate your weight on the moon.

Investigation 1 Measure Mass

Vocabulary

gram

metric system

Materials

- kitchen or mail scale marked in 100-gram or 10-gram divisions that can weigh at least 2 kg
- scale, cup measure, flour, and water

In this investigation, you will measure and calculate mass using metric units since the **metric system** measures mass. Remember that because scales rely on gravity, you can use them to measure only weight and not mass. However, on Earth, weight and mass are approximately the same.

The metric system is compatible with our decimal number system. Whether you are measuring length, mass, or volume, the relationship between the units is the same. Ten of one unit equals one of the next larger unit.

The metric system uses a prefix to indicate the size of the unit. You are probably familiar with some of these prefixes from the metric units for length and volume.

Prefix	Meaning
Milli-	Base unit ÷ 1,000
Centi-	Base unit ÷ 100
Deci-	Base unit ÷ 10
No prefix	Base unit • 1
Deca-	Base unit • 10
Hecto-	Base unit • 100
Kilo-	Base unit • 1,000

Think & Discuss

The basic unit of mass in the metric system is the **gram**. What is a situation in which you have seen grams or kilograms used as a unit of measure?

What is a situation in which you have seen milligrams used as a unit of measure?

Develop & Understand: A

1. Copy and complete this table that shows conversions among various metric units of mass. Use decimals when necessary.

Milligrams		55,000					3,589
Centigrams					195	0.4	
Decigrams	10	550	70,000	8,300	19.5	0.04	35.89
Grams	1	55					
Decagrams	0.1	5.5	700	83.0	0.195	0.0004	0.3589
Hectograms				8.30			
Kilograms	0.001		7				

Scientific notation is often used to express the magnitude of quantities that are very large or very small.

2. Use scientific notation to express the following conversions.
   a. 6 grams = ______ milligrams
   b. 5.3 hectograms = ________ grams
   c. 702 grams = ________ centigrams
   d. 45 grams = ______ kilograms
   e. 89.3 milligrams = ______ kilograms

Math Link

The kilogram was originally defined as the mass of a liter of water at 4° Celsius. Because the ratio of water's mass to volume changes at different temperatures, a specific temperature had to be designated.

3. Theo bought the following food items.

   1,300 milligrams of saffron — 68.5 grams of ground pepper
   737 grams of salt — 5 kilograms of rice

   What was the total mass of Theo's items? Express your answer in terms of one unit only.

4. The mass of one biology textbook is 1,525 grams. Suppose 300 textbooks are loaded into a truck. What is the mass of all the textbooks in the truck? Express your answer in kilograms.

Develop & Understand: B

Use a small scale marked in metric units to investigate mass. Before you begin, make sure your scale reads 0 when there is nothing on it. If it does not, you will need to adjust it.

5. Record the mass of each object or explain why you cannot find its mass with your scale.
   a. a pencil
   b. a textbook
   c. a paperback novel
   d. a full backpack
   e. a desk
   f. a calculator

6. Find an object in your classroom whose mass is approximately the following.
   a. 1 kilogram
   b. 2 kilograms
   c. 200 grams
   d. 500 grams
   e. 25 grams
   f. 5 grams

7. Find the mass of a collection of similar pens.

   a. How many pens do you need to put on the scale to get a mass of approximately 100 grams?

   b. Use your answer from Part a to estimate the mass of 1 pen.

   c. About how many pens would you need to make a mass of 1 kilogram?

   d. One metric ton equals one thousand kilograms. About how many pens would you need to make a mass of 1 metric ton?

Real-World Link

When you zero out a scientific instrument, such as a scale, you are *calibrating it.* This ensures the instruments will measure readings accurately. *Taring* refers to adjusting a scale so that a number other than 0 shows when there is nothing on the scale. This is useful if you want to weigh something inside a container.

Think & Discuss

Suppose you have two cardboard boxes with dimensions 0.5 meters by 0.5 meters by 0.5 meters. You fill one box with pocket-sized packages of tissues and one with books. Which is heavier, or are their masses the same? Explain your reasoning.

Two objects can be the same size and yet have different masses. The saying "rock is heavier than butter" means that a given volume of rock is heavier than the same volume of butter. Put another way, a large amount of butter has the same mass as a small amount of rock.

Develop & Understand: C

8. What is the mass of the measuring cup your teacher provided?

9. Put 250 milliliters of flour in the measuring cup. What is the mass?

10. Empty the cup and put in 250 milliliters of water. What is the mass?

11. What is the mass of 250 milliliters of flour? Explain how you found your answer.

12. What is the mass of 250 milliliters of water?

13. Compare the masses of 250 milliliters of flour and 250 milliliters of water.

    a. Which mass is greater? How much greater?

    b. Which has greater volume, 1 kilogram of water or 1 kilogram of flour?

14. Based on your measurements, what would be the mass of 1 liter of water?

15. Based on your measurements, what volume of flour would have a mass of 1 gram?

16. By definition, the mass of 1 liter of water at 4°C is 1 kilogram.

   a. How does your approximation of the mass of a liter of water compare to the theoretical mass?

   b. What might account for any difference between your approximation and the theoretical mass?

17. If you used a scale that was marked in milligrams for this activity, how do you think it would affect your results?

Share & Summarize

Which metric units would be most appropriate as benchmarks to measure the mass of each item below?

book	pencil
penny	person

Investigation 2 Estimate Mass

Materials

- kitchen or mail scale
- single sheets of $8\frac{1}{2}$-in. × 11-in. paper
- an unopened ream of $8\frac{1}{2}$-in. × 11-in. paper
- ruler
- scissors

In this investigation, you will use your knowledge to estimate measures of mass. To begin, how big is a gram?

Develop & Understand: A

1. Weigh a piece of paper on a small scale. You may need to fold the paper to fit it on the scale. Does the paper's mass register on the scale?

2. About how many pieces of paper do you need to put on the scale for the scale to show a number larger than 0?

3. About how many pieces of paper have a mass of 100 grams?

4. Based on your answer to Exercise 3, what is the approximate mass of one piece of paper?

5. Based on your answer to Exercise 3, compare the mass of one gram to a piece of paper.

Math Link
1 kilogram = 1,000 grams

In the next exercise set, you will consider the mass of a kilogram.

Develop & Understand: B

6. Based on your answer to Exercise 3, about how many pieces of paper have a mass of 1 kilogram?

7. Examine a ream, or package, of paper.
   a. How many pieces of paper are in the ream?
   b. Based on your answer to Exercise 4, estimate the mass of the ream of paper.
   c. Was your estimate close to the mass you measured? If not, what might account for the difference?

Real-World Link

About ninety billion kilograms of paper are used in the United States each year. According to the American Forest and Paper Association, about 50% of the used paper is recycled.

8. Without using a scale, estimate the mass of your *IMPACT Mathematics* textbook.
   a. How many pages are in the book? How might you use this information to estimate the book's mass?
   b. Compare the book to the ream of paper whose mass you have already found. Does it look bigger or smaller? Does it feel heavier or lighter?
   c. Estimate the book's mass. Explain how you decided on your estimate.

9. Weigh the book on the scale. What is the book's mass?

10. Was your estimate close? Explain what you think might account for any difference between your estimate of the book's mass and the measured mass.

11. Based on your answer to Exercise 8, about how big is 1 kilogram compared to the mass of your Math book?

In Exercises 12–16, you will consider the mass of 100 kilograms.

Develop & Understand: C

12. Based on your answer to Exercise 9, about how many Math books would have a mass of 10 kilograms?

13. About how many Math books would have a mass of 30 kilograms?

14. About how many Math books would have a mass of 100 kilograms?

15. Suppose you stack all of the *IMPACT Mathematics* books belonging to you and your classmates.

    a. Approximately what is the mass of the stack?

    b. What is the height of the stack in meters?

16. Find three objects of different sizes. Estimate the mass of each one, using what you already know about the size of grams and kilograms. Record your estimates. Then weigh the objects on the scale. Record each object's actual mass.

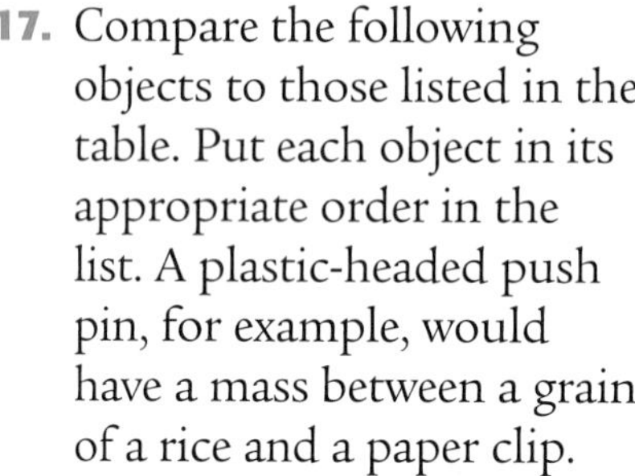

Develop & Understand: D

17. Compare the following objects to those listed in the table. Put each object in its appropriate order in the list. A plastic-headed push pin, for example, would have a mass between a grain of a rice and a paper clip.

    a. Newborn human baby

    b. Nickel

    c. Full-size refrigerator

    d. Individual sugar packet

    e. Paperback book

Object	Approximate Mass
Grain of sand	$3.5 \cdot 10^{-4}$ milligrams
Grain of rice	20–30 milligrams
Paper clip	0.5–1.5 grams
Penny	2.5 grams
Wooden pencil (unsharpened)	7 grams
1 liter of water	1 kilogram
Cat	5–7 kilograms
Dog	5–70 kilograms
Adult human	45–90 kilograms
Cow	700 kilograms
Car	800–1600 kilograms
Largest known dinosaur	80–100 metric tons
Blue whale	108 metric tons
Earth	6.0 • 1024 kilograms
Sun	2 • 1030 kilograms

18. Estimate the mass of each object from Exercise 17, using the objects whose mass is given as benchmarks to help. Choose two objects from the list. Explain how you made your estimate.

19. Think of three objects whose masses you do not know. Use the benchmarks to help you estimate the mass of each object.

Think & Discuss

Compare your estimates from Exercise 17 to the actual (approximate) masses of the objects.

Which of your estimates were close? Which were not close?

Are there any masses that surprise you? For example, was there an object that you thought was heavier or lighter than it turned out to be?

Share & Summarize

How would you estimate the mass of an object if you did not have a scale to use?

Investigation 3 Measure Weight

Vocabulary

customary system

ounce

pound

ton

Materials

- kitchen or postal scale marked in ounces
- bathroom scale marked in pounds

As you saw in Investigations 1 and 2, metric units for mass, such as the kilogram, are often used as though they measure weight. Mass is a more scientific and precise property of objects than weight. However, we can perceive weight directly and measure it easily. In science and engineering, it is important to distinguish between mass and weight. For most other situations, it usually does not make a difference.

The **customary system** of measurement is used in the United States.

The most commonly used units of weight in the customary system are **ounces**, **pounds**, and **tons**. A decimal or a fraction can be used to express a fractional portion of a pound, 0.75 or $\frac{3}{4}$ pound.

Develop & Understand: A

1. Copy and complete this table for converting customary units of weight.

Ounces	32	256000				1
Pounds	2		1.25		1	
Tons		8		1	0. 0005 = 5.00×10^{-4}	0.00003125 = 3.125×10^{-5}

2. What fraction of 1 pound is 1 ounce?
3. What fraction of 1 ton is 1 pound?
4. Write an equation that gives you weight in pounds p if you know the weight in ounces n.
5. Write a formula to convert weight in pounds to weight in ounces.
6. How are these two formulas related?
7. Write formulas to convert tons to pounds, pounds to tons, tons to ounces, and ounces to tons. Be sure to state what your variables represent.

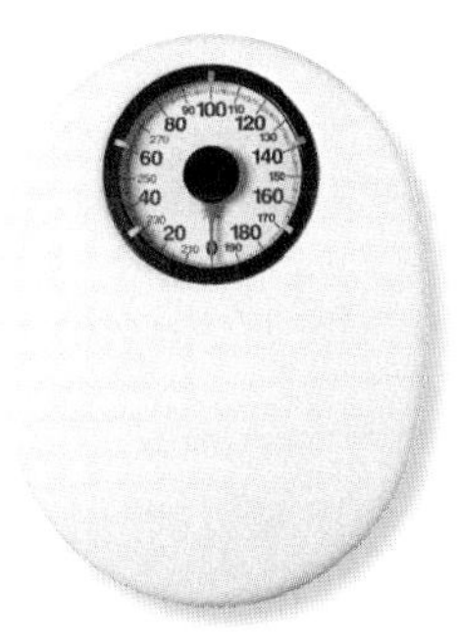

Math Link

The *ounce* is also a unit of capacity. There are eight ounces in one cup, or sixteen ounces in one pint. For clarity, *fluid ounces* describe capacity, and just *ounces* are used to describe weight.

While fractions, decimals, and mixed numbers can be used to express a quantity like $1\frac{1}{2}$ pounds, or 1.5 pounds, you can also express some of these amounts in terms of a whole number of pounds and a whole number of ounces.

For example, 1.5 pounds is equivalent to 1 pound 8 ounces.

8. Fill in the blanks to make each equation true. Write your answer to Part c as a fraction and your answer to Part d as a decimal. Write your answer to Parts e–g as a fraction and as a decimal.

a. 2.5 pounds = _____ pounds _____ ounces

b. $\frac{3}{4}$ pounds = _____ ounces

c. 38 ounces = _____ pounds

d. 10 ounces = _____ pound

e. 5 pounds 5 ounces = _____ pounds or _____ pounds

f. 3 pounds 4 ounces = _____ pounds or _____ pounds

g. 2 tons 1,000 pounds = _____ tons or _____ tons

Think & Discuss

What things are commonly measured in pounds? In ounces? In tons?

Not all scales are suitable for weighing every object.

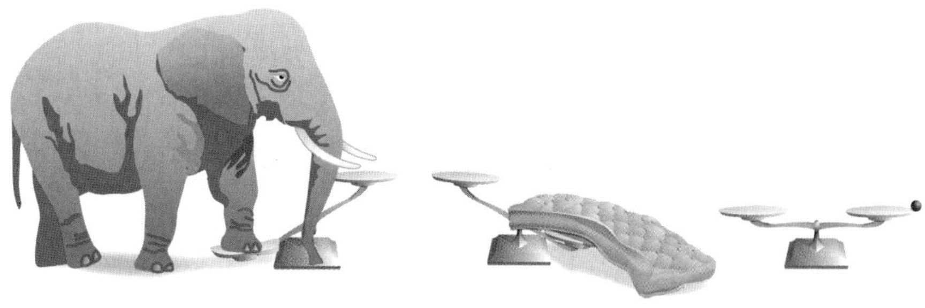

How can you tell if an object is too small to weigh on the scale you are using?

How can you tell if an object is too large to weigh on the scale you are using?

Real-World Link

Pharmacists and jewelers measure very small quantities of material in the metric unit of milligrams or in the customary unit of grains. The grain is a very small unit. There are 7,000 grains in one pound. The grain as a unit of weight was originally based on the weight of 1 grain of barley.

Develop & Understand: B

9. Find one object in the classroom that meets each of the following conditions.

   a. It can be weighed on the large scale but not on the small scale.

   b. It can be weighed on the small scale but not on the large scale.

   c. It can be weighed on either scale.

   d. It is too light to be weighed on either scale.

   e. It is too heavy to be weighed on either scale.

   f. It cannot be weighed on one of the scales for some reason other than inappropriate weight.

10. Weigh these sample objects. Copy and complete the table below. If you do not have the specified object available, substitute a similar object.

Math Link

Why don't we have a unit somewhere between pounds and tons, when the difference in size is so big? *Hundredweight* is a customary unit equal to 100 pounds, but it is hardly ever used any more.

Object	Which scale did you use to weigh it?	Weight
Shoe		
Textbook		
Stapler		
Paperback book		
Overhead projector		
Full box of paper clips		

11. About how many of the following would make one pound?

   a. staplers

   b. paperback books

   c. full boxes of paper clips

12. Based on the results of Exercise 10, what could you say if someone asked you, "How heavy is a pound?"

13. What could you say if someone asked you, "How heavy is an ounce?"

14. Approximately how heavy is one paper clip? Explain your reasoning.

15. About how many Math books would weigh the following amounts?

a. 10 pounds

b. 50 pounds

c. 100 pounds

d. 1 ton

Math Link

You will often see these abbreviations used for units of mass.

g (grams)

mg (milligrams)

kg (kilograms)

Think & Discuss

How could you use a scale meant for heavy objects to estimate the weight of a light object?

How could you use a scale meant for light objects to estimate the weight of a heavy object?

Name some objects you could weigh on a bathroom scale.

Name some objects you could not weigh on a bathroom scale.

Share & Summarize

How are mass and weight different?

What are appropriate customary units to measure the weight of each of the following?

apple

bag of apples

truck

dog

Practice & Apply

1. Convert each quantity to grams. Write your answers in scientific notation.

a. 25 kilograms
b. 29,000 centigrams
c. 34,683 milligrams
d. 4.2 milligrams
e. 4,532 kilograms
f. 0.5 kilograms

2. Which of the following metric units is heavier?

a. 350 milligrams or 200 grams?
b. 21 grams or 2 kilograms?
c. 5.4×10^3 milligrams or 5.03×10^{-3} kilograms?
d. 4×10^5 centigrams or 4×10^5 hectograms?

3. Which metric units would be most appropriate to measure the mass of each of the following?

a. a car
b. a china plate
c. a cat
d. a pinch of salt

4. Which metric unit is most appropriate to measure the mass of each of the following items?

a. a person
b. an apple
c. a bulldozer
d. a blade of grass

5. Do you think you could lift 100 grams? 10 kilograms? 100 kilograms? 1000 kilograms?

6. Name an object whose mass might be between 1 and 10 kilograms.

Refer to the table on page 246 to answer Exercises 7–13.

7. Which of these are plausible masses for an adult human?

a. 74 kilograms
b. 89 kilograms
c. 6.9×10^4 grams
d. 5.4×10^{-2} tons

8. How does the mass of an adult compare to the mass of a car?

9. How does the mass of a grain of rice compare to the mass of a grain of sand?

10. About how many cows have the same mass as a blue whale?

11. About how many grains of rice are in a 5-kilogram bag?

12. How many pennies have the same mass as 6 liters of water?

13. A mouse has about the same mass as 3 pencils. What is the approximate mass of a mouse?

Fill in the blanks to make each equation true.

14. $\frac{3}{16}$ pounds = _____ ounces

15. 9 pounds 6 ounces = _____ ounces

16. 2.6 tons = _____ pounds

17. 4.25 pounds = _____ pounds _____ ounces

18. 6,400 ounces = _____ ton

19. Jasmine needed to send a report to a co-worker in another state. She knew that she needed to put one 41-cent stamp on a letter that weighed one ounce or less. For each additional ounce of weight, she needed to add an extra 17-cent stamp. (So, a 1.5-ounce letter would cost 58 cents to mail.) She does not have a scale, but the label on her printer paper said that 500 sheets of paper weighs 5 pounds.

a. If she only used one stamp, how many pages could she send?

b. If the report was 20 pages long, how much will it cost to mail?

c. If Jasmine spends 75 cents on postage, what is the maximum number of pages she can send?

20. A queen-sized waterbed mattress is 5 feet wide, $6\frac{1}{3}$ feet long, and 8.3 inches tall. A cube of water 1 inch on each side weighs about 0.36 pounds. If the mattress is entirely filled with water, how much does it weigh?

Connect & Extend

21. The elevator in the Grand Hotel can safely carry a maximum of 1,820 kilograms. The elevator is 1.5 meters wide, 2.8 meters long, and 4 meters high.

a. What is the volume of the elevator?

b. If the elevator were filled with water, what would be the mass of the water?

c. Could the elevator safely carry that much mass?

d. An adult human typically has a mass between 45 and 90 kilograms. How many people can the elevator safely carry?

e. Assume each person is 1.65 meters tall and needs an area of about 0.5 by 0.7 meters. How many people can fit into the elevator?

f. Can the elevator safely carry as many people as can fit into it? Explain your reasoning.

22. The United States Department of Agriculture (USDA) recommends the following Daily Values for various food ingredients.

Total Fat	Less than 65 grams
Saturated Fat	Less than 20 grams
Cholesterol	Less than 300 milligrams
Sodium	Less than 2,400 milligrams
Potassium	3,500 milligrams

Packages of processed food list how much of each ingredient is in one serving. Ramona bought a box of chocolates with this nutritional information on the back.

Nutrition Facts

Serving size 4 pieces (32 grams)
Servings per container About 6

Amont per serving	% Daily value
Total Fat 15g	23%
Saturated Fat 12g	60%
Trans Fat 0g	
Cholesterol 0 mg	0%
Sodium 25 mg	1%
Total Carbohydrate 13g	4%
Dietary Fiber 1g	6%
Sugars 12g	
Protein 2g	

a. How many milligrams of dietary fiber are in one piece of chocolate?

b. How many chocolates contain 1 gram of sodium?

c. How many grams of chocolate is that? How many kilograms?

d. Is 1 gram of sodium more or less than the maximum recommended daily value?

e. What is the mass of the whole box of chocolates?

f. How much saturated fat is in the amount of chocolate in Part b?

g. What percent of the maximum recommended daily value is the amount of saturated fat in Part f?

h. How many grams of chocolate contain the maximum recommended daily value of saturated fat?

i. How many pieces of chocolate is that?

j. Ramona wants to follow the USDA dietary guidelines. Should she be worried about eating too much sodium if she eats some chocolate today? Should she be worried about eating too much saturated fat? Explain your reasoning.

23. This table shows the approximate weights for some familiar objects in both metric and customary units. Use the data in the table to find a rule that will let you complete the table.

Object	Approximate Mass (Metric)	Approximate Weight (U.S. Customary)
Penny	2.5 grams	.088 ounces
Nickel		0.176 ounces
Piece of paper (8.5 in. x 11. in.)	5 grams	
Wooden pencil (unsharpened)	7 grams	2.7 ounces
Apple	0.23–0.45 kilograms	0.5–1 pounds
Ceramic coffee mug		12 ounces
1 liter of water	1 kilogram	
Newborn human baby	kilograms	5–10 pounds
Cat	5–7 kilograms	11–15.4
Adult human	45–90 kilograms	
Refrigerator (full size)	91–109 kilograms	200–240 pounds
Car	kilograms	1700–3500 pounds
Largest known dinosaur	80–100 metric tons	
Blue whale	180 metric tons	198 metric tons

24. Use the data from Exercise 23 to create two rules, one for converting pounds to kilograms and one rule for converting kilograms to pounds. Since the weights in the table are approximate, your rules will also be approximate. Write a formula for each rule.

25. Create two rules, one rule for converting ounces to grams and one rule for converting grams to ounces. Write a formula that expresses each rule.

26. In Your Own Words Write a paragraph that describes the units of measure for mass and weight.

27. **Preview** In Chapter 6, you will further your understanding of probability. Consider the possibility of tossing a coin twice.

a. List all the possible outcomes.

b. What is the probability of the toss being heads both times?

Mixed Review

Expand each expression.

28. $4(3m + 1)$

29. $10(6x + 5)$

30. $8(2v - 11)$

31. $7(9k - 4)$

32. $-2(6y + 13)$

33. $-5(3t - 8)$

34. $0.5(4g + 15)$

35. $0.25(9x - 12)$

36. $\frac{1}{2}\left(8w + \frac{3}{4}\right)$

37. $\frac{3}{4}\left(\frac{1}{5}m - \frac{5}{7}\right)$

Use the distributive property to help you do each calculation mentally. Write the grouping that shows the method you used.

38. $5 \cdot 14$

39. $6 \cdot 42$

40. $7 \cdot 19$

41. $9 \cdot 45$

Combine like terms to simplify each expression.

42. $5m + 12m$

43. $10p - 7p$

44. $40x + 16 + 12x + 3$

45. $23t + 1 - 11t - 7$

46. $5g + 2(8 + 7g)$

47. $5(2a + 6) - 2(3a - 1)$

CHAPTER 5

Review & Self-Assessment

Chapter Summary

In this chapter, you worked with prisms and cylinders. You found a formula for the volume of prisms. You used cubic units to measure volume and square units to measure area and surface area.

You learned that nets are flat figures that can be folded to form a closed three-dimensional object. You also found a method for finding surface area by using nets.

You learned that mass is the measurement of the amount of matter in an object and weight is the pull of gravity on that amount of matter. You also learned that mass and weight are approximately the same on Earth. You used metric units to measure mass and customary units to measure weight. In addition, you used benchmarks to estimate the mass of objects and you used scientific notation to express extremely small masses.

Vocabulary

- base
- customary system
- cylinder
- gram
- mass
- metric system
- net
- ounce
- pound
- prism
- solid
- surface area
- ton
- volume
- weight

Strategies and Applications

The questions in this section will help you review and apply the important ideas and strategies developed in this chapter.

Finding the volume and surface area of a geometric solid

1. A rectangular prism has a base 4 cm long and 3 cm wide. The prism is 5 cm tall.

a. What is the prism's volume?

b. Suppose a different rectangular prism has the same height and volume as the original prism. What are the dimensions of its base?

c. Suppose a cylinder with height 5 cm has the same volume as the original prism. What is the area of its base? Estimate its radius.

2. Find all the rectangular prisms you can make with 24 blocks. Try to do this without building all of them.

a. Give the dimensions of each prism.

b. Which of your prisms has the greatest surface area? What is its surface area?

c. Which of your prisms has the least surface area? What is its surface area?

Find the surface area and volume of each figure.

3. 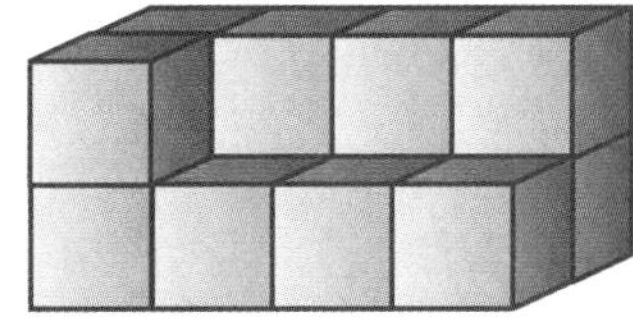

4. 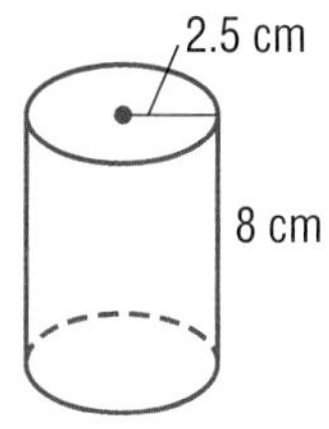

Using nets to represent geometric solids

5. The square in the figure below has sides 3 centimeters long. The triangles are isosceles, and the height of each triangle is 4 centimeters. Is the figure a net? In other words, does it fold up to form a closed solid? Explain how you know.

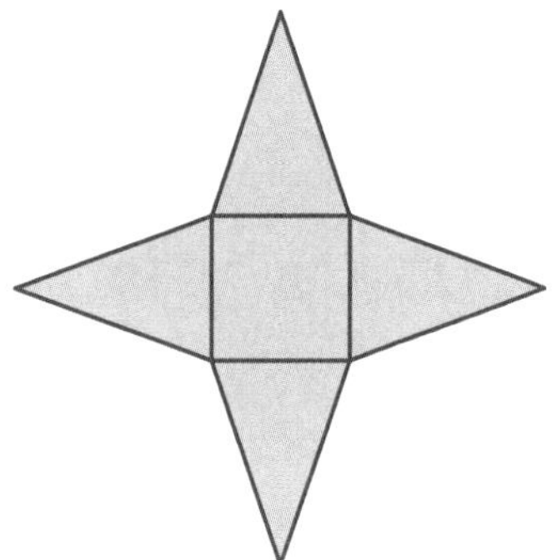

Using scientific notation and benchmarks

Use scientific notation to express the following conversions.

6. 213 milligrams __________ kilograms

7. 12 grams __________ centigrams

8. 50 grams __________ milligrams

9. 234 grams __________ kilograms

10. 5.6 hectograms __________ grams

Determine which benchmark would be closest to the given object.

11. pencil

12. a box of magazines

13. your Math book

14. an eraser

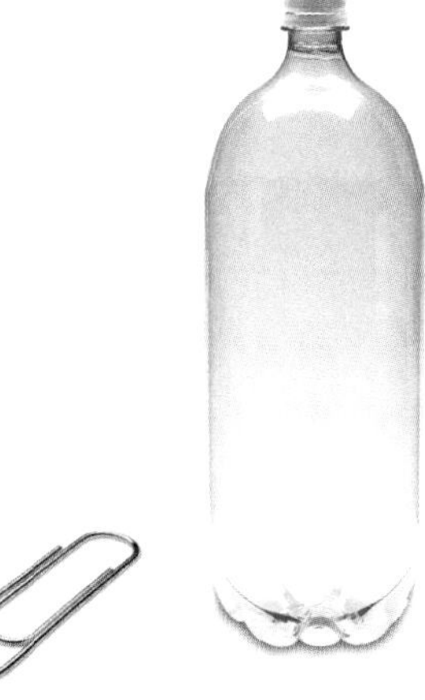

Demonstrating Skills

In Exercises 15–21, find the volume and surface area of the figure.

15.

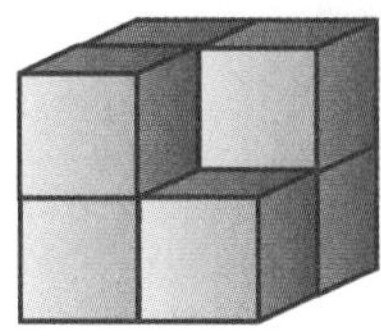

16.

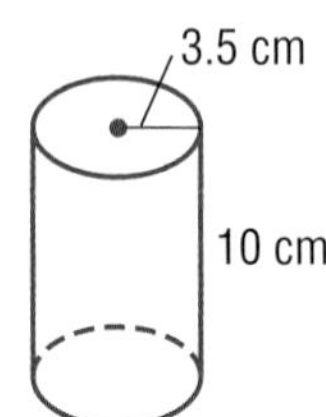

17.

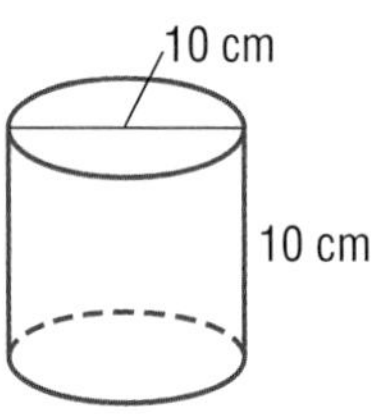

18.

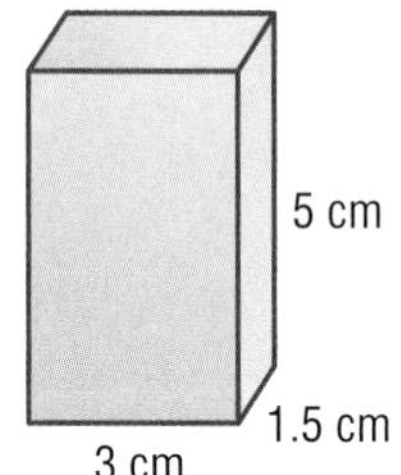

19. a cylindrical storage tank with radius 5 feet and height 12 feet

20. a can with height 10 centimeters and circumference 18 centimeters

21. a cube with side length 1.5 meters

22. Which of these three solids has the greatest surface area? Which has the least surface area?

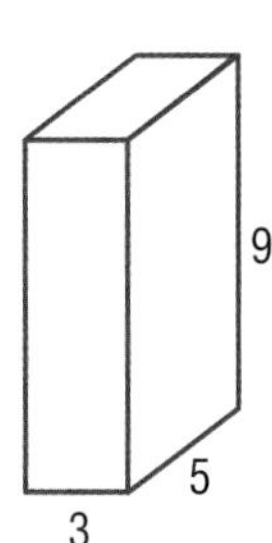

0.3

7.5

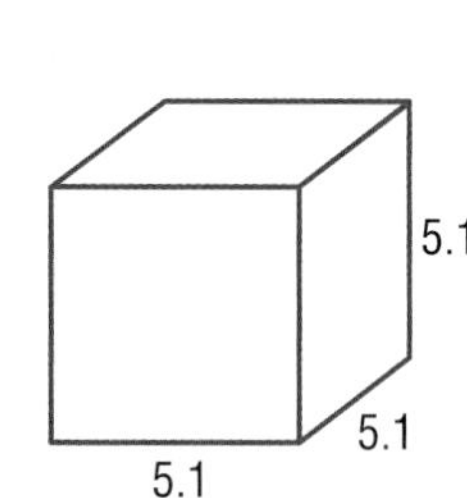

Fill in the blanks to make each equation true.

23. 3.75 pounds = ______ pounds ______ ounces

24. $\frac{1}{2}$ pounds =______ ounces

25. 12 ounces =______ pound

26. 20 ounces =______ pounds ______ ounces

27. 49 ounces =______ pounds ______ ounce

Which metric unit would be most appropriate to measure the mass of each of the following?

28. a teaspoon of sugar

29. a small dog

30. a pottery bowl

31. a shoe

Test-Taking Practice

SHORT RESPONSE

1 Meshack is building a dog house 3 feet long, 2 feet wide, and 2.5 feet high. The dog house will have a flat roof and a floor. How many square feet of wood will he need for the walls, roof, and floor of the dog house?

Show your work.

Answer ____________________

MULTIPLE CHOICE

2 A water tank measures 110 feet in diameter and is 40 feet high. How many cubic feet of water can it hold, to the nearest cubic foot?

A 13,816
B 189,970
C 379,940
D 1,519,760

3 How many pounds are equal to 64 ounces?

F 3
G 4
H 6.4
J 8

4 Which metric measure indicates the least amount of mass?

A 20 g
B 2 kg
C 200,000 mg
D 200 g

5 Which of the following measures is the best estimate for the mass of a thumb tack?

F 150 g
G 1.5 g
H 15 kg
J 1 lb

6 The mass of an object is 325 mg. What is its mass in grams?

A 0.0325 g
B 0.325 g
C 3.25 g
D 32.5 g

7 At birth, Stephen weighed 7.5 pounds. A bag of fruit weighs 70 ounces. Which weighs less?

F The fruit weighs less.
G Stephen weighs less.
H They both weigh the same.
J The answer cannot be determined from this information.

CHAPTER 6

Data and Probability

Real-Life Math

Contents in Brief

6.1	Dependence	262
6.2	Make Predictions	278
6.3	Data Graphs	294
	Review & Self-Assessment	314

What Do Most Americans Think? "Most Americans are optimistic about their futures." Have you ever wondered where statements like this orginate? After all, how would anyone know what "most Americans" think?

To uncover the American public's opinions, a group called the Gallup Organization conducts surveys on a variety of topics. This group, which has been conducting surveys for more than 60 years, polls a small sample of the American public and draws conclusions about the entire American population based on the responses. The Gallup Organization claims the margin of error in its survey results is plus or minus three percentage points.

Think About It Suppose you were to conduct a survey to determine students' favorite types of music in your class. If you were to conduct the same survey of all students in your school, do you think the results would be the same?

Math Online
Take the **Chapter Readiness Quiz** at glencoe.com.

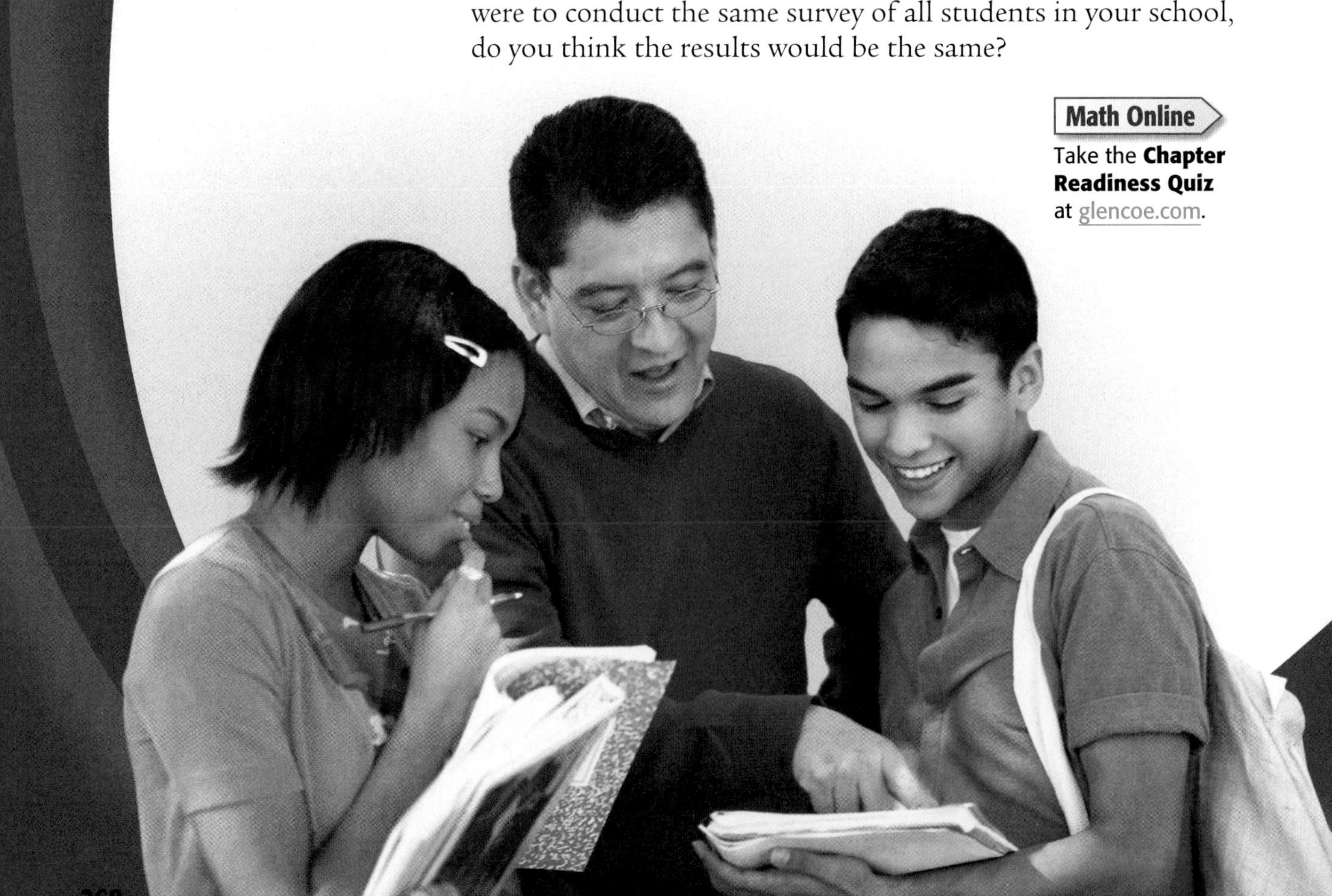

Dear Family,

Many games played at home rely on chance or probability. In the next chapter, the class will study the basics of probability. The chapter will begin by considering marbles drawn from a bag. For example, if there are six green and four yellow marbles in a bag, questions like these can be answered.

- If one marble is drawn at random, which color is it most likely to be?
- What are the chances that the marble drawn will be green?
- Suppose a yellow marble is drawn, put in a pocket, and then another marble is drawn. What are the chances of drawing a green marble?

The class will use probability to test whether games of chance are fair. A game that uses probability to develop a winning strategy will also be played.

Key Concept–Statistical Sampling

Statistical sampling is another common use of probability. In statistical sampling, a small group is chosen at random and used to draw conclusions about the entire population. Can you think of surveys or statistics which may have been based on statistical sampling? Is this a fair and accurate way to draw conclusions about a large population?

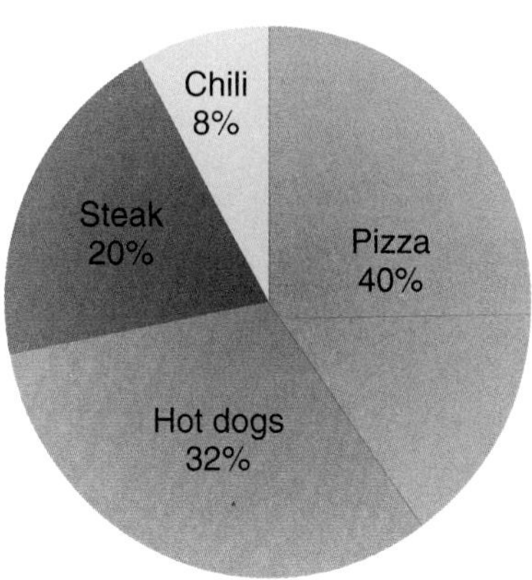

However used, finding probabilities depends on having data. The class will review and study many ways to display data, including circle graphs, and will discuss when it is most appropriate to use those displays.

Chapter Vocabulary

double-bar graph

double-line graph

population

representative sample

sample

Home Activities

- You can play different games of chance together. Your student can teach you the games that are played at school. Other games may include card games where it helps to know what cards you are likely or very unlikely to draw next.
- You can also talk about probabilities in our daily lives, such as the chance of rain.

Dependence

Many board and card games rely on *chance* or *probability* to make them fun. With some games, whether you win depends on nothing more than the cards you choose or the dice rolls you make. In this lesson, you will explore situations and games involving probability.

Inv 1	Combinations and Probability	262
Inv 2	Heads or Tails?	264
Inv 3	Is It Fair?	268
Inv 4	The Hidden Prize	270

Think & Discuss

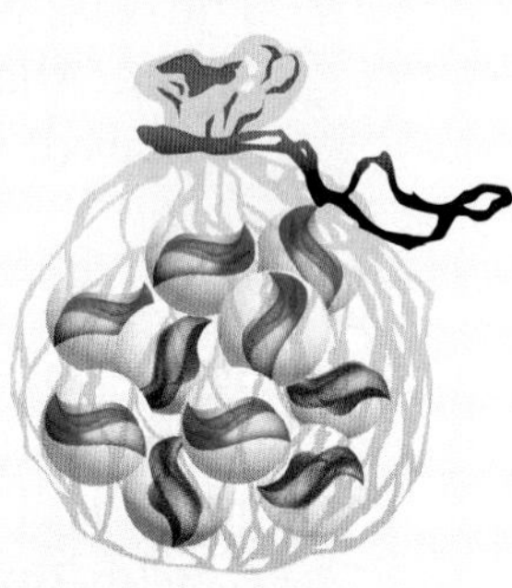

There are 10 marbles in a bag, 6 blue and 4 purple. If you draw out one marble at random, which color is it most likely to be?

The chances that the marble you draw will be purple are 4 out of 10. What are the chances that the marble will be blue?

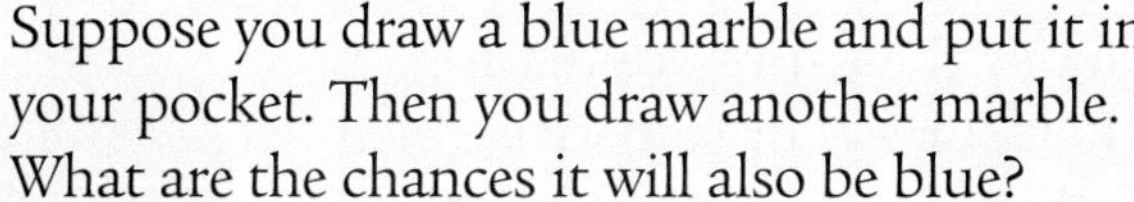

Suppose you draw a blue marble and put it in your pocket. Then you draw another marble. What are the chances it will also be blue?

Investigation 1 Combinations and Probability

You have probably heard people talk about probabilities in many ways. In the above Think & Discuss, for example, it states that the chances of the marble being purple are 4 out of 10. You might also say that the chances of choosing purple are 4 in 10, 4:10, $\frac{4}{10}$, 0.4, or 40%.

Develop & Understand: A

A bag holds six blocks, numbered 1 to 6. Imagine drawing one block from the bag.

1. What are the chances of drawing the number 2? Explain how you found your answer.
2. What are the chances of drawing an odd number? Explain.
3. What are the chances of drawing 2, 3, or 5?
4. What are the chances of drawing an odd number that is 3 or 5?

1. First, see what happens when the contestant decides *not* to change his or her mind. Play the game ten times, always staying with the first choice. Record the number of wins in the middle column of a table like the one shown below. (Your table should have one row for each game.)

Win or Lose?

Games	Do Not Change Your Mind	Change Your Mind
1		
2		

2. Now see what happens when the contestant *always* changes his or her mind. Play another 10 times, always changing to the unchosen paper in Step 4. Record the results in the last column of your table.
3. Based on your results, do you think the contestant should or should not change his or her mind?
4. Does your initial guess agree with the result found through your experiment?
5. Explain why the advice that you would now give Camila is correct.

What Did You Learn

6. When deciding which strategy to use when playing a game, why is it useful to conduct an experiment?
7. Will an experiment always give you the best strategy? Explain.

On Your Own Exercises

Lesson 6.1

Practice & Apply

1. Alejandro and Oliver have two chips. One is red on both sides. The other is red on one side and blue on the other. Each boy tosses a chip. Oliver scores one point if the chips match, and Alejandro scores one point if they do not match.

   a. Is the game fair? Explain how you know.

   b. The boys get a third chip, red on one side and blue on the other. Make up a fair game that involves tossing all three chips.

2. Suppose you turn the spinner below and then toss a cube that is numbered 1, 2, 3, 4, 5, and 6. You and another player add the two results. You score two points if the sum is 5, and your friend scores three points if the sum is 3. Is this a fair game? Why or why not?

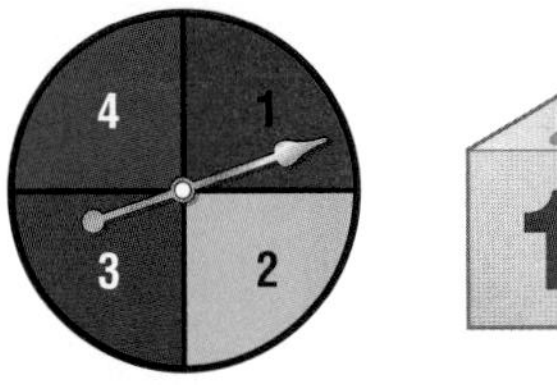

3. Rashard and Kellie are getting ready to play checkers. They devise a competition to decide who will make the first move. They will put three red and two black checkers in a bag and then each draw one. If the two checkers are the same color, Rashard wins. If they are different colors, Kellie wins.

   a. Design an experiment that could help you decide whether the competition is fair. Run the experiment 20 times. Describe your experiment and your results.

   b. List all the possible pairs of checkers that could be chosen. (Hint: There are ten possible pairs, all equally likely.) Use R1, R2, and R3 to represent the red checkers and B1 and B2 for the black checkers. For example, one possible pair is R1, B2.

   c. In how many of the ten possible pairs are the two colors the same? Does this agree with your experimental results?

   d. Is the game fair? Explain. If it is not fair, who has the advantage?

4. Arturo and Jill play a game with eight marbles numbered 1 to 8. They mix the marbles in a cup and then draw two.

a. List all the pairs of numbers that are possible to draw from the cup. (Note: Drawing 1 and 2 is the same as drawing 2 and 1.)

b. What are the chances that the two numbers are 4 and 7?

c. What are the chances that the two numbers have a sum of 11?

d. What are the chances that the two numbers include an even number?

e. What are the chances that the two numbers have an even sum?

5. Mr. Richards, a math teacher, works as a circus clown in the summer. His clown shirts come in three colors, which are yellow, green, and red. He also has four pairs of clown pants, which are yellow, purple, orange, and green.

a. List all 12 possible combinations of Mr. Richards' shirt and pants.

b. Mr. Richards chooses his clothes at random so that each outfit has the same chances of being selected. What is the probability that he will select a yellow shirt and orange pants?

c. What is the probability that at least one item will be yellow?

d. What is the probability that he will choose purple pants?

e. What is the probability that his pants and shirt will be different colors?

6. Jerry has pulled all the numbered spades from a deck of cards. He now has nine cards: 2, 3, 4, 5, 6, 7, 8, 9, and 10 of spades. He shuffles them and then holds them out for Alanah to choose one. After Alanah chooses her card, Jerry holds them out for Marquez to choose one.

   a. What is the probability that Alanah chose a number less than 8?

   b. What is the probability that Alanah chose an even number?

   c. List all the possible ways Alanah and Marquez could have chosen their cards. For example, if Alanah chose 5 and Marquez chose 3, the combination is (5, 3).

   d. What is the probability that Alanah chose a higher card than Marquez?

   e. If Alanah drew the 8, what is the probability that she chose a higher card than Marquez?

7. Mrs. Cooper has three children. Assume she is just as likely to have a boy as a girl.

   a. Make a tree diagram to find probabilities in this situation, such as how likely it is that all three of her children are boys.

   b. What is the probability that all three children are boys?

   c. What is the probability that Mrs. Cooper has two girls and one boy?

   d. Suppose you know that Mrs. Cooper has at least one girl. Now what is the probability that she has two girls and one boy?

8. A bookshelf contains three books, *After the Bell Rings, Beware of the Frog,* and *Cornered.* When Diego accidentally knocked the books onto the floor, he put them back on the shelf without paying attention to their order.

   a. Label three identical pieces of paper A for *After the Bell Rings,* B for *Beware of the Frog,* and C for *Cornered.* Fold each piece so you cannot see the labels. Shuffle them and then drop them on the floor. Pick them up one by one. How often do you pick them up in alphabetical order?

   Run this experiment at least 12 times.

   b. List all the ways the books could be put on the shelf.

   c. In how many of these ways would the books be in alphabetical order? Does this agree with your experimental results from Part a?

9. Tess has a cube with numbered faces.

- Each face has a whole number.
- Two faces have the same number.
- The sum of all six numbers is 41.

Tess rolls the cube ten times with these results.

3 11 12 4 11 12 3 4 4 4

a. Obviously 3, 4, 11, and 12 are on four of the faces. How many possibilities are there for the other two faces? What are they?

b. Which of these possibilities do you think is most likely?

10. Preview A fairground entices customers with this game.

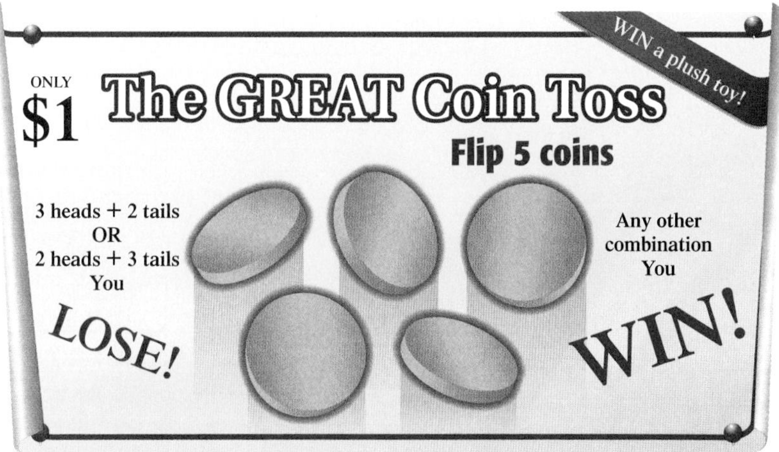

a. Draw a tree diagram to work out the chances of winning this game.

b. What is the probability that a player will win? What is the probability that a player will lose?

c. The prizes for the winners cost the fairground $2 each. If 500 people play, how much money would the fair expect to make? Explain.

11. Cary ran an experiment. Using a computer program, he examined the number of heads in various numbers of coin tosses. For example, he simulated tossing a coin 10 times and got 4 heads. For 100 tosses, he got 46 heads. He made a table of his results.

Number of Tosses	Actual Number of Heads	Expected Number of Heads	Difference |Expected – Actual|
10	4		
100	46		
1,000	513		
10,000	5,087		

Math Link

The absolute value of a number is its distance from 0 on a number line.

$|3| = 3$

$|-3| = 3$

a. Complete the last two columns of Cary's table.

b. Cary noticed that the numbers in the "Difference" column were increasing. He claimed that this shows that as the number of trials grows, the results get further from the actual probabilities of the situation. Explain why his reasoning is incorrect.

12. Kai, Steven, and Andrés are playing a dart game with this target.

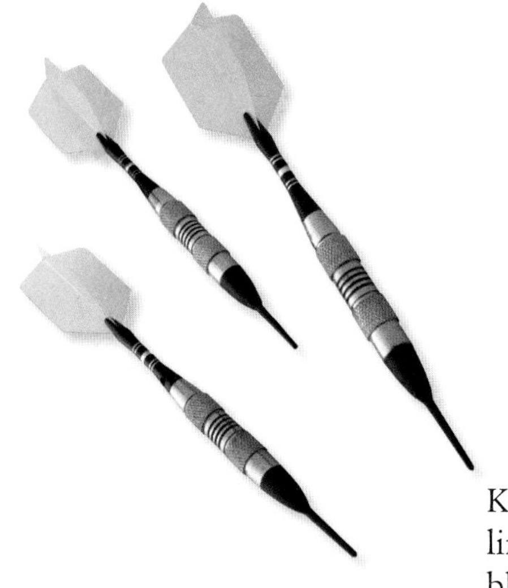

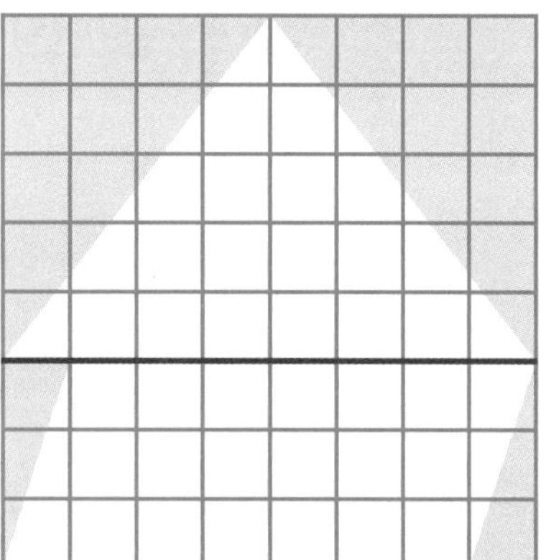

Kai scores one point if she hits the white triangle above the blue line. Steven scores one point if he hits the parallelogram below the blue line. Andrés must hit a blue region to score a point. Is this game fair? Why or why not?

13. Erika and Lani tossed a coin 20 times. Lani scored one point if it landed heads, and Erika scored one point if it landed tails. At the end of 20 tosses, there were 14 heads and 6 tails. Erika said that was impossible, since if you toss a coin 20 times, it should come out 10 heads and 10 tails. Is Erika correct? Why or why not?

Mixed Review

14. Geometry Consider this cylinder.

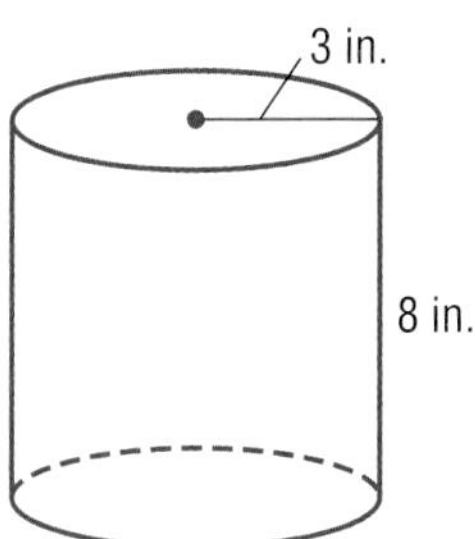

a. Find the volume of this cylinder.

b. Find the volume of a cylinder with the same height as this cylinder and twice the radius.

c. Find the volume of a cylinder with the same radius as this cylinder and twice the height.

15. A block structure and its top engineering view are shown. What are the volume and surface area of this structure?

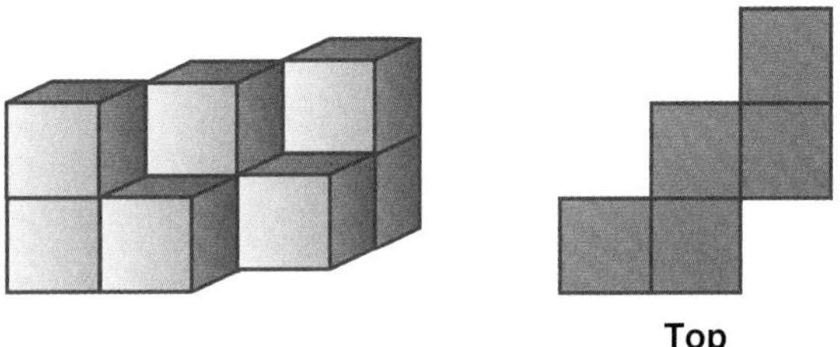

16. Here is the top-count view of a block structure. What are the structure's volume and surface area?

2	2	2	2
2	1	1	2
	1	1	

Rewrite each expression as addition, and then calculate the sum.

17. $-10 - (-9)$

18. $1.5 - (-0.14)$

19. $-3.33 - (-1.68)$

20. How long will the gummy worm be when it exits this connection? Give your answer in meters.

Make Predictions

Inv 1 Samples and Predictions 279

Inv 2 Sample Sizes 280

Inv 3 Representative Samples 283

Suppose you ask ten people in your class to name the all-time greatest professional basketball player, and seven say Michael Jordan. If your school had 1,000 students, you might predict that close to 700 of them would also name Michael Jordan. What if you were trying to predict for a group of 1,000 people in another area of the country? The seventh graders in Chicago, where Michael Jordan played for the Bulls, might not represent the opinions of people in Boston, where the Celtics play.

Understanding how to evaluate groups, and how to use that information to make predictions about larger groups, is the focus of this lesson.

Materials

- colored tiles
- paper bag

Explore

Work in groups of two or three to play *What's in the Bag*. Each group will need about 20 colored tiles. From these 20 tiles, your group should select a total of 10 tiles. Put the 10 tiles into a paper bag. Then, without revealing your colors, exchange bags with another group.

Someone in your group should draw 4 tiles from the bag you are given. Discuss how to predict, based on those 4 tiles, the number of tiles of each color in the bag. For example, if the tiles chosen are RBBY, 1 red, 2 blue, and 1 yellow, someone might say, "I think there are likely to be more blues than the others, so my guess is 5B, 2R, 2Y, 1G."

Return the tiles to the bag, shake the bag, and draw 4 tiles again. Record the tile colors in a table like the one below, adding a second prediction about the contents of the bag. As you make your new prediction, keep in mind the first group of tiles.

Attempt	Tiles Drawn	Prediction
1	RBBY	4B, 3R, 3Y, 0G
2	BBBG	6B, 2R, 1Y, 1G

Draw a third group of 4 tiles. Make another prediction. Finally, look at all the tiles in the bag. How close were your predictions?

Investigation  Samples and Predictions

Vocabulary

population

sample

Medical researchers often need to predict how well a new drug will work by testing it on part of the population. Advertisers test the effectiveness of a TV commercial on part of the viewing population before they spend large amounts of money airing it.

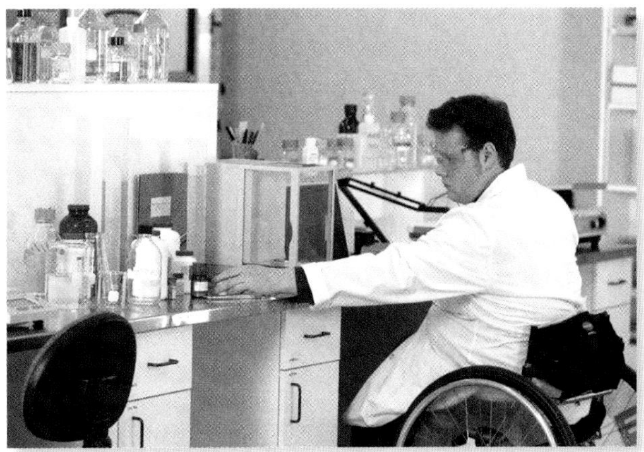

These people are all using a technique called *sampling*. The smaller group is called a **sample**. The larger group from which the sample is taken is called the **population**. In Explore, the population was the ten tiles in the bag, and you took samples of four tiles.

Develop & Understand: A

Each of the following are actual *What's in the Bag* experiments conducted by students. Look at their samples. Make your own prediction about the contents of the bags. Then check your prediction against the actual contents of the bag, which your teacher will reveal.

1. In this game, there were 20 blocks in a bag. The blocks come in yellow, blue, red, and green. Dawn and Sabina drew three samples of 5 blocks. After drawing each sample, they replaced the blocks.
   - Sample 1: YBRGG
   - Sample 2: GYYBR
   - Sample 3: RGRBY

   a. Predict the colors of the 20 blocks.

   b. Dawn and Sabina have seen three samples of 5, or 15 data items in all. Since there are 20 tiles in the bag, they may have seen some tiles twice, and some tiles were not drawn at all. Should Dawn and Sabina be fairly confident about their prediction? Explain.

   c. Check your predictions against the actual contents of the bag as given by your teacher.

2. Andrea and Tyson played a game with red, green, blue, and yellow disks. They had a bag with ten colored disks. They took five samples of four disks at a time.

- Sample 1: RRGG
- Sample 2: GYRR
- Sample 3: RBRG
- Sample 4: RYRR
- Sample 5: BRRR

a. What colors do you predict were in the group of ten?

b. Check your prediction against the actual contents of the bag.

3. Did you make a more accurate prediction in Exercise 1 or 2? In which would you *expect* to make a more accurate prediction? Why?

Share & Summarize

Kali and Kevin are playing *What's in the Bag* with 100 colored blocks. They took sample 1, returned the blocks to the bag, and then took sample 2.

- Sample 1: RRGYGBRRGR
- Sample 2: RGGRBRBGBY

Explain how you would use their sample data to make a prediction about the colors of the 100 blocks in the bag.

Investigation 2 Sample Sizes

People often use a sample to get a sense of the characteristics of a population, particularly when it would be time-consuming and costly to test the whole population. If you have a good sample, the proportion of the sample with a certain characteristic or holding a certain belief can be used to make a prediction about the population.

In a poem, for example, one characteristic you may consider is word length. The following poem has long words, such as *cheeseburgers,* which has 13 letters. It also has words with only two letters. The population in this situation is the 112 words of the poem.

Real-World Link
Although the poetry of American Emily Dickinson (1830–1886) is now known all over the world, only a couple of her almost 2,000 poems were published in her lifetime.

Love Lesson

By Kristen Pitts, 8th grade, Ohio

He is a four-year-old with blonde hair and blue eyes
He loves music about cheeseburgers and paradise

He lives each day to the fullest while he's awake
Always wanting a chocolate milkshake

He learns through his videos and hand-held games
He loves to play and play
He is a mastermind at puzzles

He reads his favorite books quite a bit
Reading them over and over before calling it quits

And when it's time for him to sleep
His blanket and warm thoughts he's sure to keep
Always such a sweet, innocent boy

He is one of the greatest boys I know
Because he taught me how his autism goes

Develop & Understand: A

1. Before continuing, estimate or guess the mean word length in the poem.

The *certain* way to find the mean word length is to count all the letters in the poem and divide by the number of words. However, your task here is to see how well you can *predict* the answer by using a sample.

2. The poem has 14 lines.
   a. For each line, count the numbers of letters and words. Record your totals in a table like the one shown here.

Line	1	2	3	4	5	6	7	8	9	10	11	12	13	14
Letters														
Words														

   b. Find the mean word length for Line 6.
   c. Find the mean word length for Line 10.
   d. Do you think you can estimate the mean word length for the poem from a one-line sample? Explain.

3. Make a table for recording the results from different sample sizes.

Lines	Letters	Words	Mean Word Length
2			
4			
6			
8			
10			

Randomly choose two lines from the poem. Add the numbers of letters for the two lines. Add the numbers of words. Calculate the mean word length. Write your results in the first row of your table.

Next, choose a sample of any four lines. Record the results. Then choose a sample of any six lines, then any eight lines, and then any ten lines. Enter the data you collect into your table.

4. Do all samples give the same mean word length? Discuss with your partner which sample you think gives the most accurate prediction.

5. Find the mean word length in the poem by dividing the total number of letters in the poem by the total number of words. How close was your prediction?

6. If you wanted to choose a one-line sample to bias the prediction toward longer words, which line would you choose? Explain.

Real-World Link

One way teachers check the "readability" of a piece of writing for their students is by using a similar sampling technique. They count the number of sentences and syllables in three randomly chosen 100-word passages and plot their results on a chart that gives them the approximate grade level of the reading material.

Your results may be different from those of other students in your class, depending on which lines you chose.

7. Create a class data set of the number of lines in the samples that gave the closest prediction. Each student should report how many lines he or she used for the sample that gave the best prediction.
   a. Count the number of times each sample size was reported.
   b. Which sample size seems to be most reliable for predicting the mean word length in the poem?

Share & Summarize

1. Count the numbers of letters in the first names of the six students sitting nearest to you. What is the mean word length for these data?
2. Do you think your group of six students is a reasonable sample for predicting the mean number of letters in the first names of everyone in your class?
3. Describe a situation for which six students is a good sample size and one for which six students is not a good sample size.

Investigation 3 Representative Samples

Vocabulary

representative sample

Materials

- protractor

For a statistics project, Alison's group wanted to predict which of the following after-school activities were the favorites of seventh graders in the four schools in their district.

- playing sports
- playing music
- listening to music
- watching TV
- going to the mall
- playing video games
- reading
- drawing or painting

Think & Discuss

How would you recommend that Alison's group collect a sample that would help them make an accurate prediction? Make a list of suggestions.

As you have learned, a good sample must be large enough to give accurate results. In addition, it must be *representative* of the population.

A **representative sample** has approximately the same proportions as the population with respect to the characteristic being studied. For example, Alison's group will want its sample to have about the same proportion as the district population of males and females. Boys might be more likely, or less likely, than girls to engage in a particular activity.

Finally, anyone conducting a survey using a sample will want a method that is practical to perform.

That gives three important questions to ask when examining whether a particular survey method is a good one.

- Is the sample *large enough* to give accurate results?
- Is the sample *representative* of the population?
- Is the survey method *practical*?

Develop & Understand: A

Alison's group proposed five strategies for conducting its survey.

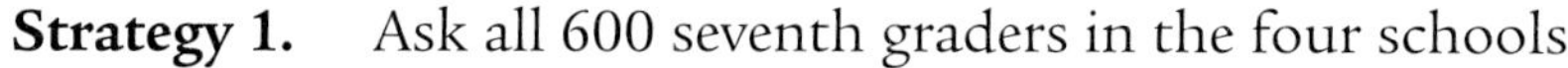

Strategy 1. Ask all 600 seventh graders in the four schools.

Strategy 2. Distribute questionnaires to the 130 seventh graders at the district band competition next week.

Strategy 3. Get a list of all the seventh graders in the district. Call 5% at random from each school.

Strategy 4. Survey every third seventh grader out of the 125 in our school as they enter their homerooms.

Strategy 5. Survey the 12 seventh graders on one student's bus.

1. Which strategy or strategies do you think would give the least representative data? The most representative data?

2. Which strategy would give the smallest sample size? The largest sample size?

3. Which strategy seems the least practical? The most practical?

4. Which strategy do you think is the best? Explain your reasoning.

Develop & Understand: B

Alison's group decided to survey every fourth student to enter each seventh grade homeroom in the school. The six homerooms have 28 students each. The survey results are displayed below in a bar graph.

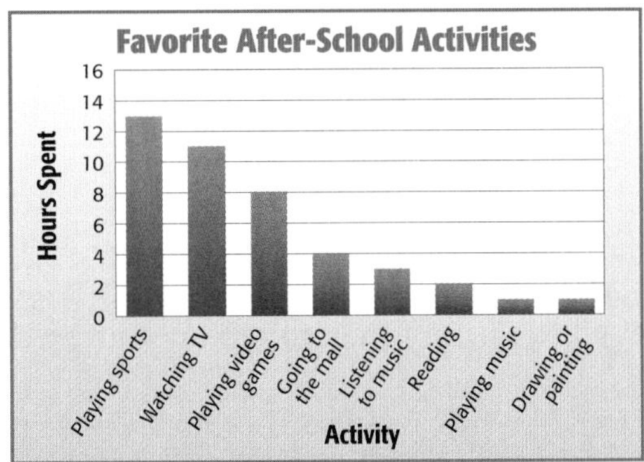

5. How large is the sample of students surveyed?

6. How many students in the sample prefer playing sports after school?

7. Based on the sample, predict how many in a district of 600 seventh graders would prefer reading.

8. Suppose you could choose one seventh grader from the 600 in Alison's school district at random. What is the probability that he or she prefers playing video games? Explain how you found your answer.

9. If Alison was predicting the favorite leisure activities of all students in the district, from kindergarten to twelfth grade, would sampling only seventh graders give an accurate picture of the population? Explain.

Share & Summarize

1. Name some situations in which you are more likely to predict from a sample rather than measure an entire population.

2. Look back over your class' suggestions for the Think & Discuss on page 283. For each suggestion, respond to these three questions.
   - Is the sample large enough?
   - Is the sample representative?
   - Is the survey method practical?

On Your Own Exercises
Lesson 6.2

Practice & Apply

1. Tobias and Kareem were playing *What's in the Bag*. From 12 hidden tiles, they took five samples of 3.

- Sample 1: BYG
- Sample 2: YBG
- Sample 3: BGB
- Sample 4: GBB
- Sample 5: RBG

What colors do you predict were in the group of 12 tiles?

2. Jemma and Sarita were playing *What's in the Bag*. There were 20 hidden colors. They drew four samples of 5.

- Sample 1: RYRGR
- Sample 2: YBGYB
- Sample 3: BYGYB
- Sample 4: RYGRY

What colors do you predict were in the group of 20?

3. Rico and Abby were playing a computer simulation of *What's in the Bag*. The software creates a group of 200 tiles in a bag with a combination of four colors. Users enter the sample size and the number of samples.

Rico and Abby selected four samples of 10.

- Sample 1: RBBGGGGYYY
- Sample 2: RRBGGGGGYY
- Sample 3: RRRGGGGYYY
- Sample 4: RRBBGGGGGY

a. From this sample, what do you think the software chose for the percentage of each color?

b. Is the sample large enough for you to feel confident about your prediction?

Investigation 1 Double-Bar and Double-Line Graphs

Vocabulary

double-bar graph

double-line graph

Material

- lined or grid paper
- colored pencils or markers

Real-World Link

Many newspapers and Internet sites present data in charts and graphs. Knowing how to read and interpret them can help you understand the article or Web site you are reading.

Mike and his twin brother each made a graph to show how they spend their weekend days. Mike wants to compare how he spends his free time to how his brother spends his free time.

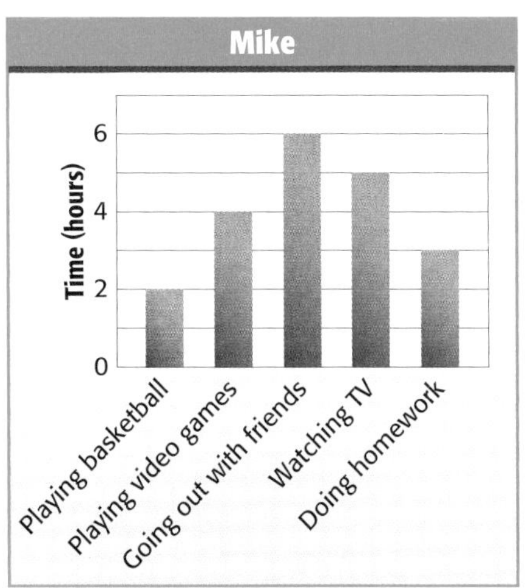

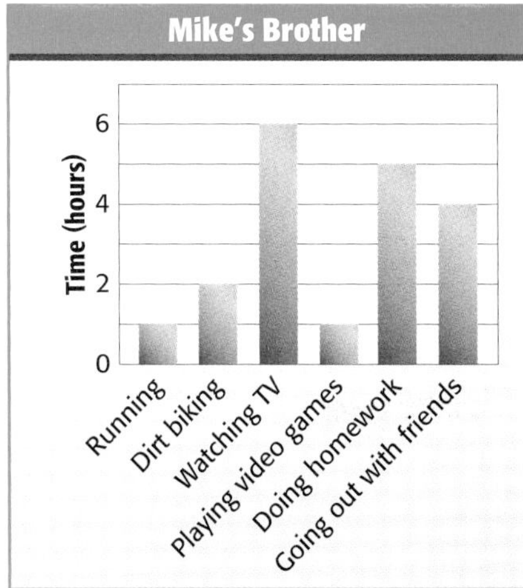

List three things you can tell about the similarities and differences in how Mike and his brother spend their days.

What would make it easier to compare the data?

Develop & Understand: A

1. Create one graph that shows the way both Mike and his brother spend their weekend time. Use the following guidelines to create the graph.

   a. Include a title at the top of a the graph.

   b. Draw a set of vertical and horizontal axes like the one that both boys used for their original bar graphs. Label the vertical axis "Time (hours)." Mark off each hour up to 7 hours.

   c. Label categories along the horizontal axis. Create a single category that can include basketball, running, and dirt biking. You should have five categories.

   d. Using blue for Mike and red for his brother, create two adjacent bars above each category name, one bar for each boy. The bars for one category will touch, but there will be space between each category.

   e. Create a key to the side of the graph to indicate that the blue bars show data for Mike and the red bars show data for Mike's brother.

2. Which boy spends more time watching TV? How much more?

3. Explain if this is an accurate statement. "Mike spends twice as much time playing video games as his brother."

4. How did the double-bar graph make comparing the brothers easier?

A bar graph with two bars for each category on the same graph is called a **double-bar graph**. A double-bar graph is used to compare two sets of data based on the same categories. For example, you could compare boys versus girls using the types of sports they play, such as football, softball/baseball, basketball, volleyball, or track.

Develop & Understand: B

5. Jennifer and Selena have collected information about students in their class. They want to compare preferences of winter activities between the girls and boys.

   Create a double-bar graph to compare the data and answer the following questions.

Activity	Girls	Boys
Skiing	6	8
Sledding	2	2
Building snowmen	5	4
Snow boarding	1	7
Snow shoeing	10	2

   a. Which activity is preferred most by girls? By boys?

   b. Which activity is peferred least by both boys and girls?

   c. If each student could pick only one activity, how many students did Jennifer and Selena interview?

Think & Discuss

In another situation, you have the following data about the use of cell phones and home Internet service in Cell Region IV. The data is shown by the number of users throughout the day.

What graph type should we use to quickly compare how many people are using each service throughout the day?

Use in the hour Ending at	Cell Phone Service	Home Internet Service
9:00 A.M.	2250	1000
10:00 A.M.	2600	900
11:00 A.M.	3000	750
12:00 P.M.	2400	700
1:00 P.M.	2100	800
2:00 P.M.	2500	700
3:00 P.M.	2600	700
4:00 P.M.	1800	1100
5:00 P.M.	1500	2500
6:00 P.M.	2900	2750
7:00 P.M.	3200	3300
8:00 P.M.	2500	3300
9:00 P.M.	2250	3200
10:00 P.M.	2000	2400

Develop & Understand: C

6. Use the same idea as the double-bar graph and put two line graphs on the same axis. Place the hour throughout the day along the horizontal axis and the number of users of each service along the vertical axis.

7. During which hour of the day is cell phone use the highest? Did you refer to the table or the graph to answer this question? Why?

8. During which hour of the day is home Internet use the highest? Did you refer to the table or the graph to answer this question? Why?

9. During what hour of the day does use of cell phone service increase the most? During what hour of the day does use of home Internet service increase the most? What factors might influence this?

10. Describe how use of cell service changes throughout the day. What factors influence use of cell service?

11. Describe how use of home Internet service changes throughout the day. What factors influence use of home Internet service?

In the last set of exercises, you created a *double-line graph.* A **double-line graph** can be used to compare how quantities of different groups change over the same time frame.

Develop & Understand: D

Use the data in the table below for the Exercises 12–14.

Sales in Hundreds of Dollars

Month	JAN	FEB	MAR	APR	MAY	JUN	JUL	AUG	SEP	OCT	NOV	DEC
Cold Beverages	6	6.2	8	5.5	7	7.2	10	10	9	8	7.5	6
Hot Beverages	5.5	7.2	4.5	3	2.5	2.2	1.5	1	1.5	2.2	2	4.5

Real-World Link

Water is the most popular beverage consumed worldwide. Tea is second.

12. Create a double-line graph that shows the sales of hot beverages (coffee and hot chocolate) and cold beverages (sodas and water) at the school concession stand throughout the year.
    a. What variable will you put along the horizontal axis? The vertical axis?
    b. How will you indicate the cold beverages and the hot beverages?

13. Write a paragraph to describe the information that is presented by the graph and table in Exercise 12. Include information about when sales of each type of beverage were highest and lowest, when one kind of beverage outsold the other, and when overall sales of beverages seemed highest. Speculate about factors that might influence beverage sales.

14. If you were to extend this graph by four more months after December, what might it look like?

Share & Summarize

1. Give a description of a situation in which you would want to create a double-bar graph to compare two things. Explain why a double-bar graph would be the best choice.
2. Give a description of a situation in which you would want to create a double-line graph to compare two things. Explain why a double-line graph would be the best choice.

Investigation  Make a Circle Graph

Materials

- protractor
- 1-meter strip of heavy paper
- 4 copies of the quarter-circle template
- scissors
- tape
- colored pencils
- straightedge

Real-World Link

Soccer's World Cup is the most-watched sporting event in the world. In 2006, an estimated 829 million North American and 26.3 billion worldwide television viewers watched World Cup broadcasts.

Data that represent parts of a whole are sometimes displayed in a circle graph, or pie chart. In this investigation, you will gather data from your class and use a circle graph to summarize your results.

Think & Discuss

Work as a class to gather student responses to this question.

Which is your favorite sport to watch? Choose one.

❑ *football* ❑ *soccer* ❑ *basketball*

❑ *baseball* ❑ *ice hockey* ❑ *other*

Copy this table. In the second column, record the number of students who voted for each choice. In the third column, record the fraction of the total votes each choice received. You will fill in the last column later.

Sport	Number of Votes	Fraction of Total Votes	Angle Measure
Football			
Soccer			
Basketball			
Baseball			
Ice hockey			
Other			

In the next set of exercises, you will create a circle graph to display the results of your survey.

Develop & Understand: A

Your teacher will give you a strip of heavy paper exactly one meter long. The strip has been divided into equal-sized rectangles, one for each member of your class. The entire strip represents your whole class, or 100% of the votes.

1. On your strip, color groups of rectangles according to the number of votes each choice received. Use a different color for each choice. For example, if five students voted for football, color the first five rectangles orange. If seven chose soccer, color the next seven rectangles green. When you are finished, all the rectangles should be colored.

Football | Soccer

Now you are ready to create a circle graph.

2. Tape the ends of your strip together, with no overlap, to form a loop with the colored rectangles inside.

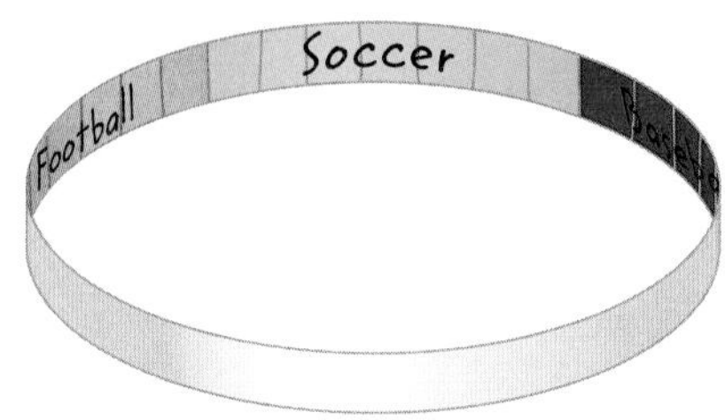

3. Next, tape four copies of the quarter-circle template together to form a circle. Find the center of the circle and mark it.

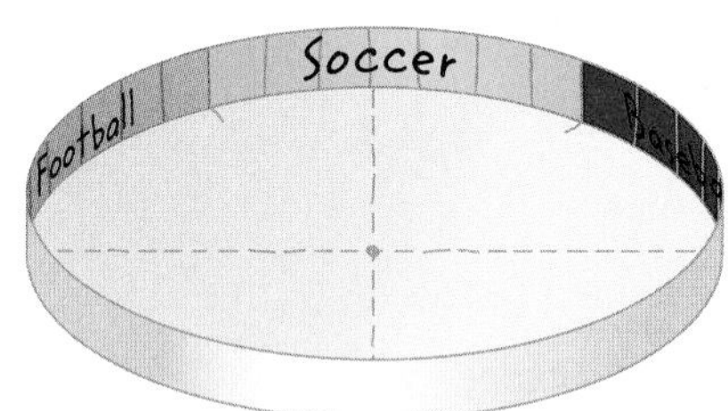

4. Place your loop around the circle. On the edge of the circle, mark where each color begins and ends.

5. Remove the loop, and use a straightedge to draw a radius from the center of the circle to each mark where there is a color change. Repeat until you have drawn six radii, one for each change in color.

6. Measure each section with your protractor. Choose one of the six radii previously drawn and line up the zero on the straight edge of the protractor with one of the sides of the angle. Find the point where the second side of the angle intersects the curved edge of the protractor. Read the number written on the protractor at the point of intersection. This is the measure of the angle in degrees.

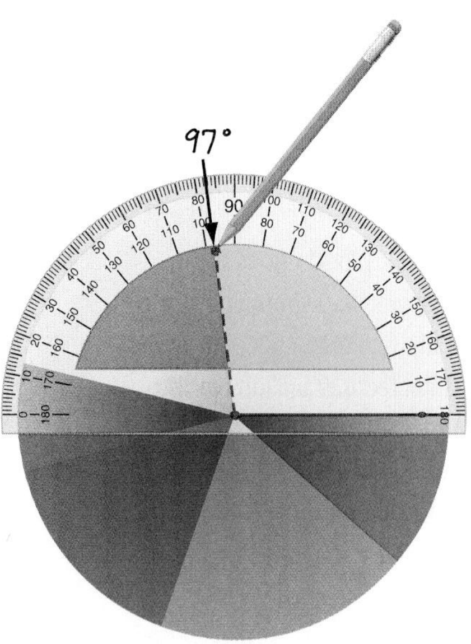

7. Fill in the *Angle Measure* column of the chart on page 299 with the angle measurements. There are 360 degrees in a circle, so your angles should have the sum of 360 degrees.

Real-World Link

Professional football is the most-watched sport in the United States, followed by figure skating.

8. Color the sections of your graph. Label each section with the sport name and the fraction of votes that sport received. For example, your circle graph might look like this.

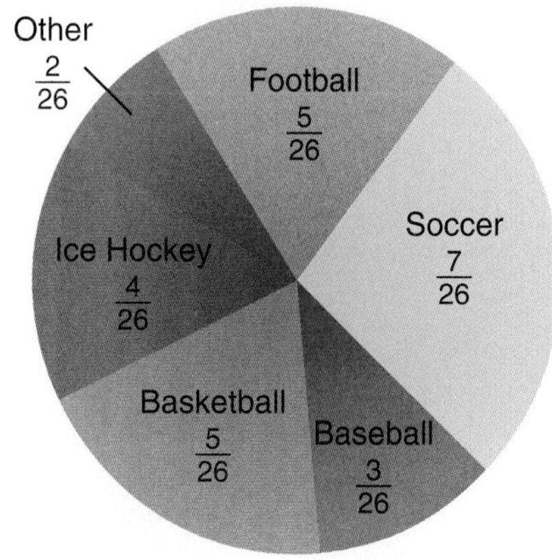

Develop & Understand: B

In Exercises 1–8, you learned to make a circle graph using a strip of paper. You will now make a circle graph using a protractor.

9. Work as a class to gather student responses to this question.

   What is your favorite season of the year? Choose one.

   - ❑ *Winter*
   - ❑ *Spring*
   - ❑ *Summer*
   - ❑ *Fall*

10. Construct a table to record your data. You can set up a table like the one on page 299.

11. Calculate the fraction of the class in each category. A circle is 360 degrees. If you multiply the fraction representing the part of the circle times $360°$, you will then have the angle measure of the respective part.

12. Construct a circle using the quarter-circle template and mark the center.

13. From the center point, draw a radius. Use this radius as a starting point to measure and draw the first angle with your protractor. Draw the remaining radii on the circle graph. You can check your calculations by making sure that the sum of all of your angles is $360°$.

Share & Summarize

Give a description of a situation in which you would want to create a circle graph. Explain why a circle graph would be the best choice.

Investigation 3 Stem-and-Leaf Plots

Vocabulary

stem-and-leaf plot

Materials

- homework—time data for your class

You have seen that line graphs, bar graphs, and circle graphs are useful for showing the distribution of numerical data. However, for some data sets, creating a line graph, bar graph, or circle graph may not be practical.

Think & Discuss

Have each student in your class estimate how many minutes he or she spent doing homework yesterday. Your teacher should record the data on the board.

The students in Ms. Washington's class kept track of how many minutes they spent doing homework one evening. Their results are listed below.

Student's Initials	Time Spent Doing Homework (minutes)
AF	42
JB	5
RC	60
HE	30
JL	45
MM	47
DL	0
SK	25
FR	67
CO	51
DW	56
PG	20

Student's Initials	Time Spent Doing Homework (minutes)
RL	90
DD	39
AG	30
RT	49
CB	58
MC	55
FB	75
JM	45
TK	44
MG	37
LK	62
EL	65

Think about what a circle graph of these data would look like. Do you think a circle graph is a good way to display these data? Why or why not?

When a data set contains many different values, or when the values are spread out, a **stem-and-leaf plot** (also called a *stem plot*) may give the best representation of the data's distribution.

Example

To make a stem-and-leaf plot of the homework-time data, think of each data value as being made up of two parts: a tens digit and a ones digit.

Write the tens digits, from least to greatest, in a column. Draw a vertical line to the right of the digits. The values in this column are the *stems.*

Stem	Leaf
0	
1	
2	
3	
4	
5	
6	
7	
8	
9	

To add the *leaves*, write the ones digit for each data value to the right of the appropriate tens digit. For example, to plot the first time of 42 minutes, write a 2 next to the stem value 4. The first five data values have been plotted here. Those values are 42, 5, 60, 30, and 45.

Stem	Leaf
0	5
1	
2	
3	0
4	2 5
5	
6	0
7	
8	
9	

After you have plotted all the data values, redraw the plot, listing the ones digits for each stem in order from least to greatest.

Stem	Leaf
0	0 5
1	
2	0 5
3	0 0 7 9
4	2 4 5 5 7 9
5	1 5 6 8
6	0 2 5 7
7	5
8	
9	0

Key: 4|2 = 42

Stem plots and histograms both group data into intervals. However, unlike a histogram, a stem plot allows you to read individual values.

For example, if you made a histogram that grouped the data above into the intervals 0–9, 10–19, and so on, it would show that there are four values between 50 and 59, but it would not show that these values are 51, 55, 56, and 58.

Develop & Understand: A

1. Look at the completed stem-and-leaf plot in the example on page 303.
   a. Explain how you can use this graph to find the range of the homework times.
   b. Use the stem plot to find the mode homework time. How did you find it?
   c. Use the stem plot to find the median homework time. Explain how you found it.
   d. Describe the shape of the graph. Explain what the shape tells you about how the data are distributed.
2. In the Think & Discuss at the beginning of this lesson, you collected homework times for your class. Make a stem plot of these data. If your data include values of 100 minutes or more, your plot will need to include two-digit stem values. For example, a value of 112 would have a stem of 11 and a leaf of 2.
3. Find the range, median, and mode of your class data.
4. Describe the distribution of your class data.
5. Write a few sentences comparing your class data to the data from Ms. Washington's class.

In the plots in Exercises 1–5, the tens digits are used as stem values and the ones digits are used as leaves. The stem and leaf values you choose for a given situation depend on the minimum and maximum values in the data set.

Real-World Link
A player's batting average is the number of hits divided by the number of times the player is officially at bat.

Develop & Understand: B

6. This table shows the batting averages for the 2007 Cleveland Indians baseball team.

2007 Cleveland Indians

Batter	Average	Batter	Average
Barfield	.243	Gomez	.283
Blake	.270	Gutierrez	.266
Cabrera	.283	Hafner	.266
Choo	.294	Lofton	.283
Dellucci	.230	Marte	.193
Francisco	.274	Michaels	.270
Garko	.289	Sizemore	.277

Real-World Link

Runs batted in, or RBI, is the number of people who score a run as a result of a player's hitting the ball.

a. The batting averages range from .193 to .294. In this case, you can use the tenths and hundredths digits as stem values and the thousandths digits as leaves.

Make a stem-and-leaf plot using the stem values shown at right.

b. Describe the distribution of data values.

c. Find the mode and median batting average.

2007 Cleveland Indians Batting Averages

Stem	Leaf
.19	
.20	
.21	
.22	
.23	
.24	
.25	
.26	
.27	
.28	
.29	

Key: 27|5 = .275

7. This table shows the American League RBI leaders for each season from 1978 to 2007:

American League RBI Leaders

Year	Player	RBI
1978	Jim Rice	139
1979	Don Baylor	139
1980	Cecil Cooper	122
1981	Eddie Murray	78
1982	Hal McRae	133
1983	C. Cooper / J. Rice	126
1984	Tony Armas	123
1985	Don Mattingly	145
1986	Joe Carter	121
1987	George Bell	134
1988	Jose Canseco	124
1989	Ruben Sierra	119
1990	Cecil Fielder	132
1991	Cecil Fielder	133
1992	Cecil Fielder	124

Year	Player	RBI
1993	Albert Belle	129
1994	Kirby Puckett	112
1995	A. Belle / M. Vaughn	126
1996	Albert Belle	148
1997	Ken Griffey, Jr.	147
1998	Juan Gonzalez	157
1999	Manny Ramirez	165
2000	Edgar Martinez	145
2001	Brett Boone	141
2002	Alex Rodriguez	142
2003	Carlos Delgado	145
2004	Miguel Tejada	150
2005	David Ortiz	148
2006	David Ortiz	137
2007	Alex Rodriguez	156

a. Make a stem-and-leaf plot of the RBI data.

b. Describe the distribution of data values.

c. Find the minimum, maximum, mode, and median of the RBI data.

Share & Summarize

1. In what types of situations would you use a stem-and-leaf plot, rather than a line plot, to display a set of data?
2. What information can you get from a stem plot that you cannot get from a histogram?

Investigation 4 Misleading Statistics and Graphs

Many times, people do not trust statistics and data because they have been misled by people using statistics. If you know how data can be distorted, you know what to look for in charts and graphs to accurately interpret the data.

People can be misled by graphs or by statistics such as "average," or mean and median. This investigation will help you become aware of how the same data can be shown differently and determine what representation may be most appropriate.

Think & Discuss

The world record for the men's 100-meter sprint has been broken six times since 1991. Below are two bar graphs that represent the times for the 100-meter sprint.

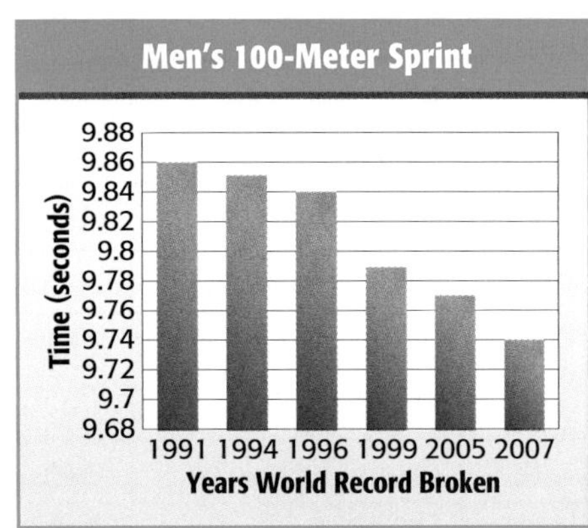

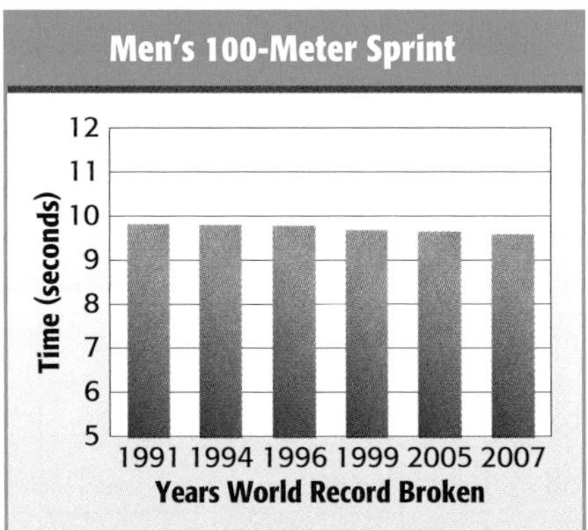

Zoe, Darnell, and Maya discussed the graphs.

Do these graphs represent the same data? Why do they look different?

5. **Measurement** Two classes of elementary school students measured their heights in centimeters. Here are the results.

   Ms. Cho's class: 117, 117, 119, 122, 127, 127, 114, 137, 99, 107, 114, 127, 122, 114, 120, 125, 119

   Mr. Diaz's class: 130, 147, 137, 142, 140, 135, 135, 142, 142, 137, 135, 132, 135, 120, 119, 125, 142

   a. For each class, make a stem-and-leaf plot of the height data.

   b. Find the range, mode, and median for each class.

   c. The two classes are at two different grade levels. Which class do you think is the higher grade?

   d. What percent of the students in Mr. Diaz's class are as tall or taller than the median height of the students in Ms. Cho's class?

6. The two circle graphs below show two college roommates budgets for how they spend their money each month.

Rafael's Budget	Amount of Money
Rent and bills	50%
School expenses	10%
Food	15%
Entertainment	10%
Savings	15%

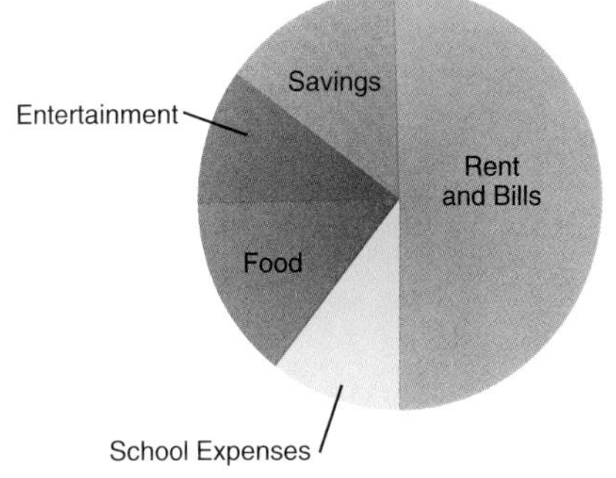

George's Budget	Amount of Money
Rent and bills	50%
School expenses	10%
Food	20%
Entertainment	15%
Savings	5%

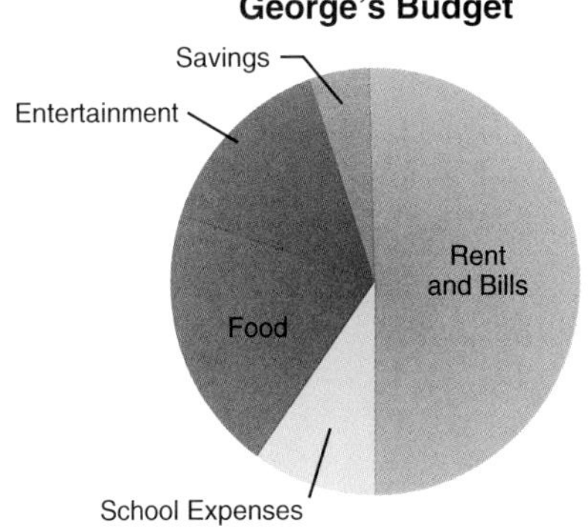

   a. Create a double-bar graph to display the data from these two circle graphs.

   b. Who is more responsible with his money? Explain.

7. The number of admissions to movie theaters went from 1.14 billion in 1995 to 1.49 billion in 2005.

a. Make a graph that shows a small change in admission.

b. Make a graph that shows a large change in admission.

c. What is the difference between your two graphs?

8. Sports You can create a *back-to-back stem plot* to display and compare two data sets. This plot shows the batting averages for the 2007 New York Yankees and the 2007 New York Mets.

Leaf values for the Yankees are given to the left of the stem, and leaf values for the Mets are given to the right of the stem. For example, the leaves 2 and 5 in the tenth row indicate a Yankee batting average of .322 and a Met batting average of .325.

Yankees and Mets 2007 Batting Averages

Yankees	Stem	Mets
6	.23	
	.24	
	.25	8
	.26	
7 3 0	.27	2 2 6
5 3	.28	0 0
2	.29	1 5 6
6	.30	
4	.31	
2	.32	5
8	.33	
	.34	1

Key: 21|5 = .215

a. Write a few sentences comparing the two teams' batting averages. Be sure to discuss the distribution of values and the mean and mode.

b. These data give the total number of home runs hit by American League and National League teams in 2002. Make a back-to-back stem plot comparing the number of home runs for the two leagues.

National League

Team	Home Runs
Arizona	165
Atlanta	164
Chicago	200
Cincinnati	169
Colorado	152
Florida	146
Houston	167
Los Angeles	155
Milwaukee	139
Montreal	162
New York	160
Philadelphia	165
Pittsburgh	142
San Diego	136
San Francisco	198
St. Louis	175

American League

Team	Home Runs
Anaheim	152
Baltimore	165
Boston	177
Chicago	217
Cleveland	192
Detroit	124
Kansas City	140
Minnesota	167
New York	223
Oakland	205
Seattle	152
Tampa Bay	133
Texas	230
Toronto	187

Mixed Review

Find the area of each trapezoid.

9.

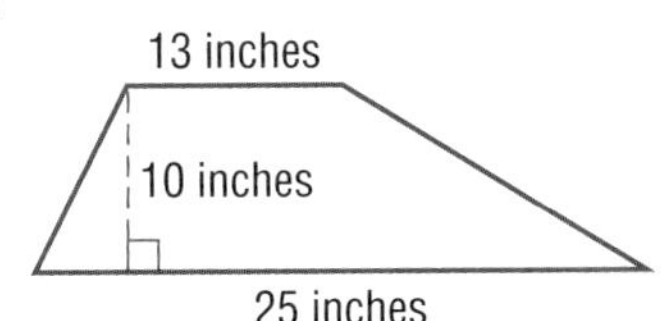

10.

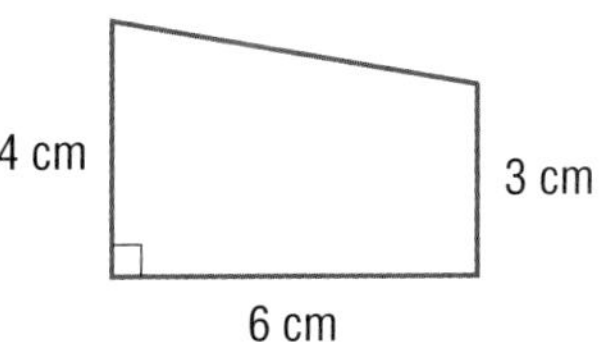

Solve.

11. $\frac{x}{12} = \frac{18}{24}$

12. $\frac{3}{4} = \frac{x}{16}$

13. $\frac{2}{6} = \frac{x}{36}$

CHAPTER 6

Review & Self-Assessment

Chapter Summary

In this chapter, you learned more about probability. You studied situations in which outcomes were influenced by previous outcomes and situations in which each outcome was independent of the others. You also determined the fairness of several games of chance and devised strategies, based on probabilities, to increase your chances of winning.

The basics of sampling a population and how to make predictions from samples were introduced in this chapter. You learned that the size of a sample and whether it is *representative* are important factors in sampling.

Finally, you encountered graphs that help you make comparisons, such as double-bar graphs and double-line graphs. You learned that circle graphs are useful for representing parts of a whole, or summarizing results.

Vocabulary

double-bar graph
double-line graph
population
representative sample
sample

Strategies and Applications

The questions in this section will help you review and apply the important ideas and strategies developed in this chapter.

Recognizing when previous outcomes influence later ones

1. Yago has two quarters, one dime, and four pennies in a pouch of his backpack. He needs a quarter for a vending machine. Suppose he reaches into his pack without looking and pulls out the first coin he touches.

a. What is the probability Yago will select a quarter?

b. The first coin Yago pulled out was a penny, so he will reach in again and pull out the first coin he touches. Which would give him a better chance of choosing a quarter, putting the penny back into the pack first or leaving it out? Explain your reasoning.

c. Yago put the penny back before pulling out a second coin. He pulled out another penny. He put it back, then pulled out yet another penny. He put it back and tried once more, muttering, "This time, I will get something other than a penny!"

Is this true? Is Yago more likely to get a quarter or dime? Explain.

Identifying whether a game is fair

2. Ben and Gabriela are playing a game with two dimes and two nickels. They toss the four coins. If one of the pairs of coins lands heads up, Ben scores 1 point. If both pairs land heads up, Ben scores 2 points. On every toss for which Ben does not score, Gabriela is awarded 1 point. They play to 5 points.

Determine whether this game is fair. Identify who has the advantage. Explain.

Using probability to make decisions and create strategies

3. Three boards and two spinners are used to play the game *Cover Up*.

Board A

1	4	2
2	9	2
9	2	1

Board B

3	9	4
1	3	9
3	4	3

Board C

2	2	3
3	6	2
2	3	6

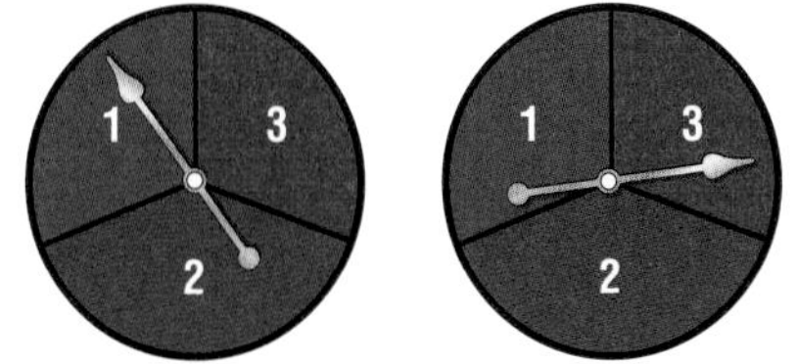

To play the game, each of two players chooses a board. They take turns. On each turn, a player spins both spinners, computes the product of the two numbers, and then covers a space on the board with that product, if there is one. The winner is the first player to cover all the spaces.

a. Which board gives the best chance of winning *Cover Up*? Explain your choice.

b. Of the remaining two boards, which is better? Explain.

Analyzing the appropriateness of a sample or a sampling process

4. A home-decorating magazine wanted to determine whether Americans think a living room should be painted or wallpapered. The people in charge of the survey considered different ways to gather this information.

For each suggestion, comment on whether the sample would be a good one and whether the process seems practical. Explain your answers.

a. Run a poll in the magazine with two toll-free numbers. If you prefer wallpaper, you call one number. If you prefer paint, you call the other number.

b. Get a listing of home purchases over the past month in 100 cities across the country, chosen at random. The surveyors would call 20% of the new homeowners and ask them what they prefer.

c. Call a random sample of 100 telephone numbers from across the country and ask the people who answer.

d. Call a random sample of 10,000 people across the country for their opinions.

Demonstrating Skills

Eddie conducted a survey to determine the favorite school lunch. His results are shown in the table to the right.

Lunch	Boys	Girls
Hamburger	12	10
Salad	2	5
Grilled cheese	10	20
Hot dog	13	13
Pizza	20	18

5. Create a double bar graph of the information given.

6. Which lunch is preferred most by boys? By girls?

7. Which lunch do the same number of boys and girls prefer?

8. If each student could only choose one lunch, how many students did Eddie survey about school lunches?

The table shows the results of a survey run by the student council.

Color for School Shirts	Number
White	20
Grey	40
Black	10
Green	20
Pink	5
Yellow	5

9. Create a circle graph using the data.

10. How many students were surveyed?

11. Which color received the most votes?

Use the circle graph to answer Questions 12–14.

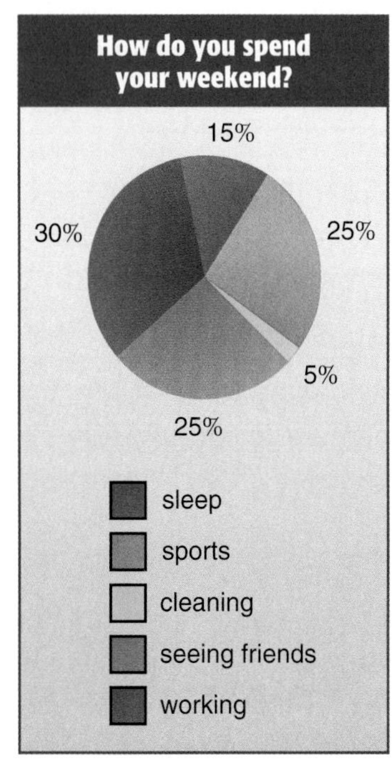

12. If 200 students were surveyed, how many spent the weekend cleaning?

13. How did most of the students spend the weekend?

14. If 300 students were surveyed, how many spent the weekend working?

15. Sports Caroline had a bowling party for her birthday. Each person at the party bowled three games. The stem-and-leaf plot shows their scores.

a. What is the lowest score? What is the highest score?

b. Find the mode and median scores.

c. Describe the distribution of scores.

Bowling Party Scores

Stem	Leaf
5	2
6	
7	3 7
8	1 6
9	2 9
10	0 3
11	0 1 4 4 4 9
12	3 3
13	2 6 6
14	
15	
16	
17	
18	5

Key: 9|2 = 92

Test-Taking Practice

SHORT RESPONSE

1 For a class project, Elise questioned five families about how many basketball or football tickets were bought in the past four months. She wrote the information in the table below. Create a double-bar graph to represent the data in the table.

Family	Basketball	Football
A	5	3
B	8	6
C	4	11
D	10	5
E	6	7

Show your work.

MULTIPLE CHOICE

2 There are 10 tiles, numbered 1 through 10, turned face down. If the first tile turned over is 3, what is the probability that an odd numbered tile will be turned over as the second tile?

A $\frac{1}{2}$

B $\frac{5}{9}$

C $\frac{4}{9}$

D $\frac{2}{5}$

3 Julia wants to predict the outcome of the sheriff election. Which of the following methods could she use to collect the best data for a prediction?

F Ask the parents of her friends for whom they plan to vote.

G Ask the students in her class who they prefer.

H Ask adults at the grocery store which candidate will receive their support.

J Ask her adult neighbors which candidate they support.

4 Chris has 2 dimes and 3 nickels in his pocket. He pulls out two coins. What is the probability that both coins will be dimes?

A $\frac{1}{7}$

B $\frac{1}{10}$

C $\frac{1}{3}$

D $\frac{2}{3}$

5 The table below shows the number of students with different pets.

Dog	5
Cat	6
Fish	4
None	8

Which category would take up about $\frac{1}{3}$ of the circle of a circle graph?

F Dog

G Cat

H Fish

J None

Real Numbers

Real-Life Math

Follow the Bike Path Malina lives three miles from her school. On Saturdays, she bikes to school to participate in track practices and meets. She travels two miles east along McKinley Avenue and one mile north along Roosevelt. Because there is an abandoned factory on the land between McKinley and Roosevelt, Malina cannot bike directly from her house to the school.

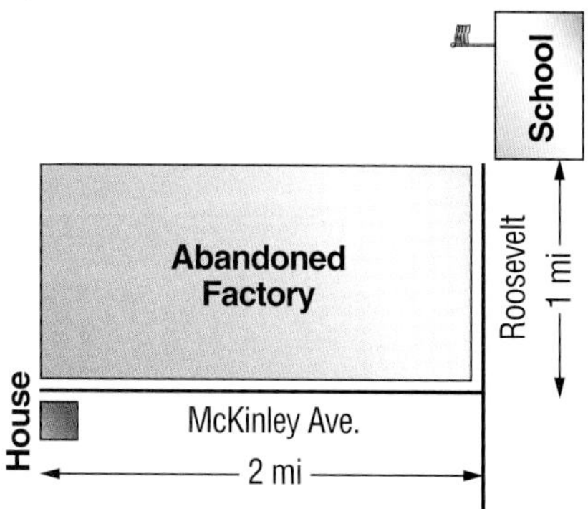

Think About It The mayor has proposed tearing down the factory and developing a park with bike paths in its place. Suppose a bike path runs from Malina's house to the school. How would this bike path affect her commute?

Contents in Brief

7.1	Rational Numbers	320
7.2	Irrational Numbers	331
7.3	The Pythagorean Theorem	343
	Review & Self-Assessment	361

Math Online
Take the **Chapter Readiness Quiz** at glencoe.com.

Dear Family,

In the elementary grades, students study the set of whole numbers, {0, 1, 2, 3, ...}. They also start working with fractions and decimals. In the middle grades, students expand their knowledge of numbers to include integers, {−3, −2, −1, 0, 1, 2, 3, ...}, and other signed numbers, such as $-2\frac{1}{2}$ and −0.75.

Any number that can be written as the ratio of two integers is called a **rational number**. In this chapter, your student will compare and order rational numbers.

Key Concept–Ordering Rational Numbers

When ordering numbers, your student may find it helpful to first write numbers as whole numbers or decimals.

$$\{\sqrt{4}, -2\tfrac{1}{2}, 1^3, -0.75\} = \{2, -2.5, 1, -0.75\}$$

The number located furthest to the left on the number line has the least value. The number located furthest to right has the greatest value. The set {−2.5, −0.75, 1, 2} lists the numbers from least to greatest.

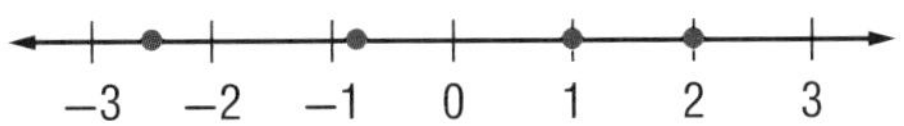

The class will also study **irrational numbers**, such as $\sqrt{5}$. Finally, your student will use rational and irrational numbers when applying the Pythagorean Theorem.

Chapter Vocabulary

irrational number	**rational number**
Pythagorean Theorem	**real number**

Home Connections

- Order numbers that you encounter in daily situations. Consider using grocery store or department store circulars to compare and order prices.
- Working with a map, determine the horizontal and vertical distances between two towns. Then, determine the diagonal distance between the same two towns. Use the Pythagorean Theorem to verify your measurements.

Rational Numbers

Integers and fractions such as $\frac{1}{2}$ and $\frac{7}{10}$ are rational numbers. The word *rational* comes from the word *ratio*, which should help you remember the definition of a rational number. A **rational number** is a number that can be written as the ratio, or quotient, of two integers.

Inv 1 Number Sets 320

Inv 2 Compare and Order Rational Numbers 324

Vocabulary

natural numbers

rational numbers

whole numbers

Think & Discuss

Is 0.5 a rational number? Why or why not?

Is $0.\overline{3}$ a rational number? Why or why not?

Is 0 a rational number? Why or why not?

You have already worked with special sets of numbers, such as odd and even numbers, prime numbers, and negative numbers. In Investigation 1, you will explore the relationship between number sets.

Natural numbers are counting numbers and can be represented as {1, 2, 3, ...}. **Whole numbers** are the counting numbers plus zero and can be represented as {0, 1, 2, 3, ...}. Integers are the counting numbers, their opposites, and zero, {..., −3, −2,−1, 0, 1, 2, 3, ...}.

Investigation 1 Number Sets

In this investigation, you will explore the relationship among four number sets.

Develop & Understand: A

Consider the four sets of numbers: natural, whole, integers, and rational. To which set(s) does each of the following numbers belong?

For example, 0 belongs to the wholes, integers, and rationals.

1. -50
2. $2.\overline{6}$
3. $-\frac{3}{4}$
4. 8^2

Math Link

The number 6 is a rational number because it can be written as the ratio of 6 and 1, $\frac{6}{1}$. The decimal -1.8 is a rational number because it can be written as the ratio of -18 and 10, $-\frac{18}{10}$, or $-\frac{9}{5}$.

For Exercises 5–11, indicate all the number sets (natural, whole, integers, and rational) for which the statement is always true. To be true for a set, the statement must be true for *all* the numbers in the set.

5. If you add two numbers in this set together, the sum must be greater than either of the numbers with which you started.

6. If you add two numbers in this set together, the sum must be greater than or equal to 0.

7. If you multiply three numbers in this set together, the product will be the same no matter the order in which you multiply them.

8. If you multiply three numbers in this set together, the product might be a negative number.

9. If you add two numbers in this set, you get a number in the same set.

10. If you subtract two numbers in this set, you get a number in the same set.

11. Write a statement that is true for the rational numbers but not for any of the other sets.

Think & Discuss

Is Lucita's claim correct?

Would the same claim be true for integers? For whole numbers? For natural numbers?

Develop & Understand: B

Math Link

Remember that in a Venn diagram, you need to be able to represent the intersection between each combination of sets, such as pairs, trios, or quartets of circles, as well as the part of each set that does not intersect the other.

12. The following Venn diagram shows the relationship between the set of odd numbers and the set of natural numbers. Copy the diagram. In each section of the diagram, fill in at least three more numbers.

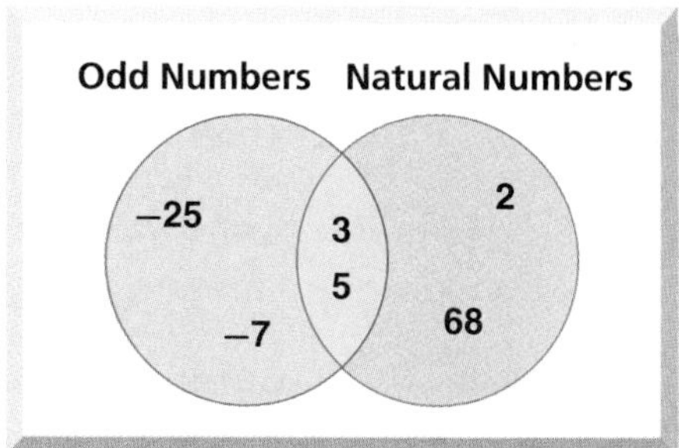

13. Which two sets could this Venn Diagram represent?

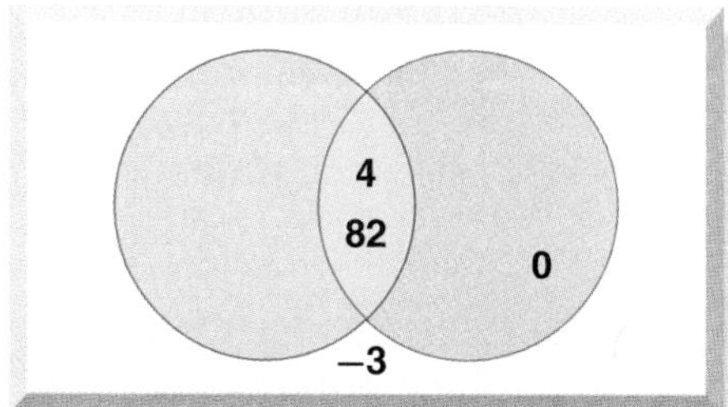

14. Draw a Venn diagram that shows the relationship between the integers and the whole numbers. Include at least three numbers in each section of the diagram.

15. Work with a partner to create a diagram with four concentric circles (circles having a common center) representing the natural numbers, whole numbers, integers, and rational numbers. Include sample numbers in each area of your diagram.

Anika, Haley, Antonio, and Kirk created number puzzles. Each student thought of a number and then gave clues so that the others could guess the secret number. Use your understanding of number sets to solve their number puzzles.

Develop & Understand: C

For Exercises 16–18, the secret number lies between −10 and 10.

16. Anika's Puzzle

- My number is not a natural number.
- My number is odd.
- If you add 5 to my number, the sum is a natural number.
- My number squared is more than my number doubled.

17. Haley's Puzzle

- My number squared is smaller than my number itself.
- My number multiplied by 4 is an integer.
- My number multiplied by 10 is larger than 7.

18. Antonio's Puzzle

- My number is a whole number.
- My number is a multiple of 3.
- My number divided by 2 is an integer.

19. Kirk started to write a puzzle, but he was not sure how to finish it. His secret number is 3.25. Write another clue that would allow Anika and Haley to find out Kirk's secret number.

- My number is a rational number.
- My number is not an integer.
- My number is evenly divisible by $\frac{1}{4}$.

Share & Summarize

Find a whole number that is not a natural number.

Find a rational number that is not a whole number.

Find an integer that is not a natural number.

Find an integer that is not a rational number.

If a number belongs to the set of whole numbers, to which other sets must it also belong?

Investigation 2 Compare and Order Rational Numbers

Think & Discuss

How many rational numbers are there between 0 and 1? How do you know?

Imagine a number line representing the set of integers. It goes on forever in both directions. Now imagine a new number line that is divided into smaller intervals. You can keep dividing the number line into smaller increments, finding numbers that are closer and closer together. No matter what two numbers you select, you can always find a number halfway between them.

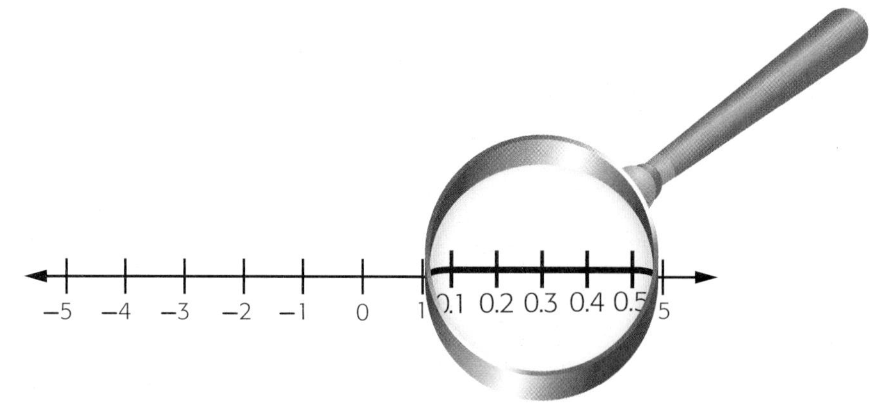

Number lines can help you compare rational numbers and arrange them in order of relative size. When you draw a number line, you have to select a scale. What are the least and greatest numbers you want to show on the line? What intervals will you mark along the line?

These answers depend on the range of numbers you need to graph and how easily you want to identify the location of the numbers. Whichever scale you use, there are always numbers between the ones you graph on the number line.

Develop & Understand: A

1. Copy the number lines shown below. On each number line, graph the following numbers.

$$\frac{1}{3} \qquad 0.338 \qquad \frac{31}{100}$$

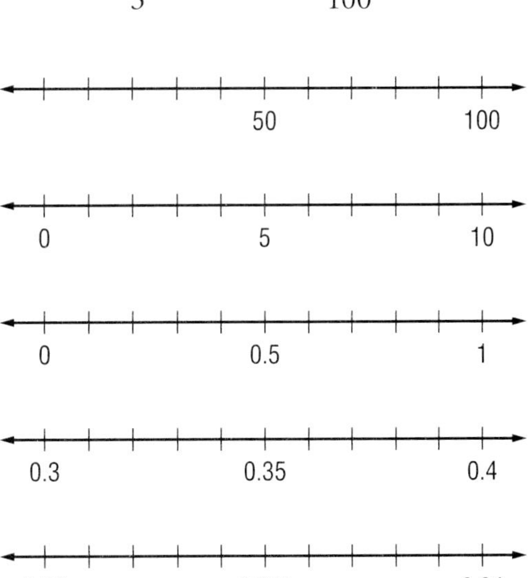

2. On which number lines can you graph all three numbers?

3. On which number line(s) can you graph the numbers most accurately?

4. How would you label a number line in order to graph $\frac{1}{3}$ exactly?

For Exercises 5–7, use the following numbers.

$$-1.09 \qquad -\frac{6}{5} \qquad -1\frac{1}{4} \qquad -0.973$$

$$-\frac{113}{100} \qquad -\frac{8}{9} \qquad -1.902 \qquad -\frac{1107}{1000}$$

5. Which numbers are less than -1 and greater than -2?

6. Which numbers are less than -1.1 and greater than -1.2?

7. Which number is less than -1.11 and greater than -1.12?

8. Draw a number line to graph the numbers shown below. Label the number line to indicate its scale.

$$10^{-2} \qquad 2^{-3} \qquad 10^{-1} \qquad \frac{4}{25} \qquad \frac{1}{5}$$

$$\frac{2}{20} \qquad \frac{3}{100} \qquad 0.18 \qquad 0.09$$

Think & Discuss

What fraction, written in simplest form, is located halfway between $\frac{3}{4}$ and $\frac{5}{6}$?

Can you find a general rule for finding a fraction that is halfway between two other fractions?

Can you find two more fractions between $\frac{3}{4}$ and $\frac{5}{6}$?

How many fractions are there between $\frac{3}{4}$ and $\frac{5}{6}$?

Ana, Marcus, and Simon use different strategies for estimating the value of a rational number and comparing two rational numbers. Consider their strategies, which are shown below.

Which strategy appeals to you? What other strategies might you use?

In the exercises on page 327, you will graph rational numbers on a number line.

Develop & Understand: B

For these exercises, draw a number line that shows -1 and 0, like the one below. Graph $-\frac{1}{3}$ and $-\frac{2}{3}$.

Without folding or measuring, use your number line for Exercises 9–13.

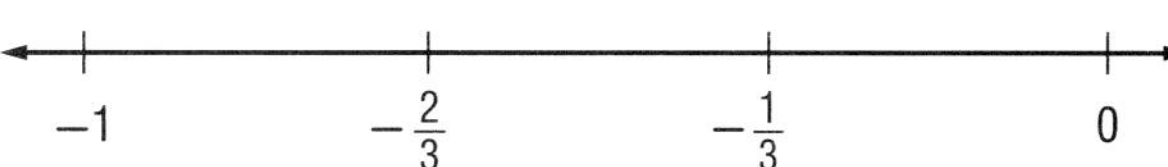

Math Link

To locate $-\frac{1}{3}$ and $-\frac{2}{3}$ precisely, you can measure with a ruler or you can fold the number line into three equal parts and mark where the folds are.

9. Graph $-\frac{1}{2}$.
   a. How did you estimate the location of $-\frac{1}{2}$?
   b. Is $-\frac{1}{2}$ closer to $-\frac{1}{3}$ or to $-\frac{2}{3}$? Explain your reasoning.
10. Mark the approximate location of $-\frac{1}{6}$ and $-\frac{5}{6}$ on the number line. Explain how you estimated the location of these numbers.
11. Mark the approximate location of $-\frac{1}{4}$ on the number line.
    a. Explain how you estimated the location of $-\frac{1}{4}$.
    b. Do you think your approximation of $-\frac{1}{4}$ is as accurate as your approximation of $-\frac{1}{6}$?
12. Consider the locations of $-\frac{1}{10}$ and $-\frac{1}{8}$.
    a. Mark the approximate location of $-\frac{1}{10}$ on the number line.
    b. Mark the approximate location of $-\frac{1}{8}$ on the number line.
    c. When you tried to approximate $-\frac{1}{8}$, did that change your idea of where $-\frac{1}{10}$ ought to be?
13. Mark the approximate location of -0.3 and -0.4 on the number line. Explain your reasoning.

Share & Summarize

Consider the following numbers.

$$-1.3 \qquad \frac{17}{20} \qquad -\frac{7}{8} \qquad 5^{-2} \qquad \frac{123}{100} \qquad 1.92$$

If you were going to put them on a number line, what range of points could you choose so that you could show all of the numbers?

What scale would you need to use for the number line if you wanted to place each of the numbers exactly?

If you graphed these numbers on a number line divided into tenths, which ones would you have to approximate? Would it be easy to show the order of the numbers?

On Your Own Exercises
Lesson 7.1

Practice & Apply

1. Consider the numbers listed below.

$$-10^2 \quad -21.3 \quad -\frac{1}{2} \quad -4 \quad 0 \quad \frac{1}{3}$$

$$12.6 \quad \frac{7}{4} \quad 5 \quad 6^{-3} \quad 17^4 \quad 0.004$$

Copy the table shown below. Write each number in all number set columns that apply.

Natural Numbers	Whole Numbers	Integers	Rational Numbers

For Exercises 2–11, write a number that fits the conditions given. If no such number exists, write "impossible."

2. a rational number that is not an integer
3. a natural number that is not a rational number
4. an integer that is not a rational number
5. an integer that is not a natural number
6. a rational number that is also a natural number
7. a rational number less than the least whole number
8. a natural number between -3 and 0
9. a rational number between 3 and 4
10. a whole number between 6 and 7
11. an integer between -10 and -12

12. Draw a Venn diagram that shows the relationship between the prime numbers and the natural numbers. Write at least three numbers in each section of your diagram.
13. Draw a Venn diagram that shows the relationship between integers and negative numbers. Write at least three numbers in each section of your diagram.
    a. Stephanie thought of a whole number. Without knowing her number, can you tell where it belongs in your diagram? Explain.
    b. Boris wrote a rational number on a piece of paper. Without knowing his number, can you tell where it belongs in your diagram? Explain your reasoning.

For Exercises 14–17, identify which number is greater. Explain how you found your answer.

14. $\frac{1}{6}$ or 0.11

15. $-\frac{6}{5}$ or -1.3

16. -0.426 or $-\frac{17}{30}$

17. 4^{-2} or 0.143

18. Graph the following numbers on a number line. Label intervals to indicate your scale.

$$\frac{43}{100} \quad \frac{3}{8} \quad 0.39 \quad \frac{2}{5} \quad \frac{7}{16} \quad \frac{365}{1000}$$

19. Copy the number line shown below. Determine the scale that was used to create the line. Label the plotted points.

For Exercises 20–22, use the puzzle clues to determine the secret number. There is more than one secret number for the puzzle in Exercise 22.

20. Puzzle A
Clue A: The number is a whole number.
Clue B: The number doubled is less than 10.
Clue C: The number is evenly divisible by 4.
Clue D: The square root of the number is a whole number.

21. Puzzle B
Clue A: The number is an integer but not a whole number.
Clue B: The number squared is less than 10.
Clue C: The number is evenly divisible by 2 but not by 4.
Clue D: The sum of the number and 5 is a natural number.

22. Puzzle C
Clue A: The number is a rational number but not an integer.
Clue B: The product of the number and 5 is an integer.
Clue C: The number is less than -1 and greater than -2.

a. List all possible secret numbers.

b. Explain how you know there are no other possible answers.

c. Write one or more additional clues that would eliminate all but one of the possible answers. Write the answer that your clues indicate.

23. Maurk and Larisa played a number game. Maurk drew a Venn diagram and kept the labels secret. Larisa chose one number at a time, and Maurk put each number in the area of the diagram where it belonged. This is what the diagram looked like after Larisa had chosen three numbers.

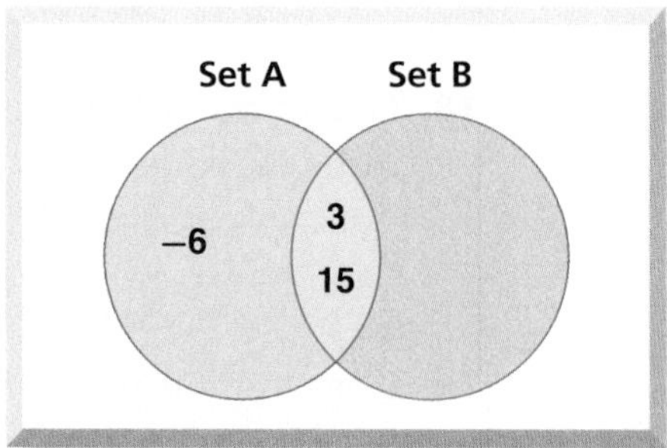

For Parts a and b, choose from the number sets listed below.

integers	natural numbers	whole numbers
rational numbers	odd numbers	even numbers
multiples of 3	prime numbers	

a. Which of the following number sets could set A represent?

b. Which of the number sets could set B represent?

c. Assume that the number sets you listed in Parts a and b are the only possible labels for set A and set B. Suggest three numbers that Larisa could choose next. For each number you suggest, explain how it will help her rule out the labels that are incorrect.

24. In Your Own Words Explain a strategy for ordering a set of rational numbers from least to greatest.

Mixed Review

Write each expression using a single base and a single exponent.

25. $x^3 \cdot x^2$

26. $(x^3)^2$

27. $y^6 \div y^2$

28. $\dfrac{(t^5)^2}{(t)(t^6)}$

Find each sum or difference.

29. $-2 + (-3)$

30. $8 + (-10)$

31. $7 - (-5)$

32. $-8 - (-3)$

Find each product or quotient.

33. $5 \cdot (-2)$

34. $-6 \cdot \frac{1}{2}$

35. $-10 \div (-5)$

36. $12 \div (-\frac{1}{4})$

Irrational Numbers

In Lesson 7.1, you worked with rational numbers. In this lesson, you will explore a new number set, the irrational numbers.

Inv 1 Squares and Square Roots 331

Inv 2 Approximate Irrational Numbers 336

Think & Discuss

Think of a number you can multiply times itself to get a product of 25. Is there more than one such number?

Next, think of two numbers you can multiply times themselves to get a product of $\frac{16}{25}$. How are these numbers different from your previous answers?

Now, what is a number you can multiply times itself to get a product of 5?

In this lesson, you will work with numbers like $\sqrt{5}$.

Investigation  Squares and Square Roots

Vocabulary

square root

Math Link

Remember that multiplying a number by itself is called *squaring* the number. The expression 7^2 can be read "7 squared."

As you remember from previous lessons, area is found by multiplying the length and width of a two-dimensional figure. Area is measured in square units.

The exponent 2 is often used to represent square units of measurement. For example, square centimeter and square inch can be represented as cm^2 and in^2, respectively.

5 cm | $A = 25\ cm^2$ | 5 cm

3 in. | $A = 9\ in^2$ | 3 in.

x units | $A = x^2$ units | x units

Develop & Understand: A

1. What is the area of the square shown below?

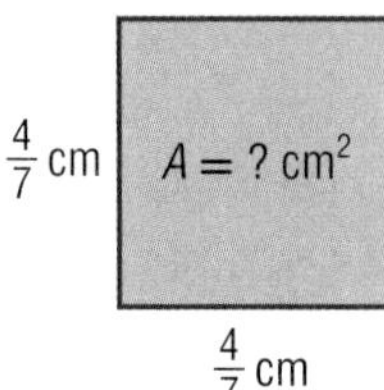

2. How long is each side of the square shown below?

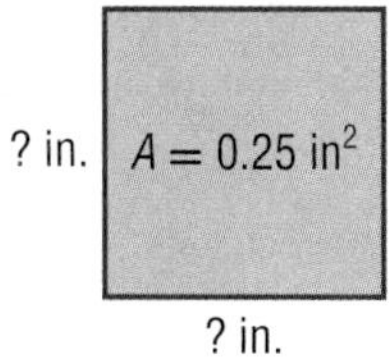

Consider the following squares with the given areas.

a.

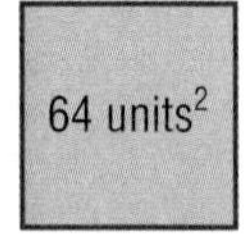

$S =$ ____ units

b.

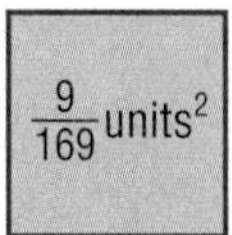

$S =$ ____ units

c.

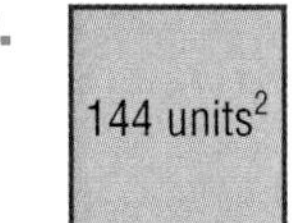

$S =$ ____ units

d.

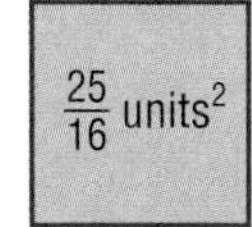

$S =$ ____ units

3. For each square, find the length of its sides.
4. Which of the squares have side measures that are whole numbers?
5. Which of the squares have side measures that are rational numbers but not whole numbers?
6. What can you say about the sides of a square with an area of 10 units^2? An area of $\frac{3}{7}$ units^2?

Think & Discuss

If you know the area of a square, how can you tell if its sides are whole numbers?

How can you tell if side lengths are rational numbers?

Can you always tell?

A number is a *perfect square* if it is equal to a whole number multiplied by itself. Geometrically, a perfect square is the area of a square with whole-number side lengths.

For example, 64, 9, and 169 are perfect squares.

Two operations that "undo" each other are called inverse operations. Addition and subtraction are *inverse operations.*

Add 12 to 15 to get 27. To undo the addition, subtract 12 from 27 to get 15.

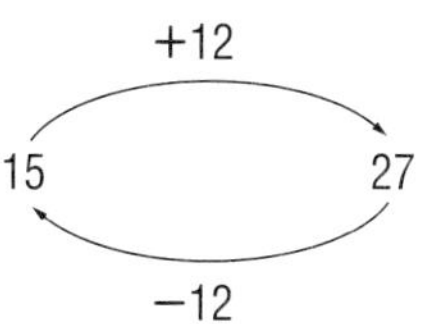

Subtract 12 from 27 to get 15. To undo the subtraction, add 12 to 15 to get 27.

Similarly, multiplication and division are inverse operations.

You will now work with the operation that undoes squaring.

Luke squared some numbers on his calculator. His results are shown below. In each case, find the number with which he started.

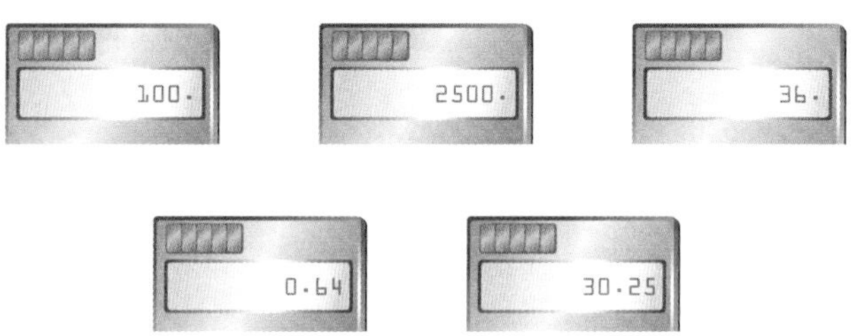

In each part of Think & Discuss, you found the number you need to square to get a given number. The number you found is the **square root** of the original number. For example, the square root of 36 is 6.

Math Link

Every positive number has both a positive and a negative square root. For example, the square roots of 36 are 6 and −6. In this lesson, you will focus on positive square roots, also known as *principal roots.*

The square root is shown using a *radical sign,* $\sqrt{}$. You can think of $\sqrt{36}$ in any of the following ways.

- the number you multiply by itself to get 36 $\quad 6 \cdot 6 = 36$
- the number you square to get 36 $\quad 6^2 = 36$
- the side length of a square with area 36

Squaring and taking the square root are inverse operations.

Square 6 to get 36. To undo the squaring, take the square root of 36 to get 6.

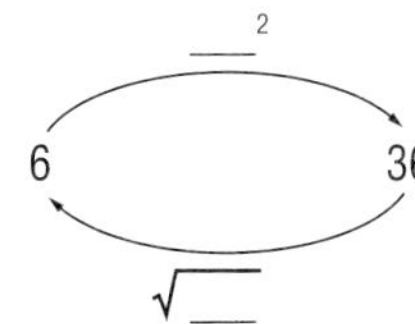

Take the square root of 36 to get 6. To undo taking the square root, square 6 to get 36.

Not all whole numbers have whole-number square roots. In fact, whole numbers that are not perfect squares have square roots that are decimals that never end or repeat. So, you can only estimate the decimal equivalents of numbers such as $\sqrt{10}$ and $\sqrt{41}$.

Math Link

Decimal numbers that never end or repeat are called **irrational numbers**. You have previously learned about the irrational number π.

$\pi = 3.14159265 \ldots$

Example

Luke estimated the decimal equivalent of $\sqrt{10}$.

Develop & Understand: B

7. Find the square of each integer from 0 to 15.
8. Draw a number line that extends from 0 to 10 with increments of 1.
9. What integers lie between 2 and 3?
10. What integers lie between 2^2 and 3^2?
11. On your number line from Exercise 8, mark the location of $\sqrt{4}$ and $\sqrt{9}$. Then mark the approximate location of $\sqrt{7}$. Explain how you decided where to put it.
12. On your number line from Exercise 8, mark the approximate location of $\sqrt{5}$. Is it greater than or less than $\sqrt{7}$?
13. Give two examples of numbers whose squares must be between 144 and 169.

Develop & Understand: C

Find the two whole numbers each given square root is between. Do not use your calculator.

14. $\sqrt{2}$

15. $\sqrt{75}$

16. $\sqrt{20}$

17. $\sqrt{26}$

18. To the nearest tenth, between which two numbers is $\sqrt{26}$? Do not use your calculator.

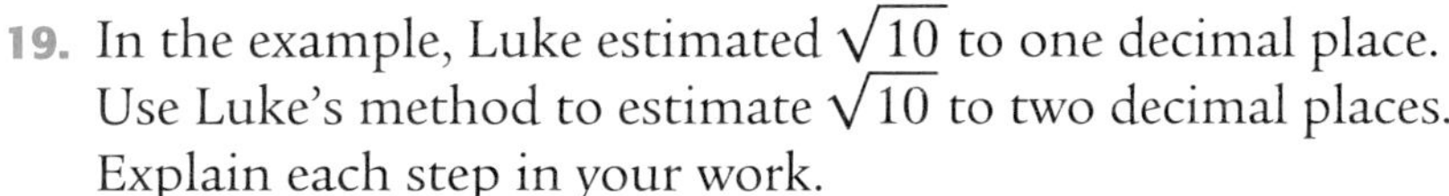

19. In the example, Luke estimated $\sqrt{10}$ to one decimal place. Use Luke's method to estimate $\sqrt{10}$ to two decimal places. Explain each step in your work.

Your calculator has a command for finding square roots accessed by pressing [2nd] [$\sqrt{\ }$]. For example, to find $\sqrt{4}$, press [2nd] [$\sqrt{\ }$] 4 [ENTER].

Develop & Understand: D

Use your calculator to approximate each square root to the nearest hundredth. Compare your results with your answers in Exercises 14–17.

20. $\sqrt{2}$

21. $\sqrt{75}$

22. $\sqrt{20}$

23. $\sqrt{26}$

24. Althea entered a number into her calculator. She then pressed [x^2] [ENTER] repeatedly until the calculator showed 43,046,721. What could have been her original number? List all the possibilities.

25. Suppose you enter a positive number less than 10 into your calculator and press [x^2] [ENTER] 5 times. Then you start with the final result and take the square root 5 times. What will happen? Explain why.

Share & Summarize

1. Describe the relationship between squaring a number and taking its square root.

2. Describe a method for approximating a square root *without* using a calculator.

Investigation 2 Approximate Irrational Numbers

Vocabulary

irrational numbers

real numbers

Remember that a rational number is a number that can be written as the ratio, or quotient, of two integers. For example, the numbers 3, 0.5, and −1.8 are all rational numbers because they can be written as $\frac{3}{1}$, $\frac{5}{10}$ or $\frac{1}{2}$, and $-\frac{18}{10}$ or $-\frac{9}{5}$, respectively.

Numbers that *cannot* be written as the ratio of two integers are called **irrational numbers**. The rational numbers and irrational numbers together form the set of **real numbers**. All real numbers can be located on the number line.

Think & Discuss

Try to decide whether each of the six numbers below is rational by looking for a ratio of integers that is equal to the number.

$3 \qquad 0.7 \qquad -2.55 \qquad \sqrt{20} \qquad 4.5678 \qquad \sqrt{25} \qquad 7.\overline{4}$

Think about numbers with one nonzero digit to the right of the decimal point, such as 0.7. Are all such numbers rational?

Think about numbers that have two nonzero digits to the right of the decimal point, such as 2.55. Are all such numbers rational?

Are numbers with n nonzero digits to the right of the decimal point rational?

Can numbers whose nonzero digits to the right of the decimal point go on forever be rational? Explain.

In the Think & Discuss above, you may not have been sure whether $\sqrt{20}$ is rational or irrational. You will now explore this number.

Develop & Understand: A

1. Using your calculator, evaluate $\sqrt{20}$. What does the calculator display for the decimal form of $\sqrt{20}$?
2. Multiply your answer to Exercise 1 by itself, using your calculator. What do you get?
3. Now imagine multiplying your answer for Exercise 1 by itself *without* using a calculator. What would be the rightmost digit of the product?
4. Could the number your calculator displayed in Exercise 1 really be equal to $\sqrt{20}$? How might your findings be explained?

Math Link

The rightmost digit of 4.36 is 6 and the rightmost digit of 121.798 is 8.

Math Link

A bar written over one or more numerals means those numerals repeat. For example,

$2.1\overline{5} = 2.1555 \ldots$

$9.\overline{32} = 9.323232 \ldots$

Calculators can be deceptive because they can only show a certain number of digits. If a number has more digits than a calculator can handle, some of the digits at the end simply do not appear on the screen.

So, what about a number like $\sqrt{20}$? Use your calculator to find the value of $\sqrt{20}$. You will discover that it has more digits than a calculator will display. That means it could be one of three types of numbers.

- It could have a set number of nonzero digits to the right of the decimal point.

  Such numbers are called *terminating decimals* because the digits eventually stop, or terminate. Since you can find a ratio of integers equal to a terminating decimal, these numbers are rational.

- It could have nonzero digits to the right of the decimal point that continue forever in a repeating pattern; $0.3\overline{24}$ is an example of such a number.

  Such numbers are called *nonterminating, repeating decimals.* They are also rational, although this fact will not be shown here. For example, $0.\overline{3}$ is equal to $\frac{1}{3}$.

- It could have nonzero digits to the right of the decimal point that continue forever without ever repeating.

  Such numbers are called *nonterminating, nonrepeating decimals.* As it turns out, *any* such number is irrational.

The number $\sqrt{20}$ is a nonterminating, nonrepeating decimal, so it is irrational.

In fact, any whole number that is not a perfect square has an irrational square root.

Develop & Understand: B

5. Draw a Venn diagram that shows each relationship. Write three numbers in each area of the diagram. If no numbers belong in an area, label the area "empty."

a. The relationship between rational and irrational numbers

b. The relationship among rational numbers, irrational numbers, and real numbers

6. Could there be a number that is both an integer and an irrational number? Explain your answer.

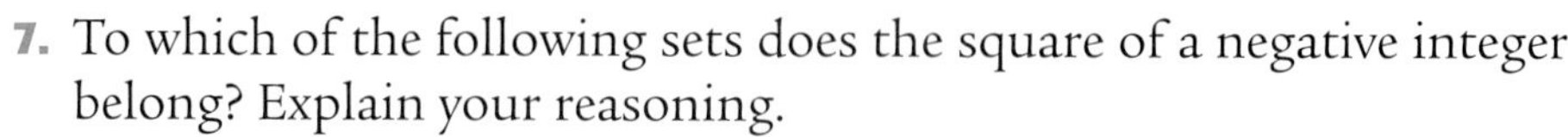

7. To which of the following sets does the square of a negative integer belong? Explain your reasoning.

   rational numbers   irrational numbers   integers

   natural numbers   whole numbers   real numbers

8. What is meant by the expression $-\sqrt{25}$?

9. What is the value of $\left(-\sqrt{25}\right)^2$?

In the next exercise set, you can use a number line to approximate the value of irrational square roots.

Develop & Understand: C

For Exercise 10, consider the following numbers.

$$\sqrt{35},\ \sqrt{50},\ \sqrt{60},\ \sqrt{90},\ \sqrt{115},\ \sqrt{125},\ \sqrt{150},\ \sqrt{200}$$

10. Draw a number line that extends from 5 to 15. Mark the approximate location of each square root.

For Exercises 11 and 12, consider these numbers.

$$\sqrt{125},\ \sqrt{126},\ \sqrt{130},\ \sqrt{137},\ \sqrt{140}$$

11. Use a number line to mark the approximate location of each square root to the nearest tenth.

12. Without using your calculator, find a square root that you estimate to be between 11.9 and 12. Then calculate to check your estimate. Were you correct?

13. **Challenge** Not all irrational numbers are roots. One famous irrational number with which you are already familiar is π.
    a. What is π to five decimal places?
    b. You may recall that π is the ratio of the circumference to the diameter of a circle and is an irrational number. Can both the circumference and the diameter be rational? Why or why not?

Real-World Link

Mathematicians have been approximating π for a very long time. In the 19th century B.C., the Babylonians approximated π as $\frac{25}{8}$, or 3.125. An Egyptian document from before 1500 B.C. approximates π as $\frac{256}{81}$, which is a little greater than 3.16. Today, mathematicians use computers to calculate π to over a trillion decimal places.

Share & Summarize

Tell whether the members of each group are *always rational, sometimes rational,* or *never rational.* Explain your reasoning or give examples.

a. integers
b. square roots
c. decimals
d. expressions that use a radical sign in their simplified form

The Pythagorean Theorem

In this lesson, you will learn about a famous mathematical concept known as the *Pythagorean Theorem*. The Pythagorean Theorem expresses a remarkable relationship between **right triangles**, which are triangles that have one 90° angle, and squares.

Inv 1 Right Triangles and Squares 343

Inv 2 Use the Pythagorean Theorem 347

Inv 3 The Distance Formula 351

Before you investigate the theorem, you will find the area of a square drawn on dot paper.

Vocabulary

right triangle

Materials

- copy of the square

Explore

Find the exact area of this square. Describe the method you use.

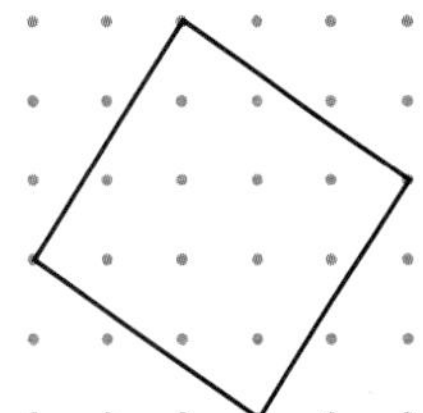

Investigation  Right Triangles and Squares

Vocabulary

hypotenuse

leg

Pythagorean Theorem

Materials

- copies of the figures
- dot paper
- scissors

Every right triangle has one right angle. The side opposite the right angle is called the **hypotenuse**. The other two sides are called the **legs**.

The relationship expressed in the Pythagorean Theorem involves the areas of squares built on the sides of a right triangle. You will discover that relationship in Exercises 1–6.

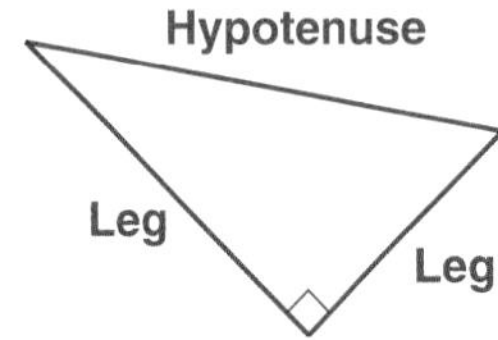

Develop & Understand: A

Exercises 1–4 show right triangles with squares drawn on their sides. Find the exact area of each square. Record your results in a copy of the table.

Exercise Number	Area of Square on Side a (units^2)	Area of Square on Side b (units^2)	Area of Square on Side c (units^2)
1			
2			
3			
4			

Real-World Link

The Pythagorean Theorem is named for Greek mathematician Pythagoras, who lived from approximately 580 to 500 B.C. Records show that the Babylonians understood the Pythagorean Theorem more than 900 years before Pythagoras was born.

1. 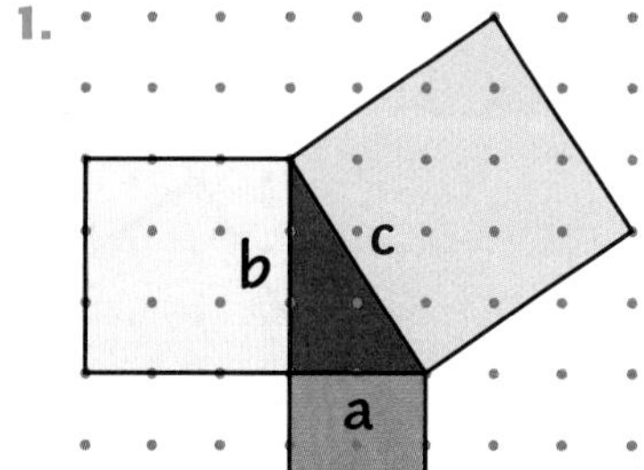

2. 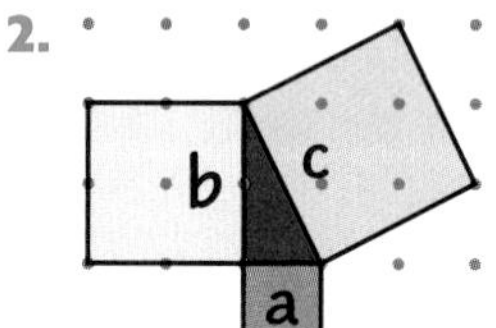

3. 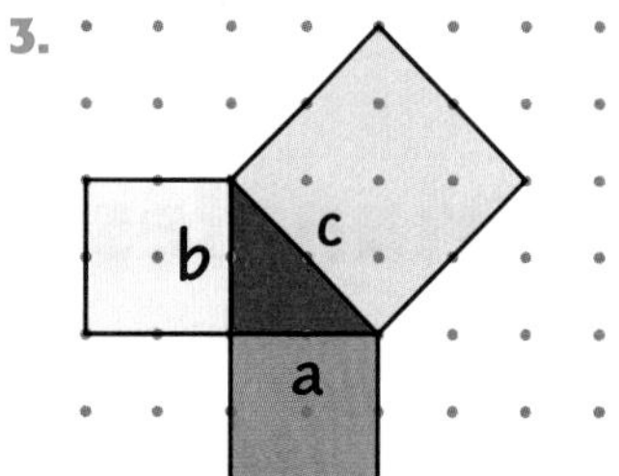

4. 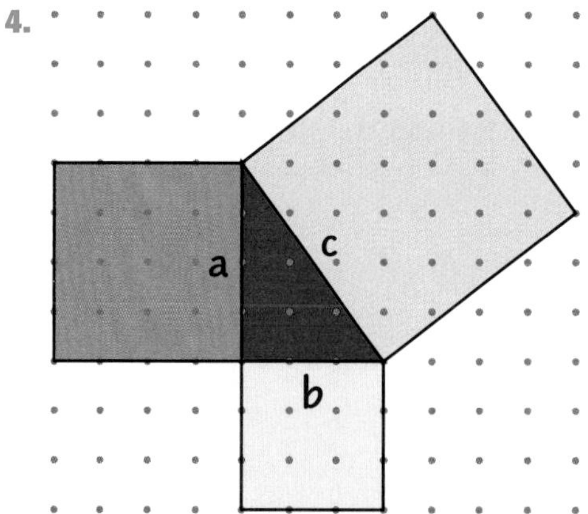

5. Look for a pattern in your table. For the cases you considered, what is the relationship among the areas of the three squares?

6. Draw your own right triangle on dot paper. Does the relationship you described in Exercise 5 hold for your triangle as well?

Exercises 1–6 illustrate the **Pythagorean Theorem**, which states the following.

The Pythagorean Theorem

In a right triangle, the area of the square built on the hypotenuse of the triangle is equal to the sum of the areas of the squares built on the legs.

The Pythagorean Theorem is often stated this way: If c is the length of the hypotenuse of a right triangle and a and b are the lengths of the legs, then $a^2 + b^2 = c^2$.

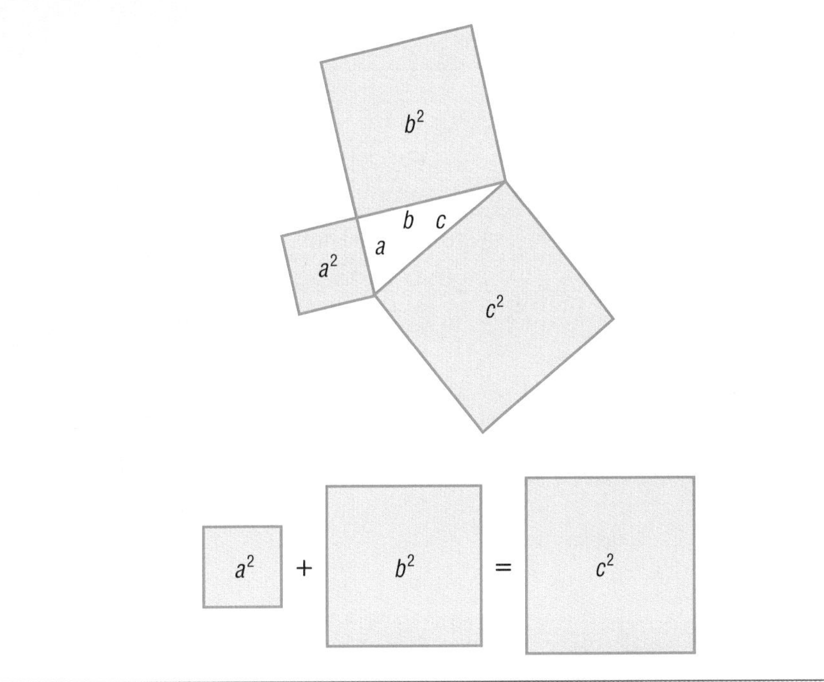

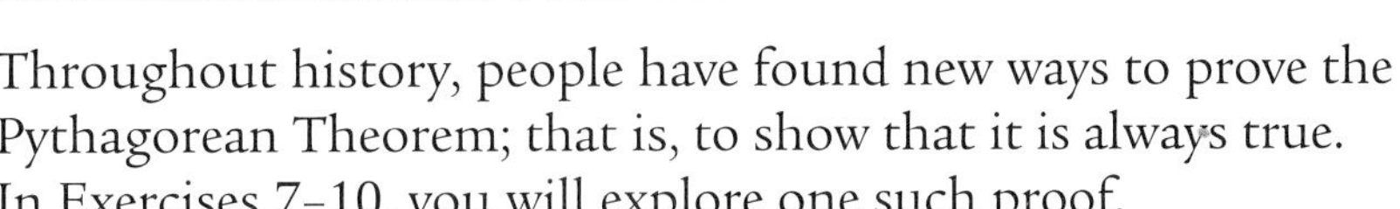

An Ionic column in Olympia, Greece

Throughout history, people have found new ways to prove the Pythagorean Theorem; that is, to show that it is always true. In Exercises 7–10, you will explore one such proof.

Develop & Understand: B

In these exercises, you will use paper triangles and squares to construct a proof of the Pythagorean Theorem.

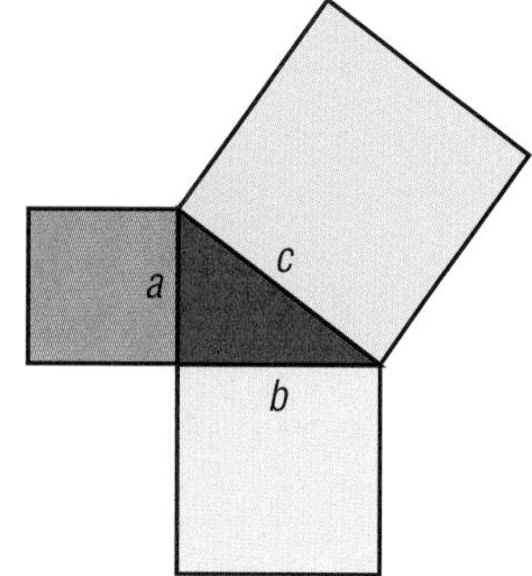

- Start with a right triangle with squares drawn on its sides.
- Carefully cut out eight copies of the triangle and one copy of each square.
- Use four copies of the triangle and the square from side *c* to make this square.

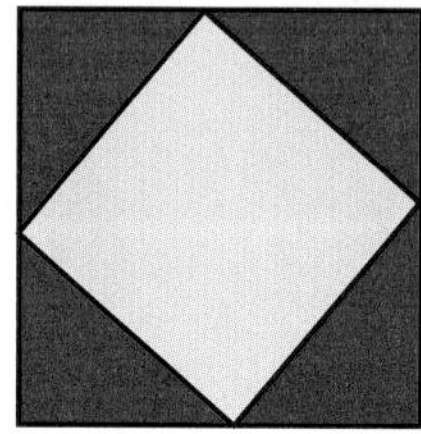

- Use four copies of the triangle and the squares from sides *a* and *b* to make this square.

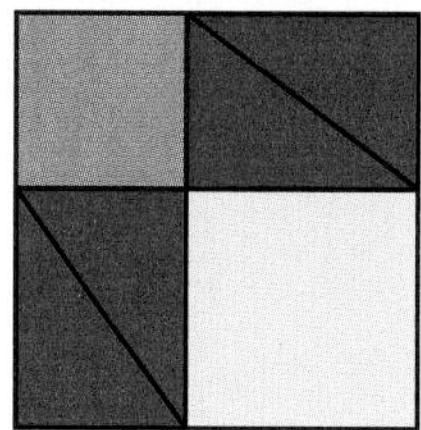

7. The two squares you made have the same area. Explain how you know this is true.

Now take the four triangles away from each square you constructed.

8. Describe what is left of each square.

9. Explain why the area of what is left must be the same for both squares.

10. Explain how your work in Exercises 7–9 shows that $a^2 + b^2 = c^2$.

Share & Summarize

State the Pythagorean Theorem in your own words. You might want to draw a picture to illustrate what you mean.

Investigation 2 Use the Pythagorean Theorem

In this investigation, you will have a chance to use the Pythagorean Theorem.

Develop & Understand: A

Find each missing area. Then use the areas of the squares to find the side lengths of the triangle.

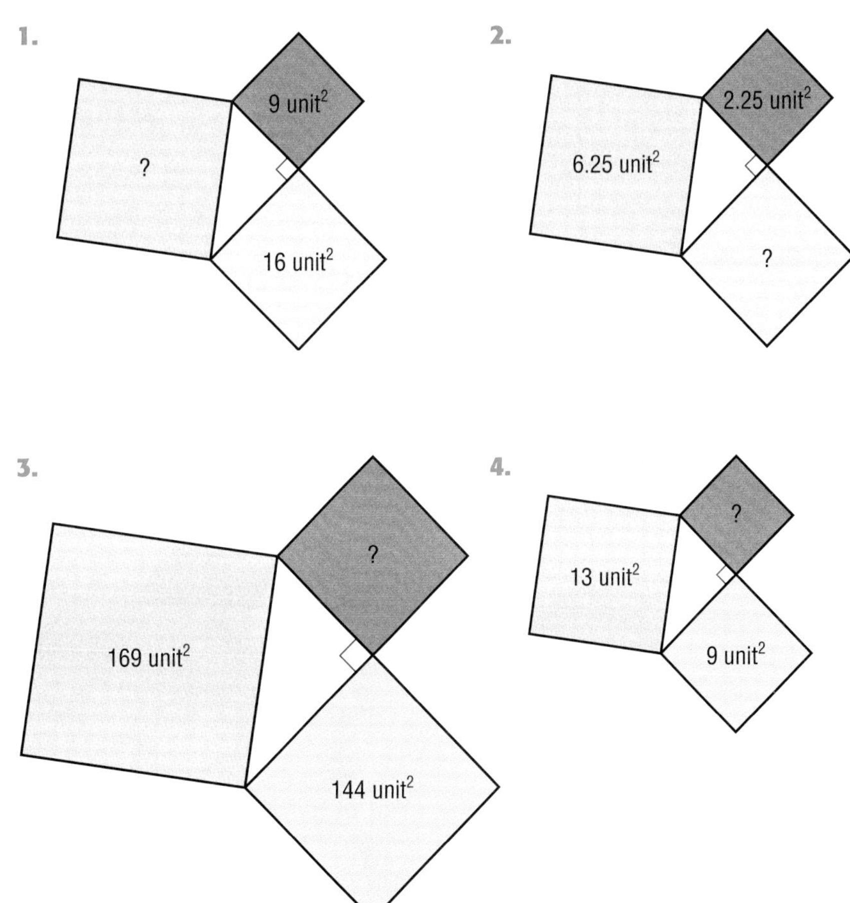

If you know the lengths of any two sides of a right triangle, you can use the Pythagorean Theorem to find the length of the third side.

Example

A right triangle has legs of length 1 inch and 2 inches. How long is the hypotenuse?

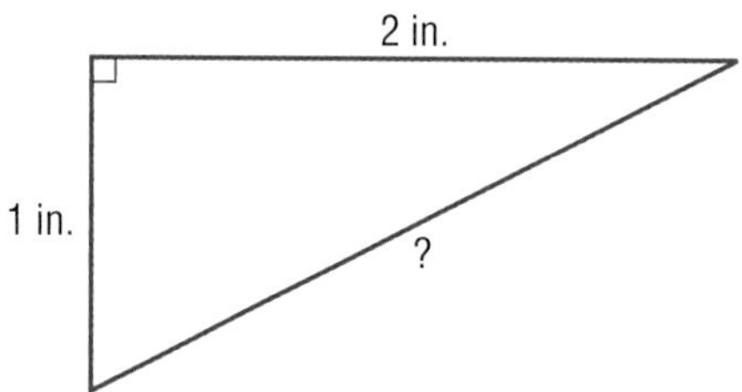

If a and b represent the lengths of the legs and c represents the length of the hypotenuse, the Pythagorean Theorem states that $a^2 + b^2 = c^2$.

In the triangle above, $a = 1$, $b = 2$, and c is the length of the hypotenuse.

$$1^2 + 2^2 = c^2$$
$$1 + 4 = c^2$$
$$5 = c^2$$

Therefore, $c = \sqrt{5}$, or approximately 2.24 inches.

Real-World Link

They may not have known the Pythagorean Theorem, but the Egyptians who built the pyramids knew that a triangle with side lengths of 3, 4, and 5 units must be a right triangle.

They used a rope device, like the one pictured here, to check that they made perfect right angles at the corners of the pyramids.

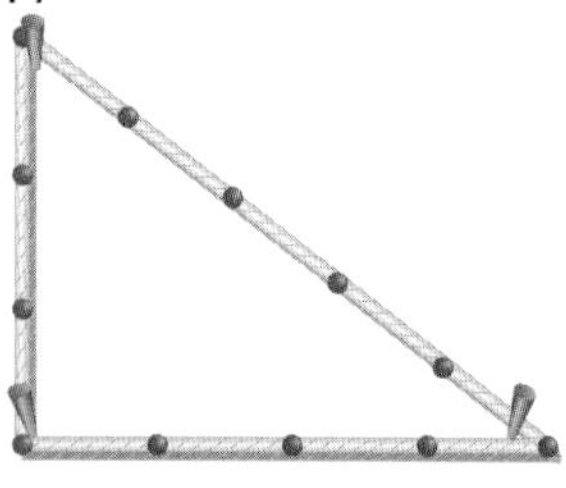

The Great Sphinx in Giza, Egypt

Develop & Understand: B

Find each missing side length. Then find the area of the triangle.

5.

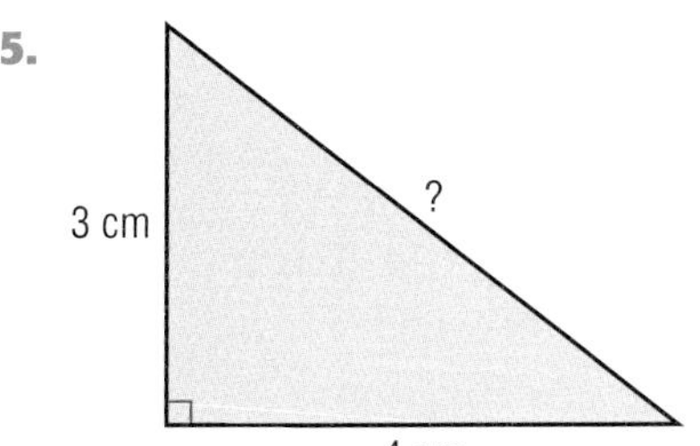

6.

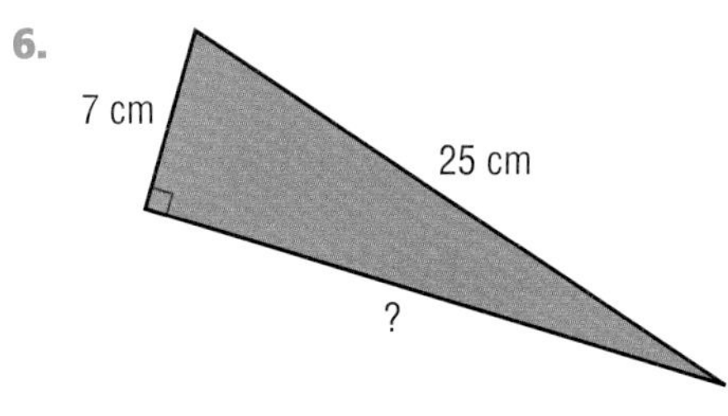

7. A rectangular lawn measures 9 meters by 12 meters. Suppose you want to walk from point *A* to point *B*.

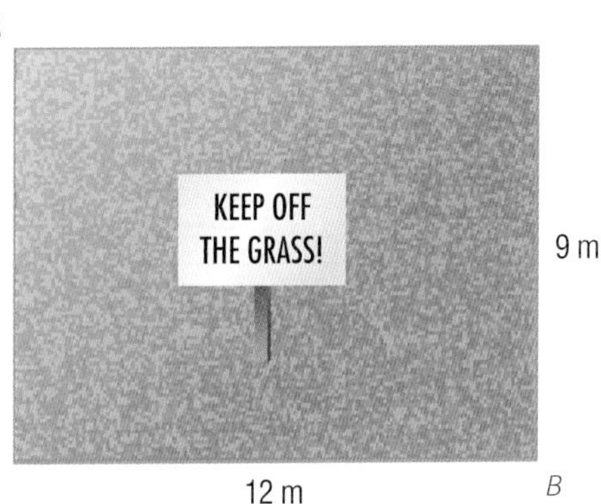

a. If you obey the sign and walk around the lawn, how far will you walk?

b. If you ignore the sign and walk directly across the lawn, how far will you walk?

8. A baseball diamond is a square measuring 90 feet on each side. What is the distance from home plate to second base?

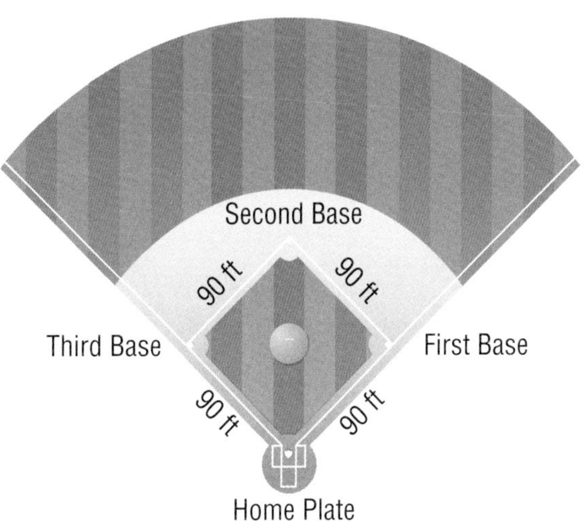

Real-World Link

In Thailand, *kite fighting* is a popular sport. A kite-fighting match involves a minimum of two heavy, star-shaped kites called *chulas* and at least four light, diamond-shaped kites called *pakpaos*. Although pakpaos are usually manipulated by a single person, it can take up to 20 people to handle a chula.

9. Caroline and Trevor are flying a kite. Caroline is holding the kite and has let out 80 feet of kite string. Trevor is standing 25 feet from Caroline and is directly under the kite.

   Caroline is holding the string 3 feet above the ground. How far above the ground is the kite? Explain how you found your answer.

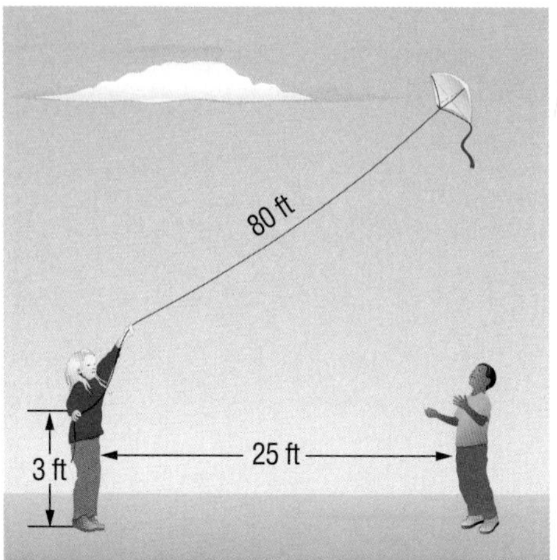

10. Safety regulations state that wheelchair ramps cannot be too steep. Suppose that, for every one foot a wheelchair ramp rises, it must cover a horizontal distance of at least 11.5 feet.

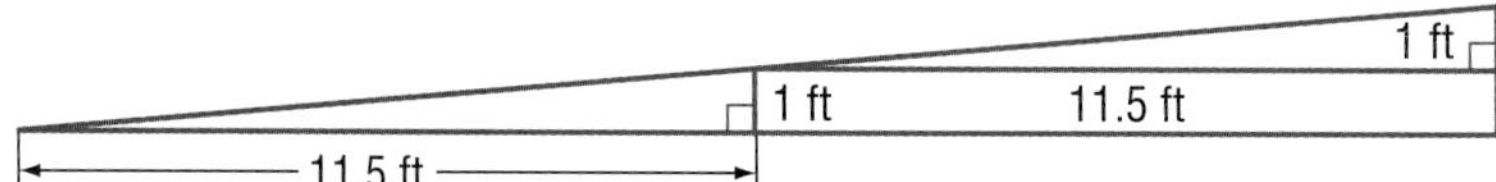

   A ramp is being built to a restaurant entrance that is 2.5 feet above the ground.

   a. How much horizontal distance does the ramp require?

   b. How long must the ramp be?

11. **Challenge** Find the values of c and h. Explain the method you used.

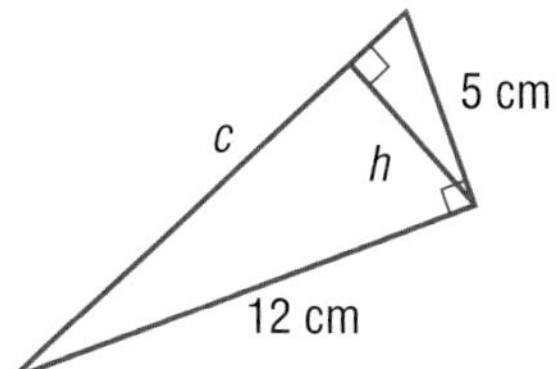

Share & Summarize

1. When do you use the Pythagorean Theorem?
2. Write your own exercise that can be solved by using the Pythagorean Theorem. Then show how to solve your equation.

Inquiry

Investigation 3 The Distance Formula

Vocabulary

distance formula

When archaeologists set up a dig, they create a grid on the ground with a string so they can record the locations of the objects they find. In this investigation, you will learn how to find distances between points on a coordinate grid.

The Situation

Dr. Davis is working on an archaeological dig. He has used string to lay out a grid on the ground. The lines of the grid are one foot apart.

So far, Dr. Davis has unearthed sections of two walls and an object he thinks might have been a toy. He drew this diagram to show the location of these objects on the grid. He labeled the corner of the walls (0, 0).

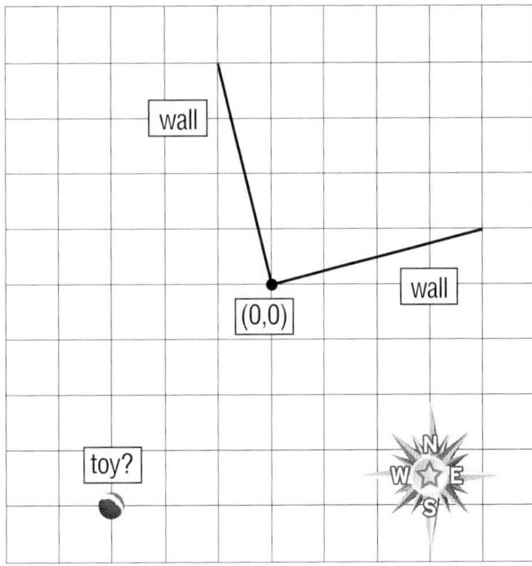

Dr. Davis wants to find the distance from the toy to the corner where the walls meet.

Try It Out

1. Estimate the distance from the toy to the corner where the walls meet.

   Dr. Davis notices that the point (0, 0) and the toy form two vertices of a triangle. He decides to use the Pythagorean Theorem to find the distance from the toy to the corner where the walls meet.

2. Determine the coordinates of the toy and use the Pythagorean Theorem to find the actual distance. How close was your estimate?

Go on

3. Len and Ava calculated the distance from the toy to the corner where the walls meet differently.

Len's Steps		Ava's Steps	
Step 1	How far left or right do I go? Find this length.	Step 1	How far left or right do I go? Find this length.
Step 2	How far up or down do I go? Find this length.	Step 2	How far up or down do I go? Find this length.
Step 3	Add the two lengths.	Step 3	Square both lengths.
Step 4	Square the sum from Step 3.	Step 4	Add the two squared lengths from Step 3.
Step 5	Find the square root of the answer from Step 4.	Step 5	Find the square root of the sum from Step 4.

a. Which student wrote the correct set of steps? Identify the mistakes the other student made.

b. How does your method for finding the distance in Question 2 compare with the correct series of steps in the table? Do you have all the same steps? Are your steps in the same order?

4. The steps for finding the distance between two points can be written in symbols as well as with words. Use (x_1, y_1) to represent one point and (x_2, y_2) to represent the other.

   Refer to the correct set of steps from Question 3. Parts a–e each show one of the steps from that set in symbols. Tell which step the symbols represent.

   a. $y_2 - y_1$

   b. $(x_2 - x_1)^2$ and $(y_2 - y_1)^2$

   c. $x_2 - x_1$

   d. $\sqrt{(x_2 - x_1)^2 + (y_2 - y_1)^2}$

   e. $(x_2 - x_1)^2 + (y_2 - y_1)^2$

Math Link

The variables x_1, y_1, x_2, and y_2 are called *subscripted* variables. They are like any other variables, just with lowered numbers called subscripts.

The symbol $\approx$ means "approximately equals."

The **distance formula** gives the symbolic rule for calculating the distance between any two points in the coordinate plane.

Try It Again

Distance Formula
If (x_1, y_1) and (x_2, y_2) represent the points, then distance $= \sqrt{(x_2 - x_1)^2 + (y_2 - y_1)^2}$.

To find the distance between two given points, first decide which point will be (x_1, y_1) and which will be (x_2, y_2).

In the following example, the distance formula is used to find the distance between $(1, -3)$ and $(-4, 5)$.

Let $(1, -3)$ be (x_1, y_1). Let $(-4, 5)$ be (x_2, y_2). Substitute the coordinates into the distance formula.

$$\begin{aligned}
\text{distance} &= \sqrt{(x_2 - x_1)^2 + (y_2 - y_1)^2} \\
&= \sqrt{(-4) - 1)^2 + (5 - (-3)^2} \\
&= \sqrt{(-5)^2 + 8^2} \\
&= \sqrt{25 + 64} \\
&= \sqrt{89} \\
&\approx 9.43 \text{ units}
\end{aligned}$$

5. Use the distance formula to find the distance between the toy and the corner where the walls meet. Let the origin be (x_1, y_1). Let the toy's coordinates be (x_2, y_2).

Use the distance formula or the Pythagorean Theorem to find the distance between the given points. If needed, round answers to the nearest hundredth.

6. (3, 1) and (6, −3)
7. (1, 15) and (6, 3)
8. (−3, −5) and (7, −1)
9. (6, 4) and (2, −3)

What Have You Learned?

10. Explain how the distance formula and the Pythagorean Theorem are related. Use drawings or examples if they help you explain.

12. Rey and Lydia have found the perfect couch for their living room, but they are not sure whether it will fit through the doorway. The doorway measures 37 inches wide and 79 inches high.

They know they can take the legs off the couch. Figure 1 is a side view of the couch with the legs removed.

The couch is too wide to fit if they carry it upright, but Rey thinks it might fit if they tilt it as shown in Figure 2.

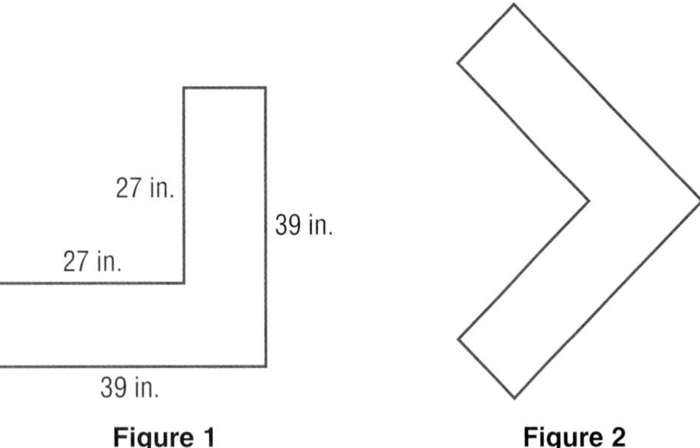

Figure 1 **Figure 2**

In this exercise, you will use the Pythagorean Theorem to figure out whether the couch will fit through the doorway.

a. For Figure 3, explain why the dashed segments shown here each have length 12 inches. Then find the length of segment *c*. Explain how you found it. Round your answer up to the nearest tenth of an inch.

b. For Figure 4, find the length of segment *d* to the nearest tenth of an inch.

c. In Figure 5, segment *b* divides segment *d* in half. Find the length of segment *b* to the nearest tenth.

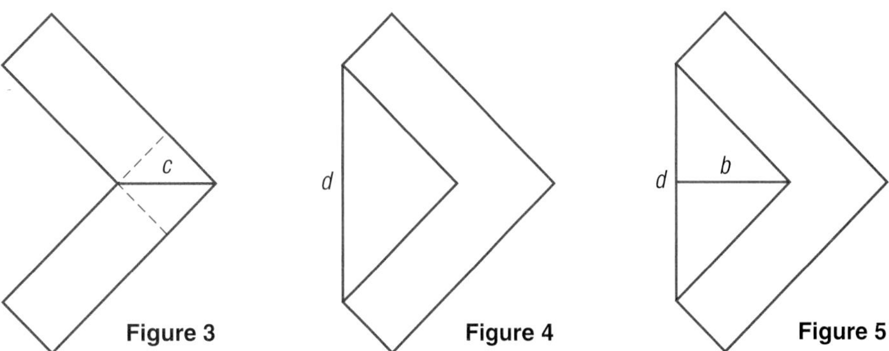

Figure 3 **Figure 4** **Figure 5**

d. Will the couch fit through the doorway? Explain how you know.

13. Mr. Mackenzie built a table. He had intended for the table to be rectangular, but he is not sure it turned out that way. He carefully measured the tabletop and found that the side lengths are 60 inches and 45 inches, and the diagonal is 73.5 inches. Is the table a rectangle? Explain how you know.

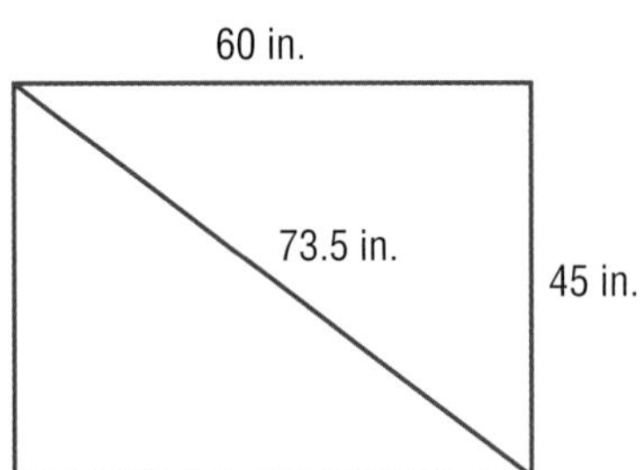

14. In Your Own Words Explain what the Pythagorean Theorem is and how it is useful for finding lengths.

15. Preview Consider the following table of values. What pattern do you see in the *x*-values? In the *y*-values?

x	1	2	3	4
y	5	10	15	20

Mixed Review

If possible, rewrite each expression using the laws of exponents.

16. $y^2 \cdot y^6$ **17.** $a^n \cdot a^n$ **18.** $2b \cdot 3b^2$ **19.** $4^5 \cdot 9^5$

20. $5a^4 \cdot 5a^4$ **21.** $a^{100} \cdot b^{100}$ **22.** $(1.1^2)^4$ **23.** $(0.9^4)^3$

24. $(a^3)^4$ **25.** $(a^4)^0$ **26.** $(x^2)^3$ **27.** $(3^n)^m$

Simplify each expression as much as possible.

28. $x^2 + x^2$ **29.** $(x^2)^3$ **30.** $x^3 \div x^2$

31. $\frac{2x^2}{x^2} - 1$ **32.** $2x^2 - x^2$ **33.** $\frac{(x^2)^3}{x^2}$

Find each indicated sum, difference, product, or quotient. Give answers in lowest terms.

34. $-\frac{3}{7} + \frac{1}{7}$ **35.** $-\frac{2}{3} + \frac{5}{8}$

36. $\frac{4}{9} - \frac{2}{3}$ **37.** $\frac{3}{8} - \left(-\frac{2}{5}\right)$

38. $\left(\frac{1}{5}\right)\left(-\frac{10}{14}\right)$ **39.** $-\frac{3}{5} \div \left(-\frac{9}{7}\right)$

CHAPTER 7

Review & Self-Assessment

Vocabulary

distance formula
hypotenuse
irrational number
leg
natural number
Pythagorean Theorem
rational number
real number
right triangle
square root
whole number

Chapter Summary

In Lesson 7-1, you were asked to take your previous understanding of types of numbers and apply it one step further to rational numbers. You used Venn diagrams to group numbers into sets including natural numbers, whole numbers, integers, and rational numbers to show your understanding of how different number sets are related. To further solidify your understanding of the relationship of these number sets, you solved number puzzles. Next, you compared pairs of rational numbers and ordered sets of rational numbers, giving justifications for your reasoning.

You were introduced to the set of irrational numbers in Lesson 7-2. You used area models to explore the inverse relationship between squares and square roots. You used Venn diagrams to explore the relationship between rational and irrational numbers. Additionally, you used calculators and number lines to show the approximate value of irrational numbers.

In the last lesson, you investigated the Pythagorean Theorem and used the theorem to find the lengths of missing triangle sides. Finally, you used the distance formula to calculate the lengths of line segments on a coordinate grid.

Strategies and Applications

The questions in this section will help you review and apply the important ideas and strategies developed in this chapter.

Identifying rational and irrational numbers

Identify all the number sets (natural, whole, integers, and rational) to which the following numbers belong.

1. -5

2. 2

3. 4.72

4. $\frac{3}{4}$

5. 5^3

6. $\sqrt{49}$

7. 2

8. 0

9. Is it always possible to think of a smaller whole number? Explain your reasoning.

10. Is the following statement true or false? Explain your reasoning.

All integers are rational numbers,
but not all rational numbers are integers.

Using Venn diagrams to show number relationships

11. Copy the following Venn diagram. In each area, write three numbers. If no numbers belong in an area, label the area "empty."

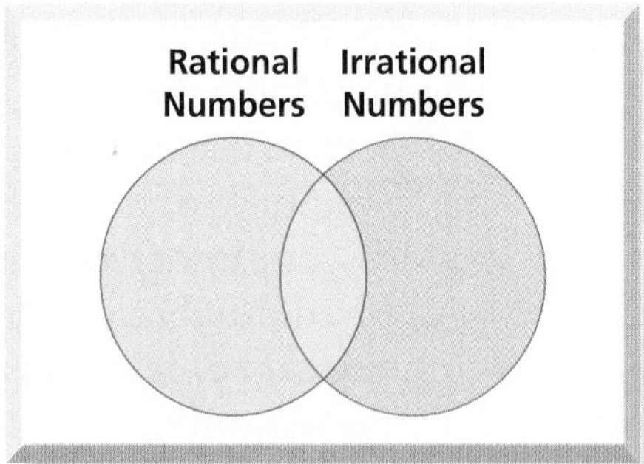

Calculating lengths using the Pythagorean Theorem

12. Before moving into her new house, Katrina is trying out furniture arrangements on graph paper. She wants to make sure there is enough room to walk around the furniture. She decides she needs at least 3 feet between pieces of furniture, except for the coffee table, which can be directly in front of the couch, and the lamp. Each square on her grid represents 1 square foot.

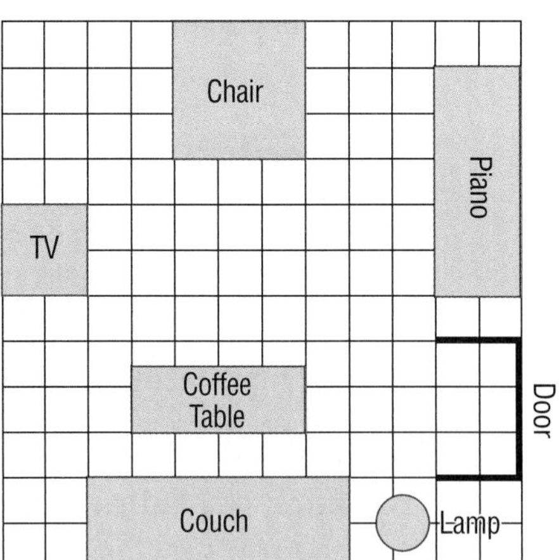

a. How many feet are there between the closest corners of the TV and the chair?

b. How many feet are there between the closest corners of the coffee table and the piano?

c. Are any pieces of furniture too close together?

13. Use the Pythagorean Theorem to find the missing side length.

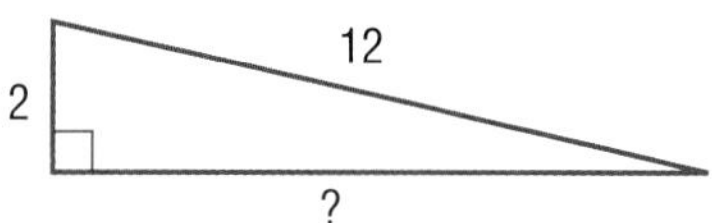

Demonstrating Skills

14. The size of a television is given in terms of the length of the diagonal of its screen. For example, a 19-inch television has a screen with diagonal length 19 inches.

Mr. Franco's television screen has a length of 21.75 inches and a width of 16.25 inches. What is the size of his television? Give your answer to the nearest inch. Explain how you found it.

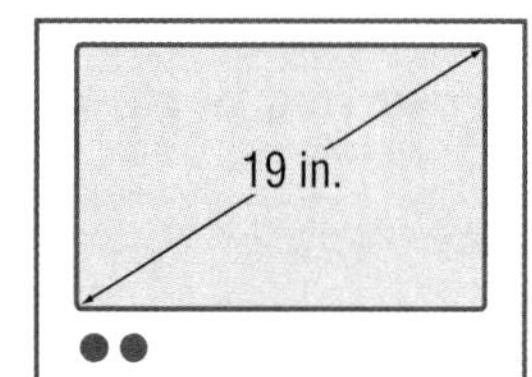

15. Order the following numbers from least to greatest.

$-1.275 \qquad -1.2 \qquad -\frac{7}{5} \qquad -1.027 \qquad -1\frac{1}{4}$

16. Which of the following numbers are less than -0.4?

$-0.48 \qquad -0.39 \qquad -\frac{9}{20} \qquad -\frac{2}{5} \qquad -\frac{1}{2}$

For Exercises 17–22, identify which number is greater. Explain how you found your answer.

17. $\frac{1}{2}$ or 0.58

18. 0.67 or $\frac{2}{3}$

19. $-\frac{3}{4}$ or -0.8

20. -1.45 or $-1\frac{2}{5}$

21. -0.745 or $-\frac{19}{25}$

22. -0.875 or -0.578

23. To the nearest tenth, between which two numbers is $\sqrt{40}$? Do not use your calculator.

24. Explain why it is impossible to have a number that is both rational and irrational.

25. Judith wants to paint a square that has an area of 361 in^2. How long must each side of her square be? Explain your reasoning.

26. Explain in your own words what an imaginary number is and why the word "imaginary" is a good explanation of these types of numbers.

27. Find five rational numbers that lie between $8\frac{11}{17}$ and $8\frac{17}{21}$.

28. Randall squared a number and the result was double the number with which he started. What was his starting number?

For Exercises 29–36, find the two whole numbers that each square root lies between without using your calculator.

29. $\sqrt{32}$

30. $\sqrt{7}$

31. $\sqrt{19}$

32. $\sqrt{83}$

33. $\sqrt{127}$

34. $\sqrt{99}$

35. $\sqrt{150}$

36. $\sqrt{200}$

37. Order the following numbers from least to greatest.

$\sqrt{16}$ $\quad -4.1 \quad -\sqrt{21} \quad \sqrt{21} \quad 4.1$

38. Order the following numbers from greatest to least.

$\sqrt{31} \quad \frac{36}{7} \quad -\frac{36}{7} \quad -\sqrt{31} \quad -5.2$

Test-Taking Practice

SHORT RESPONSE

1 Find the length of the diagonal of the quadrilateral shown below.

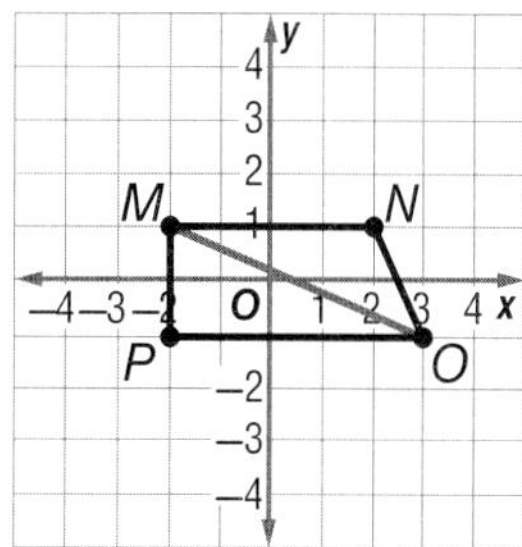

Show your work.

Answer ____________________ units

MULTIPLE CHOICE

2 Which number set does not include the number −37?

A integers

B rational numbers

C real numbers

D whole numbers

3 Which of the following numbers is an irrational number?

F $\sqrt{49}$

G 0.728

H $\sqrt{34}$

J $\frac{1}{9}$

4 Which list has the numbers ordered from least to greatest?

A $3.\overline{3}, \frac{17}{5}, 2\frac{2}{5}, 3.34$

B $2\frac{2}{5}, 3.\overline{3}, 3.34, \frac{17}{5}$

C $2\frac{2}{5}, \frac{17}{5}, 3.\overline{3}, 3.34$

D $3.\overline{3}, 3.34, 2\frac{2}{5}, \frac{17}{5}$

5 Between what two consecutive whole numbers is $\sqrt{47}$?

F 5, 6

G 7, 8

H 6, 7

J 3, 4

CHAPTER

8

Linear Relationships

Real-Life Math

At Any Rate Linear relationships always involve a constant rate. One of the most common types of rates is *speed.* The speed of an object, like a train, is a relationship between time and distance.

Imagine yourself behind the controls of the British Eurostar, the fastest train in Europe. What makes this train unique is that it runs through the Channel Tunnel, or Chunnel, an expansive tunnel drilled under the English Channel connecting Britain to France. The Eurostar, which first traveled through the Chunnel on June 30, 1993, can reach speeds of 186 mph on land and 80 mph in the Chunnel. That is more than twice the rate of an average train. People can now travel from Paris to London in just 3 hours, a journey that used to take days.

Think About It How long would it take to go 160 miles on the Eurostar through the Chunnel?

Contents in Brief

8.1	Rates	368
8.2	Speed and Slope	389
8.3	Recognize Linear Relationships	410
	Review & Self-Assessment	431

Math Online

Take the **Chapter Readiness Quiz** at glencoe.com.

Dear Family,

Chapter 8 introduces linear relationships, where a change in one variable results in a fixed change in another variable. In class, students will describe situations, make tables of data about the situations, graph the data, and write linear equations that describe the relationships.

Key Concept–Tables or Algebraic Rules

Here is an example of one kind of exercise with which the class will be working.

Three telephone companies have long-distance rates.

Company	Rate
Easy Access Company	\$1.00 for the first minute; 25 cents per minute thereafter
Call Home	20 cents per minute
Metro Communication	\$3.00 for the first minute; 15 cents per minute thereafter

Here is a question to consider.

- For each company, calculate the amount due for calls lasting 5 minutes, lasting 15 minutes, and for two other lengths of time. Show your results in a table.

Students will also learn to predict which equations have graphs that are straight lines by looking at ta bles or algebraic rules. Then, the class will be able to determine the slope and *y*-intercept.

Chapter Vocabulary

linear relationship
proportional
rate
slope
speed
velocity
***y*-intercept**

Home Activities

- Discuss with your student linear relationships and different ways linear equations appear in the world outside of school.
- Use linear equations to calculate payments for jobs using different hourly rates.

Rates

Inv 1 Understand Rates 369

Inv 2 Describe Rates 373

Inv 3 Proportional Relationships 377

Inv 4 Rolling Along 380

Vocabulary

rate

In the statement "Carlos types 30 words per minute," the rate *30 words per minute* describes the relationship between the number of words Carlos types and time in minutes. The speed of light, 186,000 miles per second, is a rate that describes the relationship between the distance light travels and time in seconds. In general, a **rate** describes how two unlike quantities are related or how they can be compared.

Here are some other statements involving rates.

When she babysits, Yoshi earns $3.50 per hour.

At the Better Batter donut shop, donuts cost $3.79 per dozen.

Franklin's resting heart rate is 65 beats per minute.

On June 24, 2003, the exchange rate from U.S. to British currency was 0.602031 pound per dollar.

Think & Discuss

Each rate above involves the Latin word *per,* which means "for each." Try using the phrase "for each" to explain the meaning of each rate. For example, the exchange rate is the number of pounds you receive for each dollar you exchange.

Work with a partner to write three more statements involving rates. Explain what your rates mean using the phrase "for each."

You can use rates to write algebraic rules relating variables. For example, the equation $d = 186{,}000t$ uses the speed of light to describe the distance d in miles that light travels during a particular number of seconds t.

In this lesson, you will explore many situations involving rates. You will also look at tables, graphs, and algebraic rules for these situations.

Investigation  Understand Rates

Vocabulary

linear relationship

Materials

- graph paper

You have probably used rates often, right in your own kitchen. Working with cooking measures will help you understand different ways of thinking about and describing rates.

Develop & Understand: A

Here is a rate expressed in words.

A teaspoon contains 5 milliliters.

This rate could also be stated this way.

There are 5 milliliters per teaspoon.

1. Suppose you double the number of teaspoons of some ingredient in a recipe. Does that double the number of milliliters? If you triple the number of teaspoons, do you triple the number of milliliters?
2. Using m for the number of milliliters, write a rule that tells how many milliliters are in t teaspoons.
3. A table of values can help you check that you have written a rule correctly. Copy this table. Without using your rule, complete the table to show the number of milliliters in t teaspoons. Use the table to check that your rule for t and m is correct.

Converting Teaspoons to Milliliters

Teaspoons, t	0	1	2	3	4	5	10
Milliliters, m							

Real-World Link

One type of spider spins an entire web in just 20 minutes. It weaves at the rate of 1,000 operations per minute.

Real-World Link

A woodpecker has a tough head. It can pound its beak against wood at a rate of 20 times per second.

4. On axes like those below, plot the data from your table.

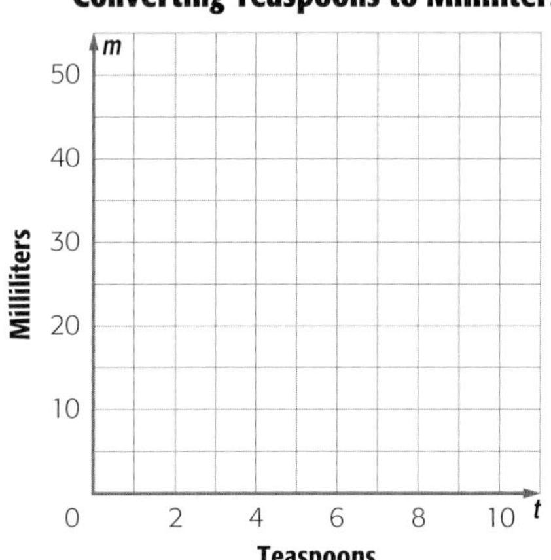

5. Does it make sense to connect the points on your graph with a line? Explain. If it does make sense, do it.

6. Does it make sense to extend your graph beyond the points? Explain. If it does make sense, do it.

Think & Discuss

Darnell and Maya wrote rules relating pints and quarts.

1 quart = 2 pints. So, my rule is q = 2p.
If I have 2 quarts, I have 4 pints. Whatever number of quarts I have, the number of pints is 2 times that amount. My rule is p = 2q.
QUART
PINT

Whose rule is correct? Explain your answer.

The relationship between teaspoons and milliliters is called a **linear relationship** because the points on its graph lie on a straight line. In a linear relationship, as one variable changes by 1 unit, the other variable changes by a set amount. The amount of change per unit is the *rate*. In the rule $m = 5t$, the 5 shows that m changes 5 units per 1-unit change in t.

Because you can have 1.5 teaspoons, 2.25 teaspoons, and so on, it is sensible to connect the points you plotted in Exercise 4 using a straight line. It is also sensible to extend the line beyond the points from the table since you can have 11, 16, 27, or even more teaspoons of an ingredient.

There are situations for which it is not sensible to connect the points on a graph. For example, this graph shows the relationship between the number of skateboards and the total number of wheels.

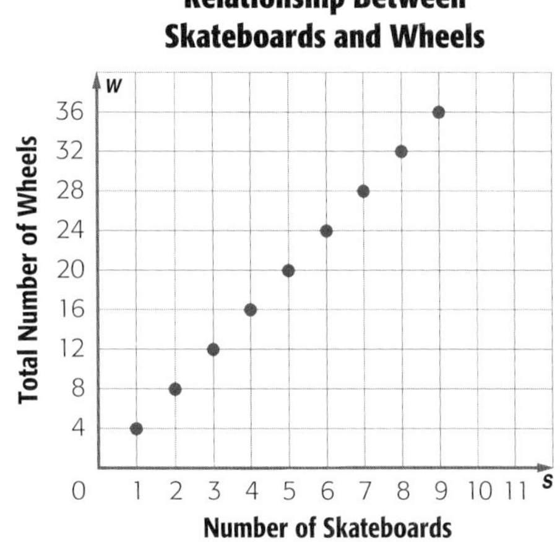

Connecting the points does not make as much sense in this case because the points in between would represent partial skateboards. For example, you would not connect the points for 5 skateboards and 6 skateboards because you could not have 5.5 skateboards or 5.976 skateboards.

Even when it does not make sense to connect the points on a graph, sometimes you may want to do so anyway to help you see a relationship. In this case, it is best to use a dashed line. You will learn more about this later.

Develop & Understand: B

Some recipes give quantities in *weight.* However, in the United States, most recipes specify quantities by *volume,* using such measures as teaspoons, tablespoons, and cups. If you know how much a cup of a particular ingredient weighs, you can write a rate statement to calculate the number of cups for a given weight. For example, it is a fact that one cup of sugar contains about a half pound of sugar.

7. Rewrite the fact above using the word *per.*

8. Write an algebraic rule relating the number of cups of sugar c and weight in pounds p. Check your rule by completing the table.

Relationship Between Cups and Pounds

Cups, c	0	1	2	3	4	5	10
Pounds, p							

9. Use the data in your table to help you draw a graph to show the relationship between the number of cups of sugar and the number of pounds. Use a set of axes like the one shown.

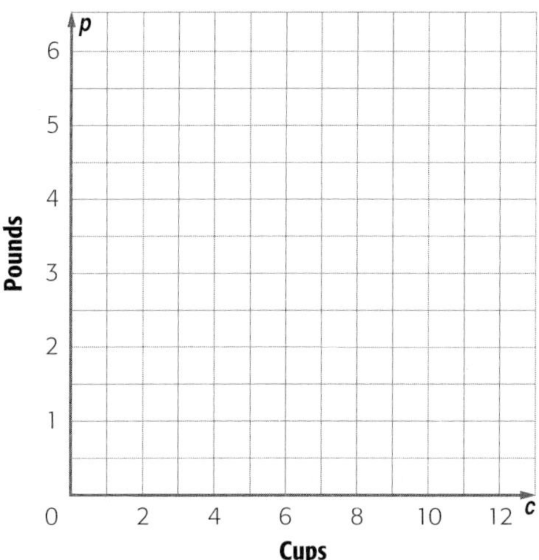

10. Does it make sense to connect the points or to extend the graph beyond the points? If so, do these things. Explain your reasoning.

11. From your graph, find the value of p when c is 7. Is that value what you would expect from your rule?

12. From the graph, find how many cups of sugar you would need if a recipe calls for 1.4 pounds. Is your answer what you would expect from your rule?

13. Do you find it easier to work from the graph or the rule? Explain.

Share & Summarize

1. You may have noticed that on many visits to the doctor, someone measures your pulse, or heart rate, to determine how fast your heart is beating. Measure your pulse by putting your fingers to the side of your neck and counting the number of beats you feel. Record the number of beats in one minute. Explain what makes the number of beats per minute a *rate*.

2. Use your heart rate to write an equation relating the number of beats b to any number of minutes timed t. Explain what your equation means.

Investigation 2 Describe Rates

Materials

- graph paper

Rate relationships can be described in many ways.

Think & Discuss

Are these three students thinking of the same or different relationships? Why do you think so?

Develop & Understand: A

Mario works in a market. His rate of pay is $10 per hour.

1. If Mario works twice as many hours one week than he does another week, will he earn twice as much? If he works three times as many hours, will he earn three times as much?
2. Copy and complete the table to show what pay Mario should receive for different numbers of hours worked.

Mario's Rate of Pay

Hours Worked, h	0	1	2	3	4	5	10	15
Pay (dollars), p								

3. Write a rule in words that relates Mario's pay in dollars to the number of hours worked. Begin your rule, "The number of dollars earned is equal to ... ." Then rewrite the rule using the variables p for pay and h for hours.
4. What part of your rule shows Mario's rate of pay?
5. Mario usually works a 35-hour week. How much does he earn in a typical week?
6. One week, Mario earned $300. How many hours did he work that week?

Alec works in a local take-out restaurant. His rate of pay is $7 per hour.

7. If Alec works twice as many hours one week than he does another week, will he earn twice as much? If he works three times as many hours, will he earn three times as much?
8. Complete the table to show what Alec would be paid for different numbers of hours worked.

Alec's Rate of Pay

Hours Worked, h	0	1	2	3	4	5	10	15
Pay (dollars), p								

9. Write a rule for Alec's pay in words and in variables.
10. What part of your rule shows Alec's rate of pay?
11. One week, Alec worked 30 hours. How much did he earn that week?
12. Compare your symbolic rules for Mario and Alec. How are they the same? How are they different?

Develop & Understand: B

In addition to variables, tables, and words, graphs are useful for comparing rates of pay.

13. Draw a graph of the data for Mario's pay. Use a set of axes like the one below.

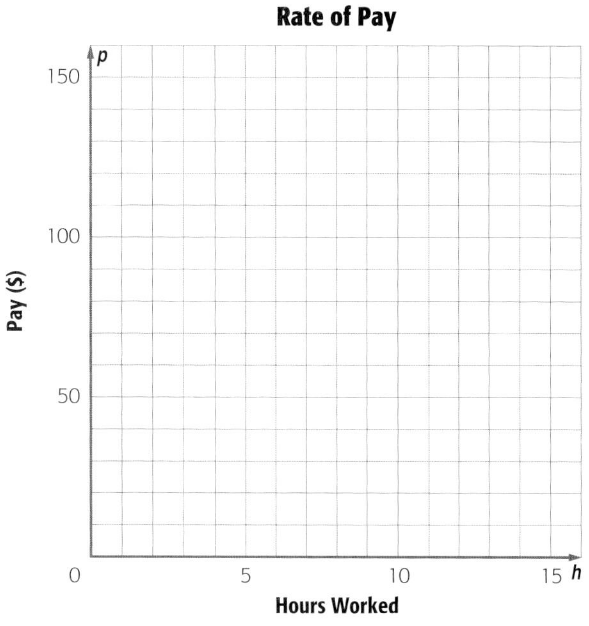

14. Does it make sense to connect the points on your graph? Does it make sense to extend the graph beyond the points? If so, do these things. Explain your reasoning.

15. On the same grid, draw a graph to show how much Alec earns. Label your graphs so you know which is for Mario and which is for Alec. Write each symbolic rule from Exercises 3 and 9 next to the appropriate graph.

16. Compare the graphs for Mario and Alec. In what ways are they the same? In what ways are they different? Which is steeper? Why?

17. Joelle earns \$15 per hour. If you added a line to your graph to show Joelle's pay, how would it compare to Mario's and Alec's graphs?

Develop & Understand: C

Tamsin works for an automobile association. Every second weekend, she is "on call." This means she must be available all weekend in case a car breaks down in her area. She is paid a fixed amount of $40 for the weekend, even if she does not have to work. If she is called, she earns an additional $10 per hour worked.

18. Complete the table to show what Tamsin would be paid for different numbers of hours worked during a single weekend.

Tamsin's Pay

Hours Worked, *h*	0	1	2	3	4	5	10	15
Pay (dollars), *p*	40	50						

19. If Tamsin works twice as many hours one weekend as she does another weekend, will she earn twice as much? Explain.

20. Write the rule for Tamsin's pay for a single weekend in words and in variables.

21. What part of your rule shows the amount Tamsin earns for being on call? What part shows her hourly rate?

22. Use the data in your table to draw a graph showing how much Tamsin earns for various numbers of hours worked. Use the same grid you used for Exercise 13 or a similar one.

Share & Summarize

1. Look again at the graphs that show how much Alec, Mario, and Tamsin earn. How do the graphs show differences in pay? How is a higher rate shown in a graph?
2. How do the symbolic rules show the differences in the rates at which the three people are paid? How is a higher rate shown in a symbolic rule?

Investigation  Proportional Relationships

Vocabulary

proportional

Materials

- graph paper

Alec and Mario, from Investigation 2, are paid at a simple hourly rate. If they work double the number of hours, they receive double the pay. If they work triple the hours, they receive triple the pay, and so on.

The word **proportional** is sometimes used to describe this kind of relationship between two variables. We can say that for Alec and Mario, the number of dollars earned is proportional to the number of hours worked. The graphs for Alec's pay and Mario's pay are shown below. They are each a straight line that begins at the point (0, 0).

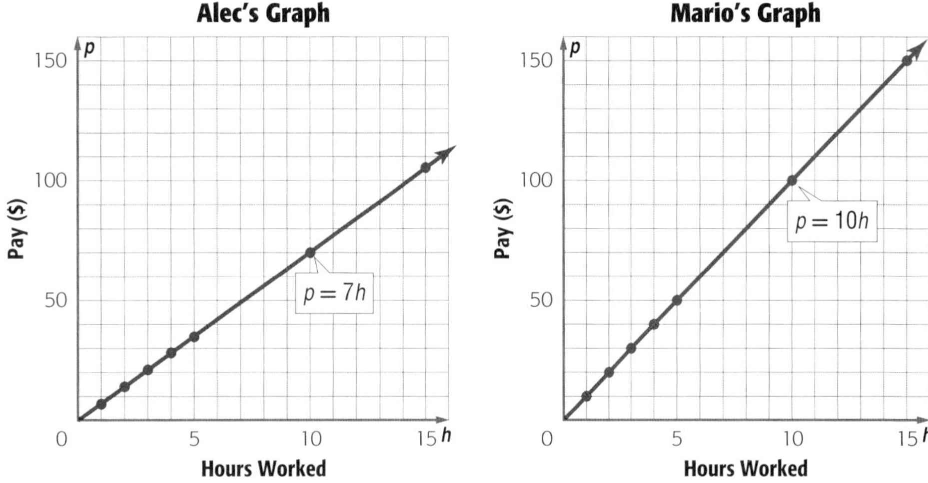

Tamsin is not paid at a simple hourly rate. She receives a fixed amount plus an hourly rate. For Tamsin, doubling the number of hours worked does not double the total pay received. The number of dollars earned is not proportional to the number of hours worked. The graph for Tamsin's pay is also a straight line, but it does not begin at the point (0, 0).

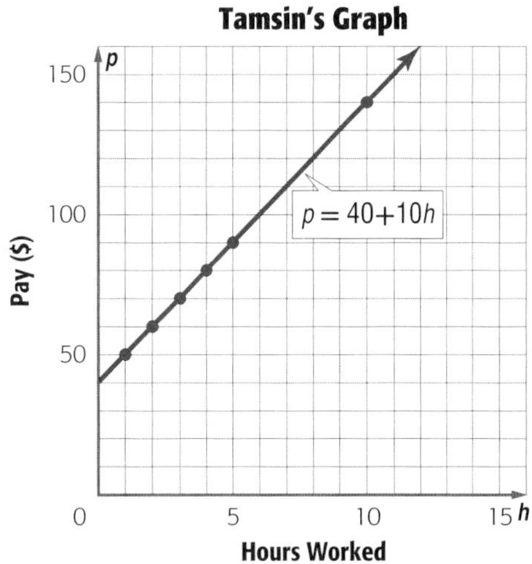

Although the relationships for Alec and Mario are proportional and the relationship for Tamsin is not, all three relationships are *linear relationships* because their graphs are straight lines.

Think & Discuss

Many companies issue cards for charging telephone calls. For which cards below are the charges proportional to the number of minutes? For which are they not proportional? Explain.

Develop & Understand: A

In Exercises 1–4, the rule describes two variables that are proportional to each other. Rewrite each rule in symbols.

1. The circumference of a circle c is 3.14 times the diameter d.

2. The perimeter of a square p is 4 multiplied by the length of a side L.

3. At a certain time of day, the length of a shadow s cast by an object is twice the length of the object L.

4. Each length h in a copy is $\frac{1}{100}$ the length k in the original.

5. Compare the four rules you developed in Exercises 1–4. In what ways are they similar? In what ways are they different?

Develop & Understand: B

Ms. Cruz gave her class a test with 40 questions.

6. Latisha got 100% correct; Kate got only 50% correct. How many of the 40 questions did each answer correctly?

7. Lupe got fewer than 20 questions correct. If she could retake the test and double the number right, would that double her percentage?

8. Does halving the number of correct answers out of 40 halve the percentage?

9. Is the number of correct answers out of 40 proportional to the percentage?

10. One evening, Ms. Cruz had to convert her class' test results into percentages. She had left her calculator at school and decided to use a graph to help her.

a. She first calculated a few values and put them in a table. Copy and complete her table.

Class Test Results

Number Correct (out of 40), n	0	10		30	
Percentage Correct, p	0		50		100

b. Ms. Cruz then drew a graph on graph paper by plotting the points from her table and drawing a line through them. She made the horizontal axis the number of correct questions out of 40 with one grid unit on the axis equal to 2 questions correct. She put percentage points on the vertical axis with one grid unit on the axis equal to 5 percentage points. She could then quickly read percentages from the graph for the various scores out of 40.

Draw Ms. Cruz's graph. Does it make sense to extend the line beyond the point (40, 100)? Explain your thinking.

11. Use your graph to convert each score out of 40 into a percentage.

a. 38 **b.** 36

c. 33 **d.** 29

e. 25 **f.** 23

12. Lucas, Opa, and Raheem scored 65%, 75%, and 85%, respectively, on the test. How many questions out of 40 did each get correct?

Share & Summarize

1. Without graphing, do you think the graphs of the relationships described in Exercises 1–4 would pass through the origin? Explain your reasoning.

2. A graph of a linear relationship goes through the point (0, 4). Is this a proportional relationship? Explain your reasoning.

3. Is $y = 5x + 7$ a proportional relationship? Explain your thinking.

Math Link

The point (0, 0) is called the origin.

Inquiry Investigation 4 Rolling Along

Materials

- 2 cylindrical pens or pencils (without edges) with different diameters
- lined paper
- hardcover book
- graph paper

Real-World Link

Historians believe that this method of rollers was used to move very heavy stones used to build the Egyptian pyramids.

In this investigation, you will conduct an experiment to understand the relationship between two variables.

Work with Your Class

Watch how a heavy box rolls on three drink cans. The rollers will keep coming out from behind the box.

1. Do the rollers go backward?
2. Does the box go faster than the rollers?
3. Why do the rollers keep coming out the back?

Try It Out

With your group, conduct an experiment that involves rolling one object on top of another. Use a perfectly round cylindrical pen or pencil, a book, and a sheet of lined paper with wide lines. Set up the paper, pencil, and book as shown.

4. Roll the book on the pen until the pen has moved two line spaces. How far has the book moved?
5. Continue to roll the book to complete the table. When you collect your data, count and measure accurately. Otherwise, you might not see the patterns that show the relationship between the variables.

Number of Line Spaces Moved

Distance Moved by Pen, p	0	2	4	6	8	10
Distance Moved by Book, b	0					

Analyze Your Data

6. Describe how the distance the book moves is related to the distance the pen moves. Write the relationship using the symbols p and b.

7. Draw a graph with the distance moved by the pen on the horizontal axis and the distance moved by the book on the vertical axis. Plot the values you found in Exercise 5. If you think it makes sense to do so, join the points with a line, and extend the line.

Apply Your Results

8. Using your rule, predict how far the book will move when the pen moves 16 line spaces. Check by doing it.

9. How many line spaces does the book move if the number of line spaces moved by the pen is 0.5? 1.5? 7.5? 1,350?

10. If the book moves 9 line spaces, how far does the pen move? Test it with your equipment.

11. Describe in words how the distance the pen moves is related to the distance the book moves. Write this rule in symbols.

Try It Again

Replace the pen with a different-sized pen or some other cylinder, and repeat the experiment.

12. Complete a new table. Are your results the same?

Number of Lines Spaces Moved

Distance Moved by Pen, p	0	2	4	6	8	10
Distance Moved by Book, b	0					

13. Is this a proportional relationship? How can you tell?

What Did You Learn?

14. Write a short report describing your experiments and the results.

On Your Own Exercises
Lesson 8.1

Practice & Apply

1. **Measurement** A kilogram is equivalent to 1,000 grams.

   a. Describe this relationship in words using the word *per.*

   b. Write a rule in symbols that relates the two measures. Use k to represent the number of kilograms and g the number of grams.

   c. Copy and complete the table without using the rule. Then use your table to check that your rule is correct.

   Converting Kilograms to Grams

Kilograms, k	1	2	3	4	5
Grams, g					

   d. Use the data in the table to help you draw a graph to show the relationship between the measures. If it makes sense to connect the points, do so. If it makes sense to extend the line beyond the points, do so.

   e. Use your table, formula, or graph to find the number of grams in 1.5 kg, 3.2 kg, 0.7 kg, and 0.034 kg.

2. **Measurement** A pound is equivalent to 0.454 kilogram.

   a. Describe the relationship in words using the word *per.*

   b. Write a rule in symbols that relates the two measures. Use p for the number of pounds and k for the number of kilograms.

   c. Complete the table without using your rule. Then use your table to check that your rule is correct.

   Converting Pounds to Kilograms

Pounds, p	1	2	3	4	5
Kilograms, k					

   d. Use the data in the table to help you draw a graph to show the relationship between the measures.

   e. Use your table, formula, or graph to find the number of kilograms in 7 lb, 40 lb, 100 lb, and 0.5 lb.

Math Link

When you plot points from a table of data, connect the points and extend the line beyond the points, if it makes sense to do so.

3. These three phone cards have different charge plans for their customers.

a. Write the rule for each card's charge plan for 1 month, using d for charge in dollars and m for minutes of calls. Underline the part of each rule that shows the rate of charge.

b. Copy the table. For each card, fill in the table to show the monthly bill for different numbers of minutes of calls.

Phone Card Charges

Minutes, *m*	0	10	20	30	40	50	100	150	200
Dollar Bill, $									
Easy Bill, $									
Fantasic Bill, $									

c. On one set of axes, draw a graph for each card's charge plan. Put minutes on the horizontal axis and dollars on the vertical axis. Label the graphs to identify which card goes with which graph.

d. Does it make sense to connect the points? Does it make sense to extend the graphs beyond the points? If so, do it.

e. How are the graphs similar? How are they different?

f. How much would the monthly bill for 85 minutes be with the Dollar card?

g. For $12, how many minutes of calls could you make in one month with the Dollar card? With the Easy card? With the Fantastic card? If your answers are more than 59 minutes, convert them to hours and minutes.

4. In Exercises 1–17 in Investigation 2, you solved exercises about Alec's and Mario's rates of pay. Suppose that Alec receives a raise to $8 per hour.

a. Write a rule in symbols relating the number of dollars p that Alec earns to the number of hours h that he works.

b. What would the graph that shows his new pay rate look like? How would it compare to the two graphs you have already drawn for Mario's pay and Alec's pay at the old rate?

c. Make a table to show what Alec would now earn for 0, 1, 2, 3, 4, 5, and 10 hours of work.

d. On your grid from Exercise 13, draw a graph to show how much Alec is paid at the new rate. Does the graph look as you predicted in Part b? If not, try to figure out where your thinking went wrong.

5. Before compact discs became popular, people listened to music and other sound recordings on record players. A record player that runs at 45 rpm turns a record 45 times per minute. The unit rpm is an abbreviation for *revolutions per minute.*

a. Complete the table to show the number of times a record turns in a given number of minutes.

Playing Records

Minutes, *t*	1	2	3	4	5	6
Revolutions, *r*	45					

b. Write in words a rule that gives the number of times a record turns in a given number of minutes. Then write the same rule in symbols. Underline the part of your rule that shows the constant rate of turning.

c. Using your rule, predict the number of revolutions in 12 seconds, or 0.2 minute.

d. Draw a graph to show how the number of revolutions is related to time. Put time on the horizontal axis.

e. Use your graph to estimate how long it would take for a record to revolve 100 times and 300 times.

6. A painter needs about 1 fluid ounce of paint for each square foot of wall she paints.

   a. Write a rule in symbols for this situation, using p for fluid ounces of paint and f for square feet. Underline the part of your rule that shows the constant rate of paint use.

   b. Complete the table.

Painting Walls

Paint (fluid oz), p	0	10	20	30	40	50	80	100	140	170	220
Area (sq ft), f											

   c. How many fluid ounces of paint are needed to paint a wall with area 120 sq ft?

   d. How many gallons of paint are needed to cover 450 sq ft? There are 128 fluid ounces in 1 gallon.

   e. Draw a graph to show how the amount of paint depends on the area to be painted. Put area on the horizontal axis.

   f. Use your graph to estimate the area covered with 1 gallon of paint, 2 gallons of paint, and 3 gallons of paint.

   g. If a painter uses twice as much paint, can she cover twice as much area? If she uses three times as much paint, can she cover three times as much area? Is the relationship between area and amount of paint proportional?

Connect & Extend

7. **In Your Own Words** How can you find the rate of a linear relationship using a graph? Using a symbolic rule? Using a table? Which do you prefer for this task? Why?

8. **Measurement** 24 ounces is equivalent to 680 grams.

   a. Describe the relationship in words using the word *per.*

   b. Complete the table. Write a rule in symbols that relates the two measures. Use your table to check that your rule is correct.

Converting Ounces to Grams

Ounces, z	0	1	2	3	4	5	6
Grams, g							

   c. Use the data in the table to help you draw a graph to show the relationship between the measures. If you think it makes sense to do so, join the points with a line, and extend the line.

Real-World Link

The metric system was first proposed in 1670 and is now used in most countries around the world.

Measurement A kilogram, kg, is equivalent to 1,000 grams, g. A pound, lb, is equivalent to 0.454 kilogram.

9. Find the number of grams in 1 lb, 2 lb, 0.6 lb, and 0.1 lb.

10. Find the number of pounds in 3 kg, 17 kg, 0.5 kg, 900 g, 300 g, and 56 g. Round to the nearest hundredth.

1 fl oz = 29.573 mL
1 oz = 28.350 g
1 gal = 3.785 L

11. Challenge Darnell bought a gallon of spring water. The label indicated that a serving size is 1 cup, or 236 mL. Use the conversions at the left to answer the questions. Round to the nearest hundredth.

a. What is the volume of the 1-cup serving size in fluid ounces?

b. The mass of 1 mL of water is 1 g. What is the mass of 1 cup of water in grams?

c. What is the weight of 1 cup of water in ounces?

d. What is the volume of 1 cup of water in liters?

e. What is the volume of 1 cup of water in gallons?

f. There are 16 ounces in a pound. What is the weight of 1 cup of water in pounds?

g. How many cups are in 1 gallon?

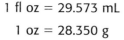

Real-World Link
The very first movie theaters were built in the late 1890s. By the 1920s and 1930s, grand "picture palaces" were springing up everywhere. The first drive-in movie opened in 1933.

12. Challenge An architect is designing a building with two movie theaters. The Green Theater will be rectangular in shape with each row holding 16 seats. The Blue Theater will be wider at the back. The first row will have 8 seats, the second row will have 16 seats, and all other rows will have 20 seats. The architect will decide how many rows to put in each theater on the basis of the number of people it is meant to hold.

a. Make a table showing how the number of seats for each theater relates to the number of rows. Assume each theater will have at least 3 rows.

b. On one set of axes, draw graphs to show how the number of seats in each theater depends on the number of rows.

c. Does it make sense to connect the points on the graphs? Explain.

d. For each theater, write a rule in symbols that tells how the number of seats depends on the number of rows.

e. How many rows should be in each theater to give the theaters an equal number of seats?

f. Which theater will have more seats if both have 11 rows?

g. Is the number of seats in each theater proportional to the number of rows? Why do you think so?

13. Carmen started to read a book at 10:20 A.M. She usually reads a page in 2 minutes. Half an hour later, her friend Yolanda started to read. She reads 15 pages per hour.

a. How many pages did Carmen read before Yolanda started to read?

b. Create a table showing how many pages Carmen and Yolanda read by 10:50, 11:00, 11:10, 11:20, 11:30, 11:40, 11:50, and 12:00.

c. Based on your table, draw graphs on the same axes showing how many pages Carmen and Yolanda read between 10:50 A.M. and noon. Does it make sense to connect the points on the graphs? Does it make sense to continue the lines beyond the 12 o'clock mark? Explain your answers.

d. Write a rule in symbols for the number of pages Carmen read since 10:50. Write a rule in symbols for the number of pages Yolanda read since 10:50. Use m for number of minutes passed after 10:50. Underline the parts of the rules that show the constant rates of reading.

e. Is the relationship for Carmen proportional? Is the relationship for Yolanda proportional?

f. How many pages would each girl read in 5 minutes? In 16 minutes? In 48 minutes?

g. At what time will Carmen finish 40 pages of her book? At what time will Yolanda finish 40 pages of her book?

h. At what time will the number of pages Carmen has read be three times the number of pages Yolanda has read?

Mixed Review

14. To simulate drawing one of three marbles from a bag, Neeraj rolled a die several times. The marbles were blue, white, and purple. Rolling 1 or 2 represented drawing the blue marble. Rolling 3 or 4 represented drawing the white marble. Rolling 5 or 6 represented drawing the purple marble.

Does this simulation represent drawing a marble and putting it back before drawing the next? Or does it represent keeping the drawn marble out of the bag before drawing the next one? How do you know?

Geometry Find the volume and the surface area of each solid.

15.

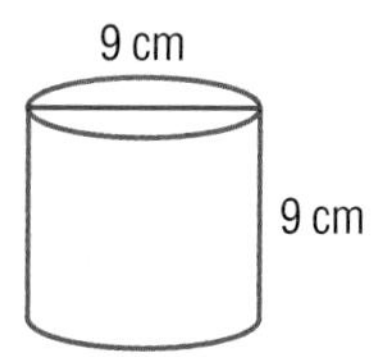

16.

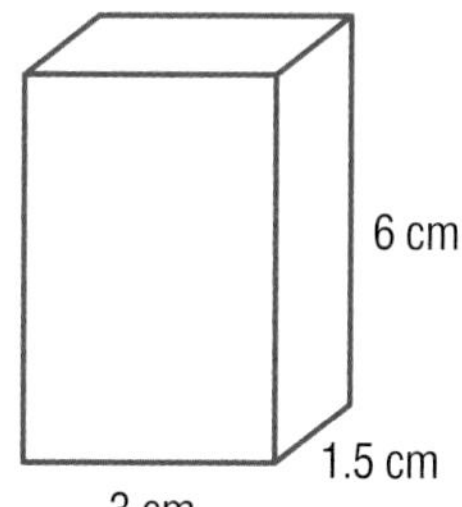

17. Calculate the area and circumference of a circle with radius 2 cm.

18. **Challenge** Calculate the area of a square with a side length of s cm and a diagonal of 1.5 cm. Show how you found your answer.

Evaluate.

19. 3^{-2}

20. $2^2 \cdot 2^3$

21. $2^2 \div 2^{-2}$

22. $0.005 \cdot 0.1$

23. $10^2 \cdot 0.1$

24. $100^2 \cdot 0.3$

25. 0.2^2

26. 0.2^{-2}

27. $\sqrt{169}$

Speed and Slope

Speed is a very common rate. It tells how the distance something travels depends on the time traveled. Speed can be measured in many ways. Here are a few common units of speed.

- miles per hour (mph)
- kilometers per hour (km/h or kph)
- feet per second (ft/s)
- meters per second (m/s)
- millimeters per second (mm/s)

Inv 1 Walk and Jog 390

Inv 2 Distance and Time 394

Inv 3 Describe Graphs 396

Inv 4 Change the Starting Point 398

Materials

- stopwatch
- yardstick

Explore

Guess an average student's normal walking speed and record your guess. Then one person in the class should walk at a "normal" speed from the back of the classroom to the front while another person accurately times him or her. Record that time and the distance the student walked, in feet.

Now here is a challenge. Several students should try to cross the classroom at a steady pace taking the amounts of time specified by the teacher, some greater and some less than it takes to cross at a "normal" speed.

You or a classmate should time how long it actually takes each student to cross the room. Then everyone in the class can compute the students' true speeds. To find the speeds, divide the distance walked by the time it took.

After you have an idea about how fast or slow the various speeds are, answer these questions.

- If a person travels 1.5 feet per second, is this a fast walk, a slow walk, or a run?
- If a second person moves with a speed of 9 feet per second, is this a fast walk, a slow walk, or a run?
- How much faster is the second person in comparison with the first?

Investigation 1 Walk and Jog

Vocabulary

slope

Materials

- graph paper

Some people walk quickly, taking several steps in a few seconds. Others walk more slowly, taking more than a second for each step. Some people naturally take longer strides than others. When you are walking with a friend, one or both of you probably changes something about the way you walk so you can stay side by side.

Develop & Understand: A

Zach's stride is 0.5 meters long. Imagine that he is walking across a room, taking one step each second.

1. At what speed is Zach walking?
2. Copy and complete the table to show Zach's distance from the left wall at each time.

Time (seconds), t	0	1	2	3	4	5	6
Distance from Left Wall (meters), d							

3. At this constant speed, is the distance traveled proportional to the time? How do you know?
4. Write a rule that shows how to compute d if you know t.
5. What would Zach's speed be if he lengthened his stride to 1 meter but still took 1 second for each step? Make a table of values, and write a rule for this speed.

6. Imagine that Maya is also crossing the room, but she is taking two steps each second. Her steps are 1 meter long. What is her speed? Make a table of values, and write a rule for her speed.

7. For all three tables, plot each set of points in the table and then draw a line through them. Put all three graphs on one grid with time on the horizontal axis. Label each graph with its speed.

Think & Discuss

Compare the three graphs you drew in Exercise 7.

- In what ways are they the same? In what ways are they different?
- What does the steepness in these distance-time graphs show?

Develop & Understand: B

Many joggers try to jog at a steady pace throughout most of their runs. This is particularly important for long-distance running.

- Terry tries to jog at a steady pace of 4 meters per second.
- Maria tries to jog at a steady pace of 3 meters per second.
- Bronwyn does not know how fast she jogs, but she tries to keep a steady pace.

8. Make tables for Terry and Maria to show the distances they travel, d meters, in various times, t seconds.

9. Write rules that show how distance d changes with time t for Terry and for Maria.

Real-World Link

The maximum speed a human being has ever run is about 27 miles per hour. The fastest animal on Earth, the cheetah, has been clocked at about 60 miles per hour.

10. A timekeeper measured times and distances traveled for Bronwyn and put the results in a table.

Time (seconds), *t*	0	5	10	15	20
Distance (meters), *d*	0	17.5	35	52.5	70

How fast does Bronwyn jog? Write a rule that relates Bronwyn's distance to time.

11. On one grid, draw graphs for Terry, Maria, and Bronwyn. Put time on the horizontal axis. Label each graph with the name of the person and the speed.

12. Explain how you can tell from looking at the graph who jogs most quickly and who jogs most slowly.

Math Link

The general rule connecting distance, rate, and time is $d = rt$, where d is the distance, r is the rate or speed, and t is the time.

All the points on each graph you drew are on a line through the point (0, 0). The steepest line is the one for which distance changes the most in a given amount of time, that is, when the speed is the fastest. The line that is the least steep is the one for which distance changes the least in a given amount of time, that is, when the speed is the slowest.

Slope describes the steepness of a line. In this case, the slope tells how much the distance changes per unit of time. More generally, the **slope** of a line tells how much the y variable changes per unit change in the x variable.

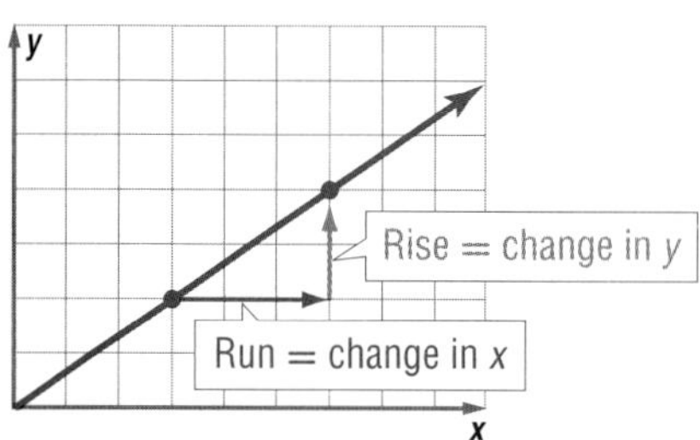

Sometimes slope is described as *rise* divided by *run*. This makes sense because y changes in the vertical direction, or "rises," and x changes in the horizontal direction, or "runs."

Example

This graph shows how Terry's distance changed over time. To find the slope, choose two points, such as (10, 40) and (20, 80). From the left point to the right point, the y value changes from 40 to 80. The *rise* between these points is 80 − 40, or 40. The x value changes from 10 to 20, so the *run* between these points is 20 − 10, or 10. The slope, the rise divided by the run, is $\frac{40}{10}$, or 4.

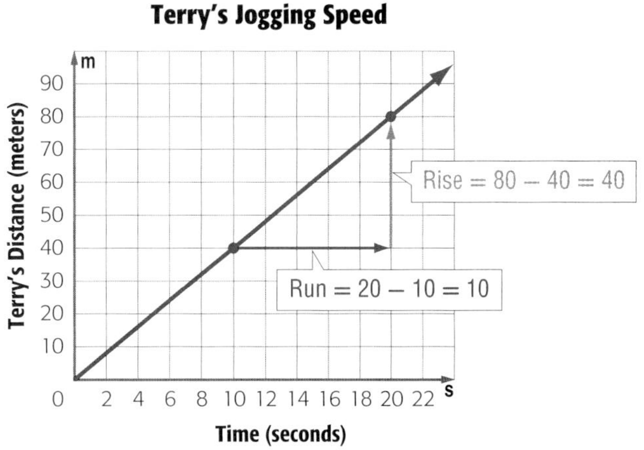

Think & Discuss

Look at your graphs for Maria and Bronwyn. What are the slopes of Maria's and Bronwyn's lines? What does the slope mean in Terry's, Maria's, and Bronwyn's graphs?

Share & Summarize

1. Javier walks at a speed of 5 feet per second. If you graphed the distance he walks over time, with time in seconds on the horizontal axis and distance in feet on the vertical axis, what would be the slope of the line?

2. Dulce walks at a speed of 7 feet per second. Suppose you graphed the distance she walks over time on the same grid as Javier's line. How would the steepness of her line compare to the steepness of Javier's line? Explain.

Investigation 2 Distance and Time

Materials

- graph paper

An airplane flies from New York to Los Angeles. There are two distances that are changing, the distance between the airplane and the New York airport and the distance between the airplane and the Los Angeles airport.

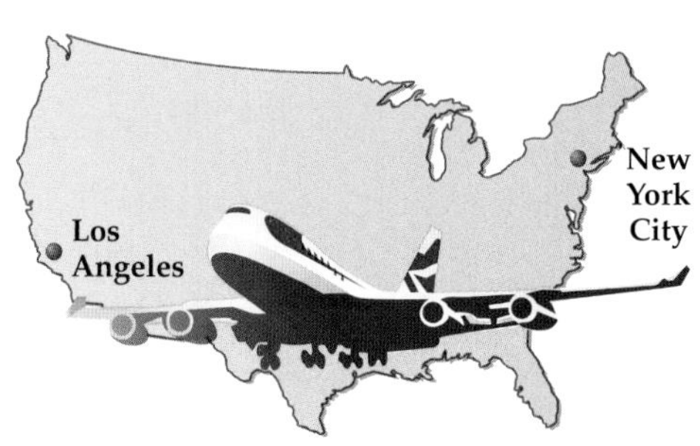

Think & Discuss

Which of the two distances described above is decreasing over time?

Think of other situations in which distance decreases over time.

Develop & Understand: A

On pages 390 and 391, Zach and Maya were walking from the left wall of a room to the right wall. You figured out how far each person was from the left wall at different points in time. Suppose instead you want to know how far the person is from the *right* wall at each point in time.

1. Is the person closest to the right wall at the beginning of the walk or at the end of the walk?
2. Suppose Maya walks at 1.5 meters per second across a room that is 10 meters wide. Copy and complete this table.

Maya's Walk

Time (seconds), t	0	1	2	3	4
Distance from Right Wall (meters), d	10				

3. Use the data in Exercise 2 to draw a graph that shows the relationship between Maya's distance from the right wall and time.
4. What is the slope of the line that you drew?
5. Use your graph to estimate when Maya would reach the right wall.
6. Explain how you can find the distance from the right wall if you know the time.
7. Write a symbolic rule that relates d to t.

Bianca and Lorenzo solved an equation on a quiz. Bianca wrote the rule $d = -2t + 20$. Lorenzo wrote the rule $d = 20 - 2t$. Can they both be right? Explain your thinking.

Create a problem that can be described by one or both of these rules.

Develop & Understand: B

Ruben and Kristen started walking away from a fence at the same time. Ruben walked at a brisk pace, and Kristen walked at a slow pace. They each measured the distance they had walked in 10 seconds. From this, they estimated how far from the fence they would have been at various times if they had continued walking. They drew distance-time graphs from their data.

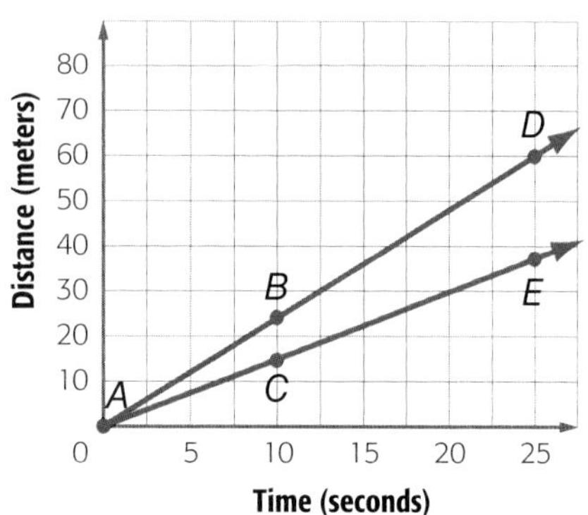

8. Which graph represents Ruben's walk, and which represents Kristen's? Explain how you know.

9. What events in the story above match points A, B, and C?

10. What do points D and E tell you about the positions of Ruben and Kristen?

11. Use the graphs to estimate each person's walking speed in meters per second. Give your answers to the nearest tenth.

12. Which line has the greater slope, Ruben's or Kristen's? Explain why.

13. What are the slopes of the two lines? How are they related to Ruben's and Kristen's speeds?

Share & Summarize

1. How are the graphs in Exercises 8–13 different from the graphs in Exercises 1–7?
2. How is the rule in Exercises 1–7 different from the rules in Exercises 8–13? How are they the same?
3. Explain how the differences in the rules relate to the differences in the graphs.

Investigation 3 Describe Graphs

Vocabulary

speed

velocity

Some rates vary. For example, if you count your pulse for one minute and then count it for another minute, you will probably get different results. It is normal for pulse rates to fluctuate, or change.

At least for a while, you would expect other rates to be fixed, or stay the same. For example, if your employer said your pay rate was $7 per hour, you would expect to earn that for each hour you work.

In this investigation, you will inspect the graphs below to find the directions, speeds, and relative locations of a group of cars along a particular highway.

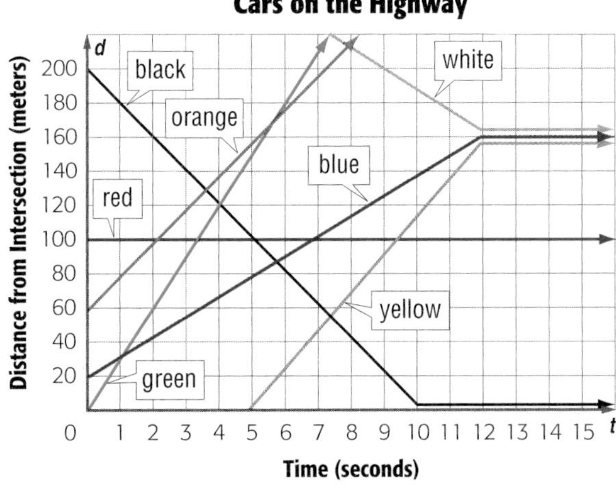

Develop & Understand: A

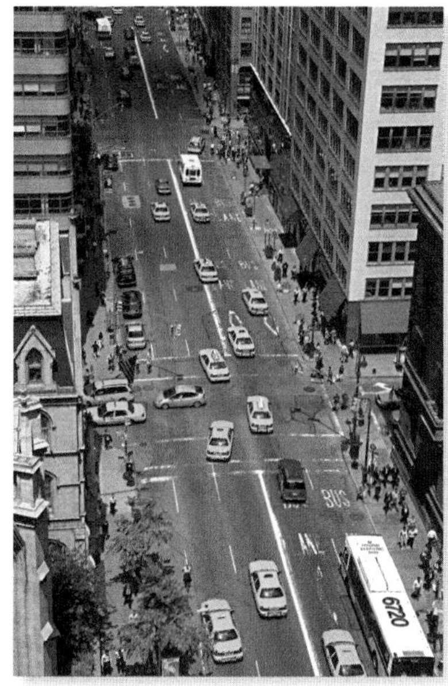

Seven cars are all near an intersection. The graphs on page 396 show the distances between the cars and the intersection as time passes.

In your group, study the graphs carefully. Then discuss the following questions. Record your group's decisions, and be prepared to talk about them in class.

1. In what direction is each car moving in relation to the intersection?
2. Compare the cars' speeds.
3. Do any of the cars stop during their trips? If so, which cars?
4. Prepare a group report for one of the cars. Imagine you are in that car, and give the highlights of your trip for these 15 seconds. Include such observations as where and when you started the trip and what you saw going on around you in front of the car, to the sides, and through the rearview mirror.

People often use one of two words when they describe how fast something is moving: *speed* or *velocity*. In fact, these two words mean different things mathematically.

Speed is always positive. It shows how fast an object is moving, but it does not reveal anything about the object's direction.

Velocity can be either positive or negative, as can slope. The sign of the velocity shows whether an object is moving from or toward a designated point. The absolute value of the velocity is the same as the speed. While the black car is moving, for example, it has a positive speed of 20 meters per second. However, its distance to the intersection is decreasing, so its velocity is -20 meters per second.

Share & Summarize

1. How can you determine from the graph whether a car is moving toward or away from the intersection?
2. How can you determine the speed of a car from the graph?
3. How can you determine from the graph whether a car is moving?

Investigation  Change the Starting Point

Vocabulary

***y*-intercept**

Materials

- graph paper

You will now explore a situation in which runners start a race at different points. You will see how rules and graphs show these differences.

Develop & Understand: A

Alita ran a race with her younger sister, Olivia. Alita let Olivia start 4 meters ahead of the starting line. Alita ran at a steady rate of 3 meters per second while Olivia ran at a steady rate of 2 meters per second.

Math Link

When you compare two things, look at both similarities and differences.

1. Make a table to show how many meters each sister was from the starting line at various times.
2. On the same grid, draw a graph for each sister showing the relationship between distance from the starting line and time. Compare the graphs.
3. Is Alita's distance from the starting line proportional to time? Is Olivia's distance proportional to time?
4. For each sister, write a rule in symbols to relate distance d and time t. How are the numbers in the rules reflected in the graphs?

Think & Discuss

In Chapter 3, you learned about elevations below sea level. For example, suppose Candace ended a hike at an elevation of -150 feet. The number -150 shows two things. First, the 150 tells Candace's distance from sea level. What does the negative sign show?

Describe some other situations in which you might use a negative sign along with a distance. Explain the meaning of the negative sign in each situation.

Develop & Understand: B

Five brothers ran a race. The twins began at the starting line. Their older brother began behind the starting line, and their two younger brothers began at different distances ahead of the starting line. Each boy ran at a fairly uniform speed. Here are rules for the relationship between distance d meters from the starting line and time t seconds for each boy.

Adam:	$d = 6t$
Brett:	$d = 4t + 7$
Caleb:	$d = 5t + 4$
David:	$d = 5t$
Eric:	$d = 7t - 5$

5. Which brothers are the twins? How do you know? Which brother is the oldest? How do you know?

6. For each brother, describe how far from the starting line he began the race and how fast he ran.

7. Which graph below represents which brother?

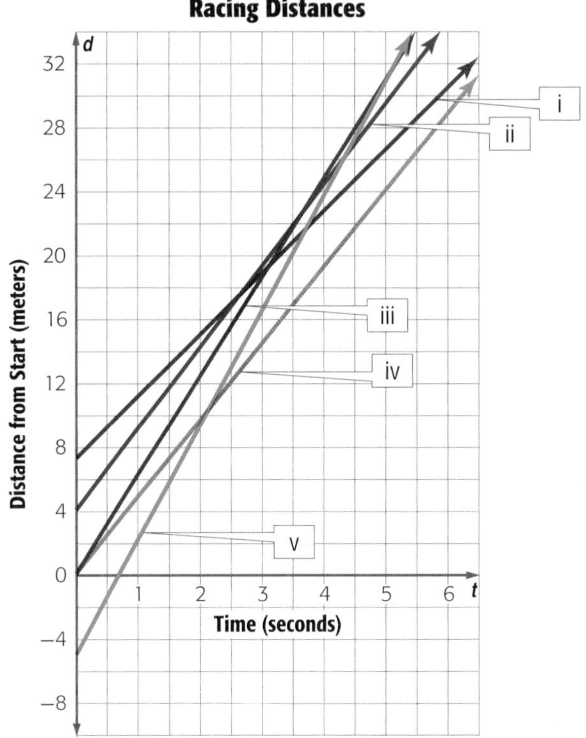

8. What events match the intersection points of the graphs?

9. Use the graphs to help you find the order of the brothers at the given times.

a. 2 seconds after the race began

b. 3 seconds after the race began

10. Which two brothers stay the same distance apart throughout the race? How do you know?

11. If the finish line was 30 meters from the starting line, who won?

12. Which brothers' rules are proportional? Which are not?

The rules for Exercises 5–12 have the form $d = rt + p$, with numbers in place of r and p. The value of r is the velocity, it gives the rate at which distance changes as time passes. The value of p tells the starting point. For example, Brett's rule is $d = 4t + 7$. His velocity is 4 meters per second. He began 7 meters ahead of the starting line.

If you graph $d = rt + p$, with t on the horizontal axis, or x-axis, and d on the vertical axis, or y-axis, the coordinates of the starting point on the graph are $(0, p)$. When $t = 0$, $d = p$. Therefore, p is the value of d at which the graph intersects the y-axis. That is why p is also called the ***y*-intercept**. For Brett's rule, the y-intercept is 7.

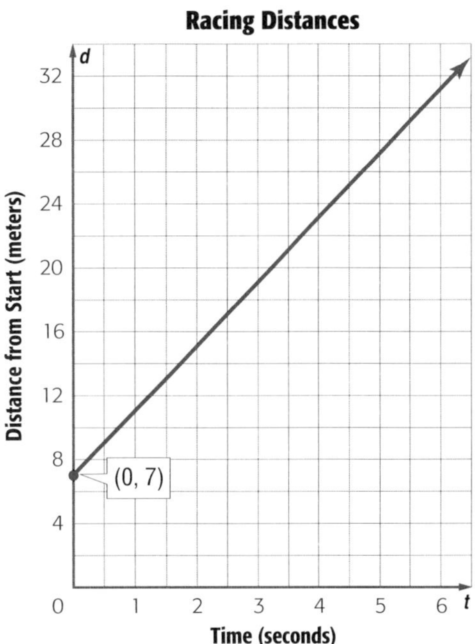

Share & Summarize

1. In Exercises 5–12, who began behind the starting line? Which part of the graph shows this? Which part of the rule shows this?
2. The brothers' dog, King, ran the race with them. His rule was $d = -15 - 2t$. What does the rule reveal about King's speed and starting position?
3. In what direction is King running?

On Your Own Exercises
Lesson 8.2

Practice & Apply

1. **Sports** Suppose you are riding a bicycle at a steady rate of 16 feet per second.

   a. Imagine that you rode at this pace for various lengths of time. Copy and complete the table.

Time (s)	0	10	20	30	35	40	50	55	60	70
Distance (ft)	0									

Math Link

1 mile = 5,280 feet

   b. Write a rule relating the number of feet you travel d to the time in seconds t. Use your rule to find how long it would take you to travel 900 feet.

   c. From your table, find how many feet you ride in 1 minute. About how many miles is this?

   d. Use the distance in feet that you travel in 1 minute to find how far you could ride in 1 hour if you kept up your pace. What is your traveling speed in miles per hour?

   e. Write a second rule that relates the number of miles you travel d to the time in hours t.

   f. Draw a graph for the distance in feet traveled in relation to time in seconds. Draw another graph for the distance in miles traveled in relation to time in hours.

   g. What is the slope of each line you graphed?

   h. Compare the two slopes. How can they be different if the speeds are the same?

2. A 90-minute cassette tape plays for 45 minutes per side. The tape is about 26 meters long and moves at a steady rate.

   a. Complete the table to show how many meters of tape play in given numbers of minutes.

Time (min), t	0	1	10	20	30	40	45
Length Played (m), L							26

   b. Write a rule relating length of tape played to time.

   c. What is the speed of the tape through the recorder in meters per minute? If you graphed the length of tape played in relation to time, what would be the slope of the line?

   d. Find the length of a 120-minute cassette tape, which has 60 minutes per side. Assume that tapes of different lengths run at the same speed.

3. **Sports** A marathon is a race of 26.2 miles. Nadia and Mark are running a marathon. Nadia's speed is 5.2 miles per hour, and Mark's is 3.8 miles per hour. Assume their speeds are steady throughout the race.

   a. Complete the table for Nadia and Mark. Distances are in miles.

Time (hours), t	0	1	2	3	4
Nadia's Distance from Start, N	0				
Mark's Distance from Start, M					
Distance between Nadia and Mark, d					

   b. On a single grid, draw graphs that show these three relationships.
      - time and Nadia's distance from start
      - time and Mark's distance from start
      - time and the distance between Nadia and Mark

   c. What is the slope of each line? Which line is the steepest? Which line is the least steep? Why?

   d. Write a rule for each of these three relationships.

   e. Think about the distance between Nadia and Mark. How is the rate this distance changes related to Nadia's speed and Mark's speed?

Real-World Link

The word *marathon* comes from the name of a famous plain in Greece. In 490 B.C., a runner was sent from this plain to Athens, which was about 25 miles away, to report the Athenians' victory in a battle against the Persians.

4. **Sports** Naia, Helena, Bryson, and Mark start their marathon run, which is 26.2 miles long. Nadia's speed is 5.2 miles per hour, Helena's is 4.1 miles per hour, Bryson's is 4.85 miles per hour, and Mark's is 3.8 miles per hour. Assume their speeds are steady throughout the run.

   a. Complete the table. Distances are in miles.

Marathon Results

Time (hours), t	0	1	2	3	4
Nadia's Distance from Finish, N	26.2				
Helenas Distance from Finish, H					
Bryson's Distance from Finish, B					
Mark's Distance from Finish, M					

b. On a single grid, draw graphs that show the relationships between time and distance from the finish for all four runners.

c. Compare the four graphs. In what ways are they similar, and in what ways are they different? What is the slope of each graph?

d. How much time will it take for each of the four runners to finish the marathon? You may want to extend your graph or table.

e. For each runner, write a rule in symbols that relates that runner's distance from finish (N, H, B, or M) to t.

5. Geography Soon after Trina's trans-Australia flight left Adelaide for Darwin, the pilot announced they were 200 kilometers from Adelaide and cruising at a speed of 840 kilometers per hour. The plane was flying north from Adelaide, toward the South Australia border. To pass the time, Trina decided to calculate her distance from Adelaide at various times after the pilot's message.

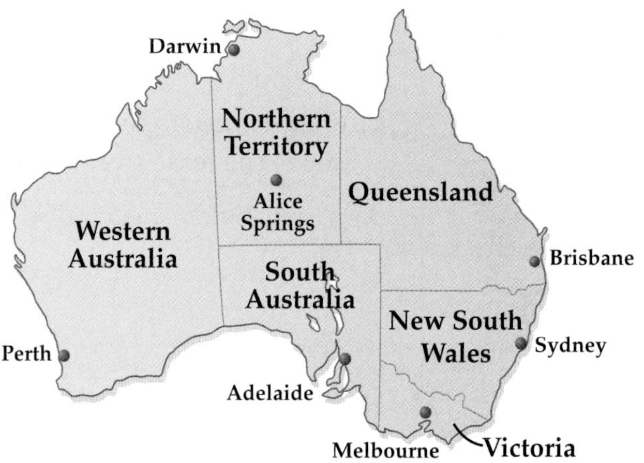

a. Complete the table to show the distance in kilometers from Adelaide at various times in minutes after the pilot spoke.

Trans-Australia Flight

Time after Pilot's Message (min), t	0	1	15	30	60	90	120
Distance from Adelaide (km), d	200						

b. Draw a graph of the data.

c. The South Australia border is 1,100 km from Adelaide. Use your graph to determine about how long after the pilot's message the plane crossed the border.

d. Alice Springs is 1,400 km from Adelaide. How long after the message would Trina expect to pass over it?

e. Explain in words how to find the distance from Adelaide if you know the time after the pilot's message. Write the rule in symbols. Explain how each number in your rule is shown in the graph.

6. One Sunday, Benito, Julie, Sook Leng, Edan, and Tia visited a park at different times. They rented bikes of different colors and rode them along the park's bike route. The bike route goes from the rental shop to a cafe. It also goes in the opposite direction, from the rental shop to a lake.

The graphs show the friends' locations from noon until 2 P.M.

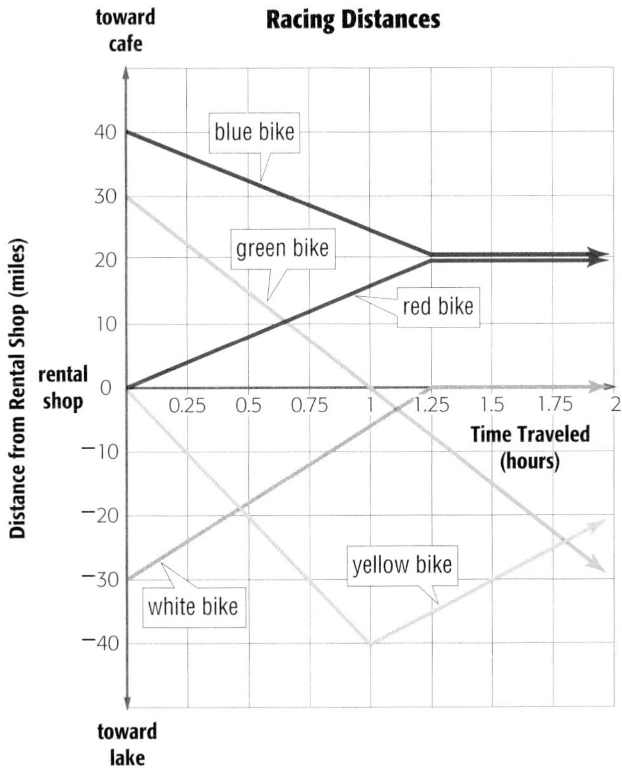

a. Describe the route taken by each bike. For example, the blue bike moves from the direction of the cafe toward the rental shop, and then stops.

b. Determine the speed of each bike. If a bike's speed changes, list all of its speeds.

c. Tia and Julie met about 20 miles from the rental shop and talked for an hour. Julie had come directly from the rental shop, and Tia had already visited the cafe and was returning her bike. What were the colors of their bikes?

d. From noon to 1 P.M., Benito had the highest speed. Sook Leng had the second highest, and Edan had the lowest. What color bikes were Benito, Sook Leng, and Edan riding?

e. Determine the slope of the line for each bike. If the slope changes, list all of them.

f. Are the bikes' speeds and the slopes of their lines always equal to each other? Why or why not?

For each equation, give the coordinates of the graph's intersection with the y-axis. In other words, give the y-intercept.

7. $y = 3x - 5$ **8.** $y = -7x + 14$ **9.** $y = 1.5x + 2.4$

10. $y = 27 - 9x$ **11.** $y = -35 + 3.1x$ **12.** $y = \frac{-3}{5} + \frac{3}{5}x$

Determine the slope of each line.

13. $y = 7x - 12$ **14.** $y = 31 - 4.6x$ **15.** $y = 8 + 3x$

16. $y = -x - 1$ **17.** $y = -0.21x + 98$ **18.** $y = \frac{1}{4}x - \frac{7}{2}$

19. In Your Own Words Explain how the speed in a distance-time situation is related to the slope of the graph for the situation.

20. Physical Science When you see lightning and then hear the clap of thunder, you are hearing what you have just seen. The thunder takes longer to reach you because sound travels much more slowly than light. Sound takes about 5 seconds to travel 1 mile, but light travels that distance almost instantly.

a. Complete the table to show the relationship between d distance in miles from a thunderstorm and t time in seconds between seeing the lightning and hearing the thunder.

Seconds between Lightning and Thunder, t	0	5	10	15	20	25
Distance from Storm (mi), d						

b. If thunder and lightning happen at the same time, where is the storm?

c. How could you calculate how far away a storm is? Write a rule that relates distance d from a storm to the time t between seeing it and hearing it.

Graph a line with the indicated slope.

21. 1

22. 12

23. -3

24. 0.37

25. -270

26. Astronomy *Pioneer 10* was launched on March 3, 1972. Designed to last at least 21 months, the spacecraft outlived and outperformed the fondest dreams of its creators. On April 25, 1983, *Pioneer 10* sent radio signals to Earth from Pluto, $4.58 \cdot 10^9$ km away.

a. Radio signals travel at the speed of light, about $3.00 \cdot 10^5$ km/s. How long did it take signals to reach Earth from Pluto?

b. Challenge What was *Pioneer 10*'s average speed, in kilometers per hour, kph, between March 3, 1972, and April 25, 1983?

In Exercises 27–31, do Parts a, b, and c for the given speed.

a. Express the speed in feet per minute.

b. Graph the distance moved over time, with time in minutes on the horizontal axis and distance in feet on the vertical axis.

c. Determine the slope of the graph.

Math Link

1 mile = 5,280 feet

27. 3 meters per minute; 1 foot is approximately 0.3 meter.

28. 15 meters per second

29. 60 kilometers per hour

30. 45 feet per second

31. 75 miles per hour

32. Write the speeds given in Exercises 27–31 from slowest to fastest.

33. Redraw the two graphs for Exercises 29 and 31 on a single grid.

a. What do you notice about the slopes of the two graphs? (Hint: Look at the angle each line makes with the horizontal axis.)

b. Explain your result in Part a.

34. A lion surveying his surroundings from a tall tree saw a horse half a kilometer to the south. He also spotted a giraffe 1 kilometer to the west. The lion jumped from the tree to chase one of these animals, and both the horse and the giraffe ran away from the lion with the maximum speed.

A lion can run 200 meters in 9 seconds, a horse can run 200 meters in 10 seconds, and a giraffe can run 200 meters in 14 seconds. The lion went after the animal that would take him less time to catch. After you solve this problem, you will know which animal he pursued.

a. Complete the table. For the distance between the lion and each of the other animals, assume the lion is running toward the animal. Distances are in meters.

Time (seconds)	0	50	100	150	200
Lion's Distance From Tree	0				
Horse's Distance From Tree	500				
Giraffe's Distance from Tree	1,000				
Distance between Lion and Horse	500				
Distance between Lion and Giraffe	1,000				

b. On a single set of axes, with time on the horizontal axis and distance on the vertical, draw these three graphs.

- the lion's distance from the tree
- the horse's distance from the tree
- the giraffe's distance from the tree

c. On another set of axes, draw these two graphs.

- the distance between the lion and the horse
- the distance between the lion and the giraffe

d. How are the five graphs similar, and how are they different?

e. What events match the points of intersection of the graphs for Part b?

f. Use the graphs to help you find how much time it would take the lion to catch the horse and how much time it would take the lion to catch the giraffe.

g. What are the slopes of the five lines? How are they different and why?

h. Write a rule in symbols for each line you graphed.

Mixed Review

35. Geometry Match each solid to its name. (Hint: For those of which you are unsure, think about what the term might mean.)

a. square pyramid

b. cone

c. cylinder

d. triangular prism

e. oblique prism

f. hexagonal prism

g. octahedron

h. tetrahedron

i. hemisphere

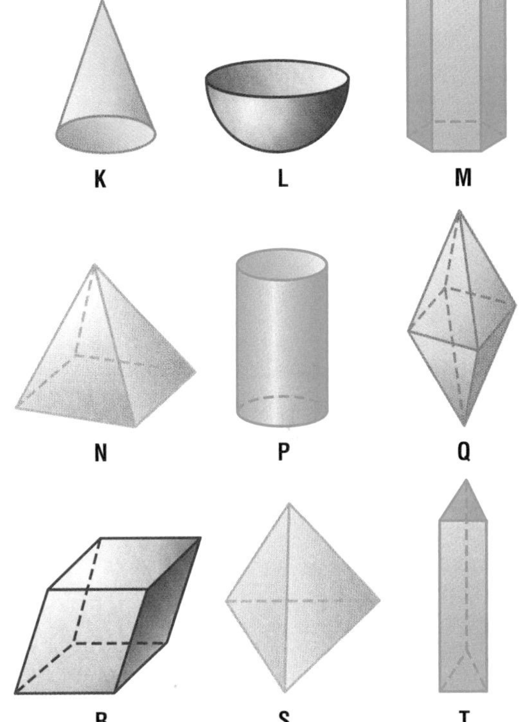

36. Mr. Ritter was curious about the copy machine in his office. He wanted to know if pressing the 200% button meant that the area of a picture would be doubled. What would you tell him?

37. Geometry A rectangular prism has a base 5 inches long and 2 inches wide. The prism is 6 inches tall.

a. What are the prism's volume and surface area?

b. Another block prism is four times the height of the original prism. What are its volume and surface area?

c. If a cylinder has the same volume as the original prism and the area of its base is 10 square inches, what is the cylinder's height?

Recognize Linear Relationships

You have learned about linear relationships between variables such as pay earned and time worked. In this lesson, you will look at both linear and nonlinear relationships, but it will not always be obvious which is which.

Inv 1	Explore and Describe Patterns	411
Inv 2	Graphs and Rules from Patterns	414
Inv 3	From Rules to Graphs	417
Inv 4	Secret Rules	420

Real-World Link

Tiles are typically made from clay that has been hardened by *firing*, or *baking*, at very high temperatures in a special oven called a *kiln*. A glaze might then be applied to the tiles.

Explore

Pat and Tillie are patio tilers who specialize in two-color rectangular patios. All of their patios are constructed from tiles measuring 1 foot by 1 foot. The border is one color, and the center is another color. The patio shown here is 9 feet by 6 feet.

Their Totally Square line of patios contains a variety of square patio designs. Below is a Totally Square patio measuring 6 feet by 6 feet.

Find a rule that tells how many border tiles will be used on a Totally Square patio of a specified size. Then find a rule that tells how many nonborder tiles there will be. Tell what each variable in your rules represents.

Pat and Tillie call their patio designs that are not square their ColorQuad line. Find a rule that expresses the number of border tiles in a ColorQuad patio with width w and length l. Then write a rule that expresses the number of nonborder tiles in a ColorQuad patio.

Decide whether each set of number pairs could describe a linear relationship.

9. (0, 20); (2, 0); (1, 10); (3, −10); (−1 , 30)

10. (−1, 21); (1, 25); (3, 29); (−3, 17); (0, 23)

Determine the slope and give the coordinates of the y-intercept for each relationship.

11. $y = 3x + 9$ **12.** $y = -15x + 39$ **13.** $y = x - 17$

Find a rule for each graph.

14.

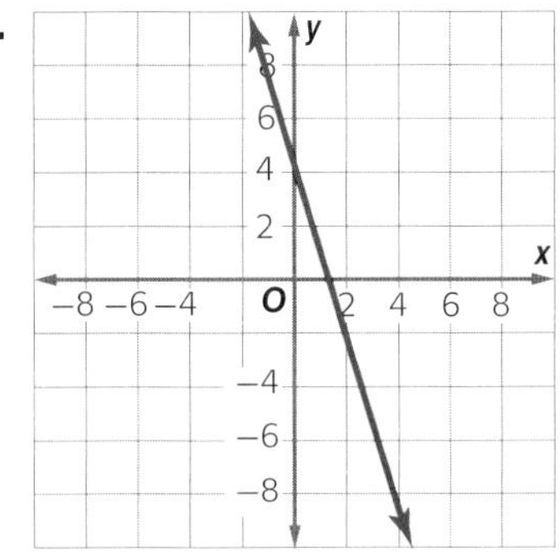

15.

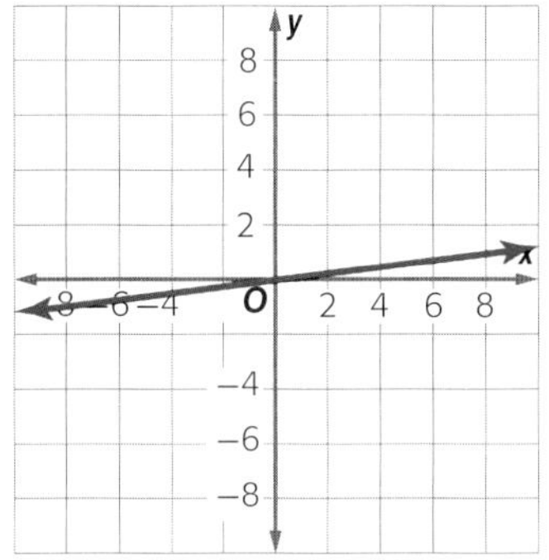

Test-Taking Practice

SHORT RESPONSE

1 Write a rule for the following pattern.

Show your work.

x	1	2	3	4	5
y	1	3	5	7	9

Answer ______________

MULTIPLE CHOICE

2 Determine the slope and give the coordinates of the y-intercept for $y = -5x - 12$.

F −5; (−12, 0)

G −5; (0, −12)

H 5; (0, 12)

J 12; (0, −5)

3 Which of the following is *not* proportional to $\frac{36}{45}$?

A $\frac{4}{5}$

B $\frac{12}{15}$

C $\frac{3}{12}$

D $\frac{32}{40}$

Equations

Real-Life Math

For Your Amusement Imagine that you and your friends are going to an amusement park. It has the most gravity-defying rides in the country.

Did you know that engineers use equations to design and build park rides? Equations are also used by park vendors to determine the cost of souvenirs and concessions.

Think About It Sandwiches cost \$5.00 for one or two for \$9.00. Lemonade costs \$3.50 for a small or \$5.50 for a large. How much would you spend to buy three sandwiches and one large lemonade? Write an equation that represents this total cost.

Contents in Brief

9.1 Find a Solution Method 436

9.2 A Model for Solving Equations 446

9.3 Solve Equations 460

9.4 Solve Equations with Parentheses 474

Review & Self-Assessment 489

Math Online

Take the **Chapter Readiness Quiz** at glencoe.com.

Dear Family,

The next chapter is about solving equations. The class will start by reviewing the backtracking and guess-check-and-improve methods of solving equations. Then a third, more efficient, solution method that involves doing the same thing to both sides of an equation will be learned. Combined with previously learned skills, your student will be able to solve a variety of equations, including equations with parentheses.

Key Concept–Solving Equations

$x + 5(x + 2) = 100$	
$x + 5x + 10 = 100$	Apply the Distributive Property.
$6x + 10 = 100$	Combine like terms.
$6x = 90$	Subtract 10 from both sides of the equation.
$x = 15$	Divide both sides of the equation by 6.

Finally, the class will use equations to solve real-world situations. For example, the equation $x + 5(x + 2) = 100$ could be used to model the following example.

Marla and her brothers, Pete and Max, collect souvenir postcards. Marla has two more postcards than Pete. Max has five times as many postcards as Marla. Pete and Max have a total of 100 postcards. How many postcards does Pete have?

Chapter Vocabulary

conjecture **model**

Home Activities

- Review the backtracking and guess-check-and-improve methods of equation solving.
- Write equations to represent daily situations, such as the cost of a family dining out for the evening.
- Look for examples of linear equations in your workplace. Work together to solve them.

The Steeplechase, Steeplechase Park, Coney Island, N. Y.
442

LESSON 9.1

Find a Solution Method

Inv 1 Choose a Solution Method 437

Inv 2 Use a Spreadsheet to Guess-Check-and-Improve 440

You already know the methods of solving equations: backtracking and guess-check-and-improve. *Backtracking,* which you used in Chapter 1, can help you solve many equations. For example, consider this equation.

$$\frac{5b+3}{3} = 6$$

To find the solution, first set up a flowchart for the expression $\frac{5b+3}{3}$. The flowchart shows the steps for calculating the value of the expression, the *output,* for any *input b.*

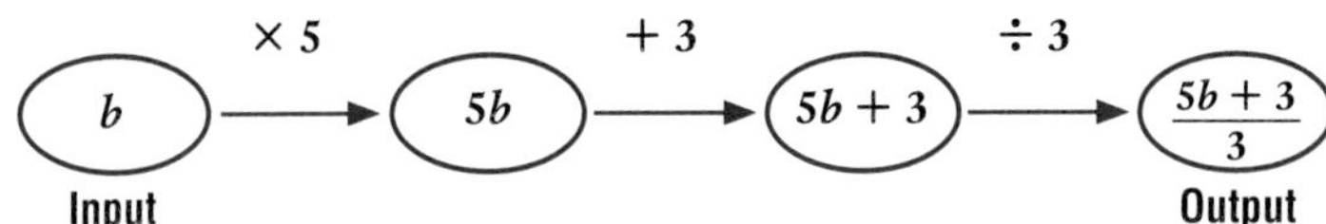

To solve $\frac{5b+3}{3} = 6$, you need to find the input that gives an output of 6. Draw the flowchart again, this time entering 6 as the output and leaving the other ovals blank.

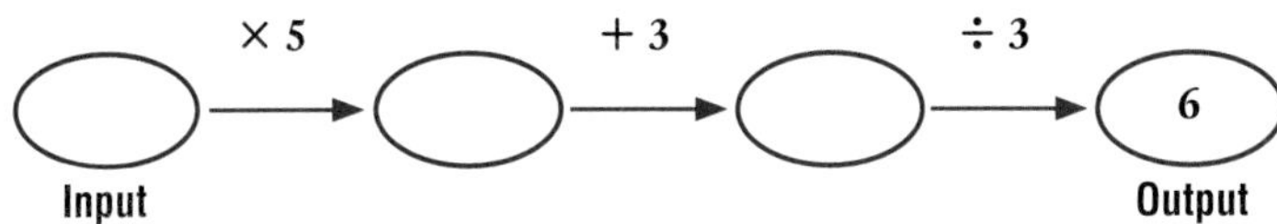

Now you can work backward, step by step, from the output to find the input. First, ask yourself, "What number divided by 3 gives 6?" Write that number in the oval to the left of the 6.

Real-World Link

Flowcharts are used in many professions. Builders use them to specify the steps in a home's construction, engineers to diagram how a part travels through the manufacturing process, and electricians to show how current flows in a circuit.

Think & Discuss

Continue backtracking from right to left to find the values for each oval in the flowchart. The number in leftmost oval is the solution of the equation.

What is the solution? Check your answer by substituting it into the equation.

You can use the *guess-check-and-improve* method to solve any equation, even equations that are difficult to solve by backtracking. To use this method, *guess* what the solution might be and then *check* your guess by substituting it into the equation. If your guess is incorrect, guess again. By comparing the results of each substitution, you can *improve* your guesses and get closer and closer to a solution.

Think & Discuss

Consider the equation $3b + 5 = b + 17$.

Suppose you begin with 1 as a guess for the solution. Substituting 1 for b gives 8 for the left side of the equation and 18 for the right side. You might enter these results in a table.

Guess	$3b + 5$	$b + 17$
1	8	18

Try 4 as your second guess. Substitute 4 for b in both sides of the equation. Record your answers in a copy of the table.

For a guess of 1, the difference between the two sides of the equation is 10. Are the two sides closer together or further apart when you substitute 4?

If 4 is not the solution, continue to adjust your guess until the two sides of the equation have the same value. What is the solution? Explain.

Investigation 1 Choose a Solution Method

Backtracking is sometimes much easier or more efficient than guess-check-and-improve. You will now practice using these two methods to solve equations. As you work on the exercises, think about how you could decide which method is easier for a particular equation.

Develop & Understand: A

Use backtracking to solve each equation. Check your solutions by substituting them into the original equations.

1. $3f - 2 = 13$
2. $3(2g + 12) = 51$
3. $\frac{2a + 1}{3} = 5$
4. $2(\frac{3n + 4}{5} - 1) = 6$

Use guess-check-and-improve to solve each equation.

5. $u + 4 = 4u - 8$
6. $2q - 2 = 4q - 7$
7. $m(m - 6) = 91$ (There are two solutions. Try to find both.)

For each equation in Exercises 8–11, discuss with your partner which solution method, backtracking or guess-check-and-improve, would be easier or more efficient. Then solve the equations.

8. $\frac{5k - 4}{3} = 7$

9. $\frac{3n + 5}{2} = \frac{5n + 3}{3}$

10. $5B + 2(B + 1) = 16$

11. $4j + 74 = 478$

12. Look at the equations in Exercises 8–11. Are any of them difficult to solve by backtracking? If so, explain why backtracking is difficult to use for these equations.

Solve each equation using any method you like. Check your solutions.

13. $4(5g - 1) = 56$

14. $\frac{3d + 1}{3} = 6d - 3$

15. $35 - 2s = 11$

16. $4x + 3 = -3x - 4$

Algebra is a powerful tool for solving real and important situations. The exercises that follow are generally much easier to solve than real situations, but they will help you develop needed algebraic skills.

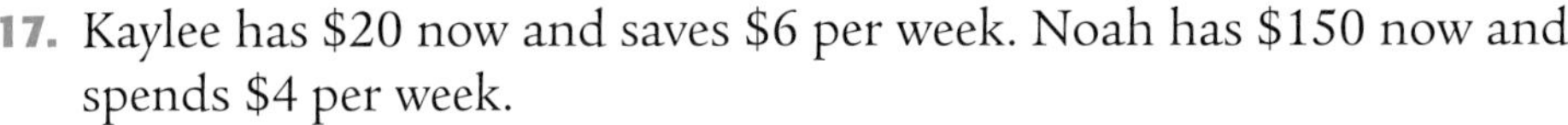

Develop & Understand: B

17. Kaylee has \$20 now and saves \$6 per week. Noah has \$150 now and spends \$4 per week.
   a. How much money will Kaylee have *n* weeks from now?
   b. How much money will Noah have *n* weeks from now?
   c. Write an equation that represents Kaylee and Noah having the same amount of money after *n* weeks.
   d. Solve your equation for *n*. What does the solution tell you about this situation?
   e. How can you check your solution?

18. Inez's family is purchasing a new stereo system from Sharon's Sound Shop. The salesperson offers a choice of payment plans.

Plan A: Make a \$100 down payment and then pay \$35 per month for 24 months.

Plan B: Make no down payment and pay \$40 per month for 24 months.

a. Which plan is more expensive? Explain.

b. For each plan, write an expression for the amount Inez's family will have paid after n months.

c. After how many months will the family have paid the same amount no matter which payment plan is selected?

In Exercises 19–23, write an equation that fits the situation. Solve the equation using whatever method you like. Check your solution.

Math Link

The sum of the measures of the angles in any triangle is 180°.

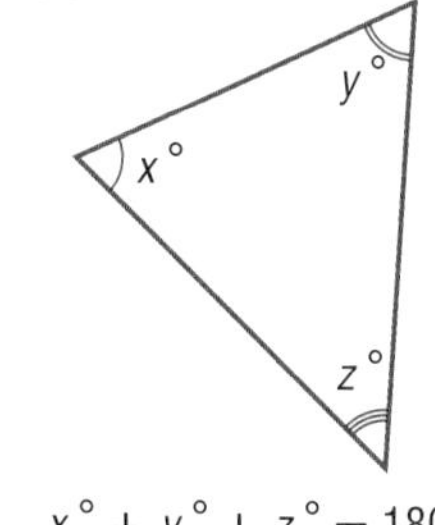

$x° + y° + z° = 180°$

19. Finding 5 more than a given number gives the same result as 3 times the sum of that number and 1. What is the number?

20. One angle of a triangle measures f degrees. One of the other angles of the triangle is twice this size, and the third angle is three times this size. Find the measures of all three angles.

21. A rectangular field is three times as long as it is wide. The fence around the perimeter of the field is 800 meters long. How wide is the field?

22. An apartment building has a TV antenna on its roof. The top of the antenna is 66 meters above the ground. The building is ten times as tall as the antenna. How tall is the antenna?

23. In a class of 25 students, there are seven more girls than boys. How many boys are in the class?

Share & Summarize

1. Write an equation that *cannot* be solved easily by backtracking. Explain why backtracking would be difficult.

2. Together, Hally and Trey have \$14. If Hally had \$1 more, she would have twice as much money as Trey.

   a. Write and solve an equation to find how much money Hally and Trey each have.

   b. Explain how you wrote your equation and how you decided which method to use to solve it.

Inquiry

Investigation 2 Use a Spreadsheet to Guess-Check-and-Improve

Materials

- computer with spreadsheet software (1 per group)

Guess-check-and-improve is a great method for solving simple equations, but it can be tedious when the equations are more complicated. The process is more efficient when you have a computer do the calculations while you do the thinking.

Sonia decided to use a spreadsheet to help her solve the following equation using guess-check-and-improve.

$$\frac{3n + 5}{2} = \frac{5n + 3}{3}$$

Setting Up the Spreadsheet

First, Sonia entered headings and formulas in the top two rows of the first four columns of her spreadsheet.

Guess-Check-and-Improve.xls

	A	B	C	D
1	Test Variable *n*	Left Side (3*n* + 5)/2	Right Side (5*n* + 3)/3	Difference
2		= (3 * A2 + 5)/2	= (5 * A2 + 3)/3	= B2 − C2
3				

Sheet 1 Sheet 2 Sheet 3

As Sonia entered guesses for *n* into her spreadsheet, she observed how the results changed so she could make a good choice for her next guess. When her guess is correct, the two sides of the equation will have the same value.

1. The formula in Cell D2 calculates the difference between the values in Cells B2 and C2. How do you think Sonia will use the difference to help make her next guess?

Sonia started guessing the solution by entering 0 in Cell A2. The following are her results.

Guess-Check-and-Improve.xls

	A	B	C	D
1	Test Variable n	Left Side $(3n + 5)/2$	Right Side $(5n + 3)/3$	Difference
2	0	2.500	1.000	1.500

Sheet 1 Sheet 2 Sheet 3

Sonia formatted the numbers in Cells B2, C2, and D2 to show three decimal places.

She decided to try 1 for her next guess. So, she entered 1 in Cell A2.

Guess-Check-and-Improve.xls

	A	B	C	D
1	Test Variable n	Left Side $(3n + 5)/2$	Right Side $(5n + 3)/3$	Difference
2	1	4.000	2.667	1.333
3				

Sheet 1 Sheet 2 Sheet 3

Try It Out

2. Set up a spreadsheet like Sonia's on your computer. Enter Sonia's first two guesses to make sure your spreadsheet works.

3. Use your spreadsheet to solve Sonia's equation. How do you know when you have found the solution?

4. Use your spreadsheet to solve each equation below. You will have to change the formulas in Cells B2 and C2 for each equation. Pay attention to the strategies you use to improve your guesses. You will be asked about them later.

a. $4S - 70 = 18S$ **b.** $4n + 1 = \frac{3n + 8}{2}$

5. What strategies did you use to help you improve your guesses?

Math Link

Answers may be positive or negative, and they may be whole numbers or decimals.

Go on 

Try It Again

Sonia's friend Teresa solved the equation by making a spreadsheet to try a series of 20 guesses at once. She started by setting up a spreadsheet like Sonia's. Teresa then used the spreadsheet's "fill down" command to copy the formulas in Cells B2, C2, and D2 into Rows 3–21 of Columns B, C, and D.

To make a series of guesses, Teresa modified Column A. She entered 0 in Cell A2 and the formula A2 + 1 in Cell A3. Then she copied this formula into Cells A4 to A21. The program created a column of numbers that increased by 1, from 0 in Cell A2, to 19 in Cell A21.

Guess-Check-and-Improve.xls

◇	A	B	C	D
1	Test Variable n	Left Side $(3n + 5)/2$	Right Side $(5n + 3)/3$	Difference
2	0	= (3 * A2 + 5)/2	= (5 * A2 + 3)/3	= B2 − C2
3	= A2 + 1	= (3 * A3 + 5)/2	= (5 * A3 + 3)/3	= B3 − C3
4	= A3 + 1	= (3 * A4 + 5)/2	= (5 * A4 + 3)/3	= B4 − C4

Sheet 1 | Sheet 2 | Sheet 3

Real-World Link

Spreadsheets range from quite simple to extremely complex. A family might use a spreadsheet to plan its monthly budget and a corporation might use one to help with financial planning.

6. Try this on your spreadsheet. How will Teresa know which value of n gives a correct answer? Where is the solution on Teresa's spreadsheet?

Use Your Spreadsheet

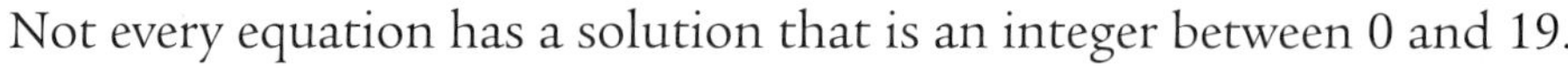

Not every equation has a solution that is an integer between 0 and 19.

7. Try to solve the equation $\frac{8x + 3}{3} = x - 3$ using Teresa's method. Make your first series of guesses the set of integers from 0 to 19.

- **a.** Notice that the differences in Column D increase as the values in Column A increase. These guesses do not get closer to a solution. How could you modify Teresa's strategy to get closer?
- **b.** Modify the formula in Column A to match your new strategy. Does 0 ever appear in Column D? If not, where does it look like the solution to this equation should be?
- **c.** Refine your guesses, if needed, and solve the equation.

Use a spreadsheet to solve the equations on page 443. Some solutions may not be positive and may not be integers. If you use Teresa's method, you may need to try more than one series of test values. Pay attention to the strategies you use to improve your guesses.

8. $15x + 32 = 3x - 16$

9. $\frac{7P + 34}{3} = \frac{13P - 4}{2}$

10. Choose one of Exercises 8–11 on page 436. Use a spreadsheet to solve the equation.

What Did You Learn?

11. When you solve equations with a spreadsheet, what strategies can you use to improve your guesses? Be sure to consider how to find negative and decimal solutions as well as positive integers.

12. Analyze this spreadsheet.

Guess-Check-and-Improve.xls

◇	A	B	C	D
1	Test Variable n	Left Side $4n + 2$	Right Side $7n + 12$	Difference
2		$= 4 * A2 + 2$	$= 7 * A2 + 12$	$= B2 - C2$

Sheet 1 | Sheet 2 | Sheet 3

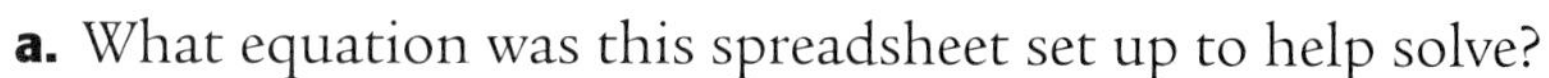

a. What equation was this spreadsheet set up to help solve?

b. What do the formulas in Cells B2, C2, and D2 mean?

c. What results will you get if you enter 1 in Cell A2?

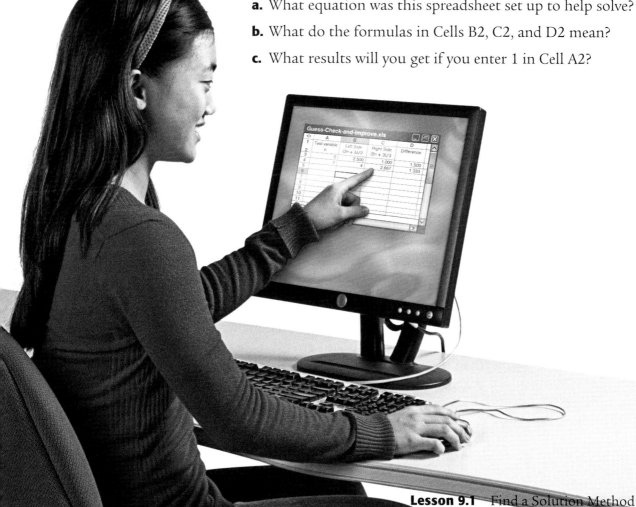

On Your Own Exercises

Lesson 9.1

Practice & Apply

Use backtracking to solve each equation.

1. $4d + 7 = -15$

2. $4(2w + 1) = 52$

3. $\frac{4x + 7}{5} = 3$

4. $3(\frac{2c + 1}{5} + 1) = 6$

Use guess-check-and-improve to solve each equation.

5. $b + 5 = 3b - 9$

6. $4t - 3 = 3t + 7$

Which would be a better way to solve each equation, backtracking or guess-check-and-improve? Use the method you think is best to find each solution.

7. $\frac{6y - 7}{3} = 5$

8. $4u + 7 = 5 + 3u$

9. $5t + 34 = 17$

10. Marcus has 20 comic books and gives 3 to his cousin Lelia every week. Mike has 10 comic books and buys 2 new ones each week.

- **a.** How many comic books will Marcus have after w weeks?
- **b.** How many comic books will Mike have after w weeks?
- **c.** Write an equation that represents Marcus and Mike having the same number of comic books after w weeks.
- **d.** Solve your equation to find after how many weeks the two boys will have the same number of comic books. Check your solution.

11. Kiran said, "Eleven more than my number is the same as 18 more than twice my number." What is Kiran's number?

12. In a box of 64 red and blue whistles, there are 16 more blue whistles than red whistles. Write and solve an equation to determine how many are red.

13. Solve this equation by guess-check-and-improve. There are two solutions. Find both.

$$m(m + 2) = 168$$

Connect & Extend

14. Preview Charles was thinking about how to solve the equation $d(d - 2) = 46$. He said, "This seems simple. I want to know two numbers that multiply to give 46. One of the numbers is 2 less than the other. I'll use 8 as my first guess. Since 6 is two less than 8, that's $8 \cdot 6 = 48$. So, 8 is not the solution."

- **a.** What should Charles choose as his next guess?
- **b.** Use guess-check-and-improve to find a solution for d to the nearest hundredth that gives a result within 0.1 of 46.

Real-World Link
The first drive-in theater opened June 6, 1933, in Camden, New Jersey. The idea was a success with close to 5,000 drive-ins operating by 1958. As the car became a part of daily life, the thrill of sitting in one to watch a movie declined. In 2007, a little more than 500 drive-ins remained in the United States.

15. At a drive-in theater, admission is \$4.00 per vehicle plus \$1.50 per person in the car.

    a. Write an equation. Use a to represent the total admission, v for the number of vehicles, and p for the number of people.

    b. Use your equation to determine how much money the theater will collect if 43 cars carrying a total of 93 people see a movie.

    c. Find the admission price for a car holding three people.

    d. The admission for one van was \$16. Use your equation to find how many people were in the van.

16. **Preview** Try to solve this equation using any method you choose. What do you discover?

$$n^2 + 3 = 2$$

17. Jim and Lillian were playing *Think of a Number.* Jim said, "Think of a number. Triple it. Add 5. Multiply your result by 10." Lillian said that her answer was 320.

    a. What equation could Jim solve to find Lillian's number?

    b. Use backtracking to solve the equation.

18. **In Your Own Words** Describe the types of equations that are better solved by backtracking, and the types that are better solved by guess-check-and-improve. Explain your reasoning.

Mixed Review

Determine the slope for each relationship.

19. $y = 2x + 1.7$

20. $y = -x - 1.2$

21. $2y = 6x + 9$

Geometry Find the value of the variable in each drawing.

22.

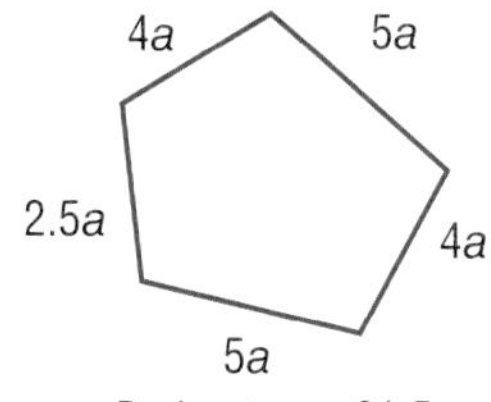

Perimeter = 61.5

23.

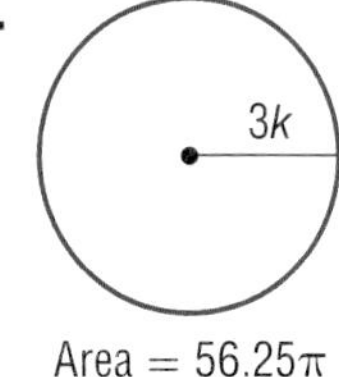

Area = 56.25π

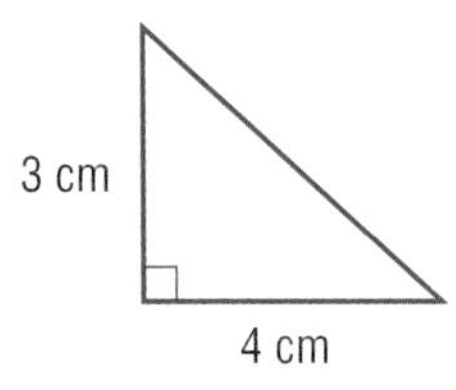

24. **Geometry** Consider the triangle at left.

    a. What is the area of the triangle?

    b. Suppose you use this triangle as the base of a prism 1 cm high. What would be the prism's volume?

    c. Suppose you use the triangle to make a prism h cm high. What would be the prism's volume?

A Model for Solving Equations

Inv 1 Balance Puzzles 447

Inv 2 Keep Things Balanced 449

Inv 3 Solve Problems with Balance Puzzles 453

Vocabulary

model

In this lesson and the next, you will learn an equation-solving method that is more efficient than guess-check-and-improve. This new method can often be used when backtracking will not work.

First, you will learn to use a model to help you think about solving equations. The word *model* has several meanings. A model can be something that *looks* like something else. For example, you have probably built or seen model cars or airplanes that look just like the real vehicles with all the parts of the model proportional to the actual parts.

In mathematics, a **model** is something that has some of the *key characteristics* of something else. In Chapter 1, you used bags and blocks as a model for algebraic expressions. Just as you can put any number of blocks into a bag, you can substitute any value for the variable in an expression.

Mathematical and scientific models help people understand complex ideas by simplifying them or making them easier to visualize. In this lesson, bags and blocks are again useful as models for algebraic expressions. By adding a balance, you can model algebraic *equations* and find more efficient ways to solve them.

Explore

On this balance, each bag contains the same number of blocks. The bags weigh almost nothing compared to the blocks.

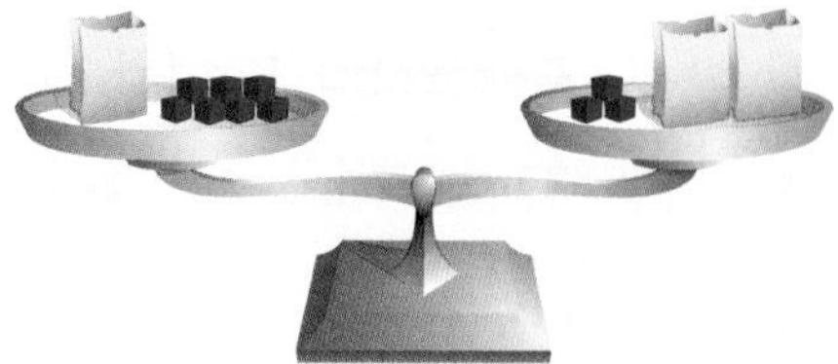

How many blocks are in each bag? Explain how you found your answer.

Check to make sure your answer is a solution to the puzzle.

Investigation 1 Balance Puzzles

Now it is your turn to draw some balance puzzles to try with your classmates. Here are the rules for creating balance puzzles.

- In any puzzle, all bags must hold the same number of blocks.
- The two sides of the balance must have the same total number of blocks with different combinations of bags and blocks on each side.

Develop & Understand: A

1. Work by yourself to draw a puzzle with no more than 10 blocks on each side of the balance, including the blocks hidden in the bags. Exchange puzzles with your partner. Solve your partner's puzzle.

2. Now draw a more difficult puzzle, using a total of 12 or more blocks on each side. Exchange puzzles with your partner. Solve your partner's puzzle.

As you work with the balance puzzles in Exercises 3–7, try to think of more than one strategy for solving each puzzle.

Develop & Understand: B

3. Melina and Sancho challenged their classmates to solve this balance puzzle.

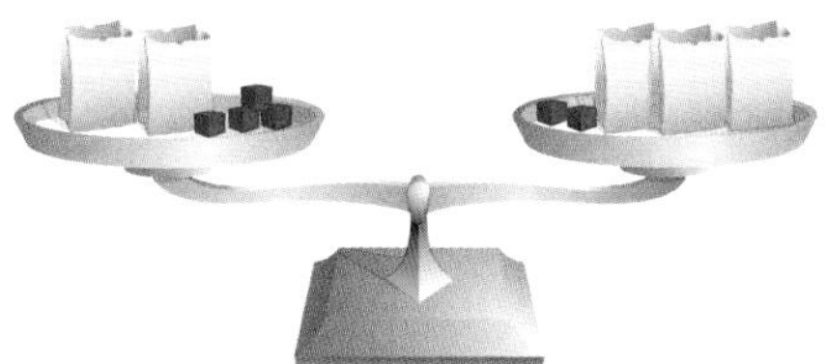

   a. How many blocks are in each bag? Check your solution.

   b. What strategy did you use to solve the puzzle?

Melina and Sancho's classmates created similar puzzles. Solve each puzzle by finding how many blocks are in each bag.

4. Tanya and Craig's puzzle

5. Margaret and Aida's puzzle

6. Uma and Ted's puzzle

7. Will and Trent's puzzle

Share & Summarize

Select one of the balance puzzles from Exercises 4–7. Explain your solution method so that someone in another class could understand it. Be sure to justify your reasoning.

Investigation  Keep Things Balanced

By now, you have probably seen and used a variety of strategies to solve exercises about balance. In this investigation, you will focus on solving puzzles by "keeping things balanced." Soon, you will see how the thinking involved in this strategy can help you solve equations.

Example

Here is how Malik solved Exercise 7 on page 448.

I'm trying to keep the pans balanced. If I remove the same thing from each side, what is left will still balance. I could remove a bag from each side and have this.

I could then take 2 more bags and a block from each side, like this.

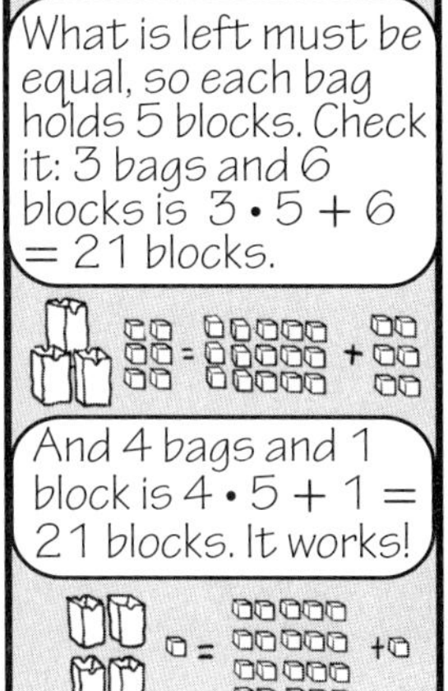

What is left must be equal, so each bag holds 5 blocks. Check it: 3 bags and 6 blocks is 3 • 5 + 6 = 21 blocks.
And 4 bags and 1 block is 4 • 5 + 1 = 21 blocks. It works!

Malik kept things balanced by taking the same number of bags or blocks from each side until he had only 1 bag on the right and 5 blocks on the left.

Develop & Understand: A

Solve each puzzle by keeping things balanced. For each step, record what you do and what is left on each side of the balance. You may draw each step if you prefer. Be sure to check your solutions.

1. 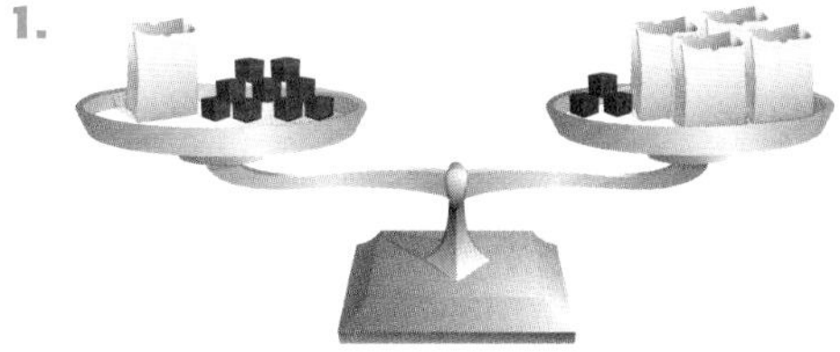

2. 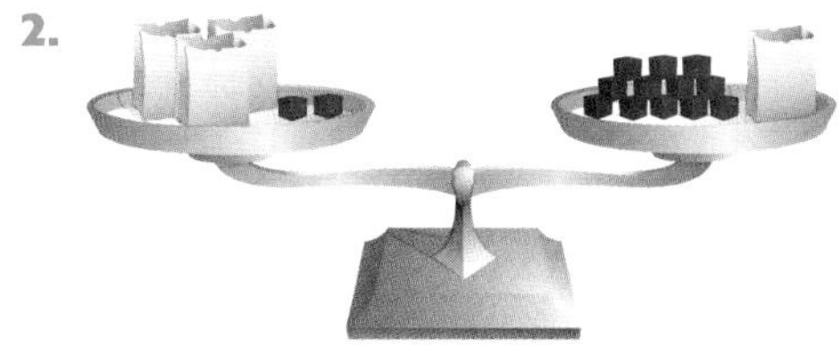

In Chapter 1, you wrote algebraic *expressions* to represent situations involving bags and blocks. In the same way, you can write algebraic *equations* to describe balance puzzles.

Example

Here is how Meg made a balance puzzle.

On this side, Meg put 2 bags of blocks and 5 single blocks. If n stands for the number of blocks in each bag, then the number of blocks in "two bags and five blocks" is $2n + 5$.

On this side, Meg put 3 bags of blocks and 2 single blocks. The number of blocks in "three bags and two blocks" is $3n + 2$.

Since the sides balance, the number of blocks on the left, $2n + 5$, is equal to the number of blocks on the right, $3n + 2$. This can be expressed with the following equation.

$$2n + 5 = 3n + 2$$

When you find the number of blocks in each bag, you have found the value of n.

Develop & Understand: B

3. Look at Meg's balance.

   a. Find the number of blocks in each bag.

   b. Check that your answer is a solution of the equation $2n + 5 = 3n + 2$.

Practice & Apply

The students in Ms. Avila's class made some balance puzzles. In Exercises 1–4, tell how many blocks are in each bag. Remember, in a balance puzzle, every bag holds the same number of blocks.

1. Carla and Toshiro's puzzle

2. Alberto and Elena's puzzle

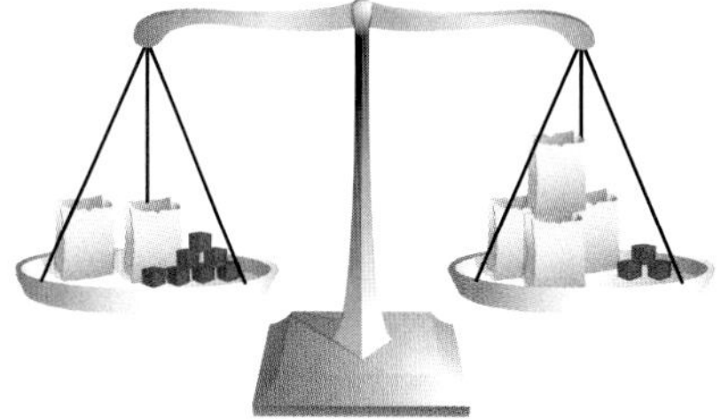

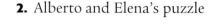

3. Benita and Ayana's puzzle

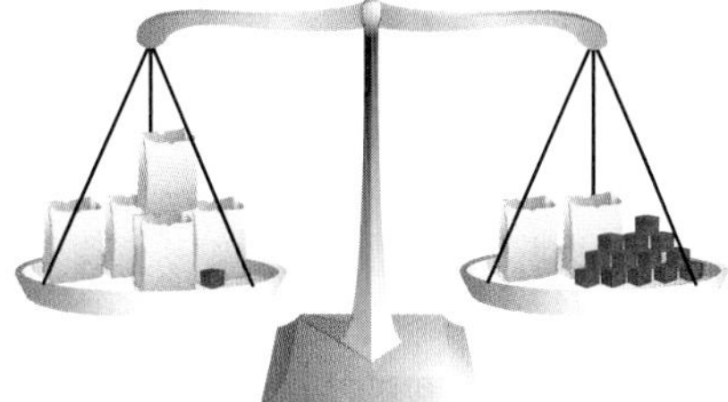

4. Maggie and Jonas' puzzle

5. Make up three balance puzzles with 11 blocks total on each side. Write an equation for each puzzle, and solve each equation.

6. Raphael and Gary made this balance puzzle. How many blocks are in each bag? Write down each step you take to solve this puzzle.

7. Consider this balance puzzle.

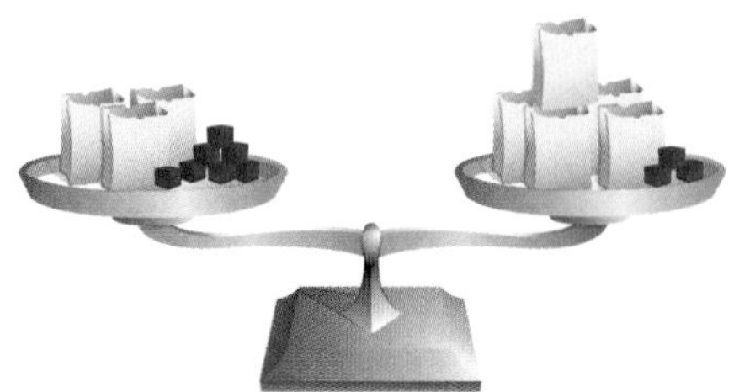

a. Write an equation to fit this puzzle. Let n represent the number of blocks in each bag.

b. Use the drawing to find the value of n.

c. Check that your answer to Part b is a solution of your equation.

8. Consider the equation $3p + 10 = 5p + 3$.

a. Draw a balance puzzle to represent the equation. Explain how you know your puzzle matches the equation.

b. Use your drawing to solve the equation. Check your solution by substituting it into the equation.

9. Nicky said, "The sum of my number and 21 is the same as 4 times my number. What is my number?"

a. Use N to represent Nicky's number. Write an equation you could solve to find the value of N.

b. Solve your equation by imagining a balance puzzle. Check your solution.

Connect & Extend

10. Mass The balances below each hold different sets of blocks, jacks, and marbles. Each marble has a mass of ten grams.

a. Find the mass of each jack and the mass of each block. Describe the steps you take.

b. This balance has blocks, marbles, and jacks from Part a, plus some toothpicks. Find the mass of each toothpick.

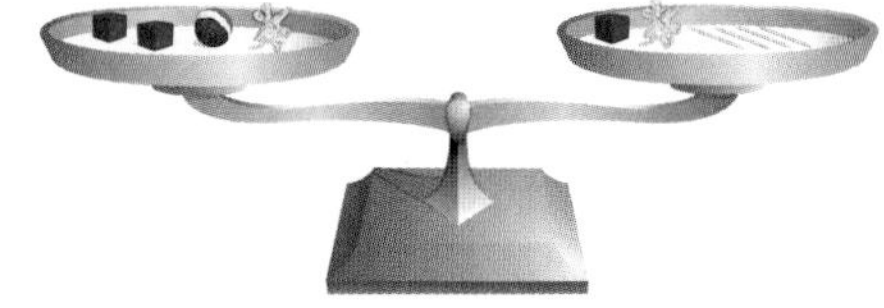

11. Sareeta and her brother Rashid put stickers in bags to give away at Sareeta's birthday party. They started with an equal number of stickers and put the same number in each bag. Sareeta filled five bags and had one sticker left. Rashid filled four bags and had seven stickers left.

    a. Let n represent the number of stickers in a bag. Write an expression that describes the way Sareeta distributed her stickers.

    b. Write an expression that describes Rashid's four bags and seven extra stickers.

    c. Because Sareeta and Rashid started with the same number of stickers, it is possible to balance their distributions of stickers. Draw a balance puzzle to show this.

    d. Write an equation that matches your balance puzzle.

    e. Use your balance puzzle to find how many stickers are in each bag.

    f. Show that your answer to Part e is a solution of the equation.

12. Read the story about Sareeta and Rashid in Exercise 11.

    a. Create a similar story that can be represented by the equation $3n + 2 = 2n + 9$.

    b. Draw a balance puzzle for this equation, and find the value of n.

    c. Show that your answer to Part b is a solution of the equation.

13. Consider the equation $2v + 50 = 7v + 30$.

    a. Describe a situation that matches this equation.

    b. Find the value of v. Check your solution.

    c. Explain what your solution means in terms of the situation you described in Part a.

14. Patterns Study the pattern of shapes.

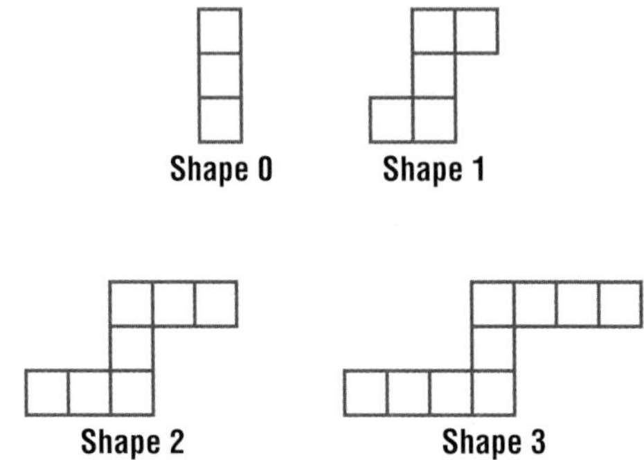

a. How many squares will be in shape 4? In shape 5?

b. Write an expression for the number of squares in shape *n*.

c. How many squares will be in shape 20?

d. Write an equation to find which shape will have 25 squares. Solve your equation.

15. Challenge Sergei and Hilary were cutting streamers of a particular length for a party. Sergei held up a strip of crepe paper and said he could cut one streamer from it and have three feet left. Hilary said that if the strip of crepe paper was one foot longer, she could cut exactly two streamers from it.

a. Find the length of the streamers cut by Sergei and Hilary. Show how you found your answer.

b. Find the length of Sergei's strip of crepe paper.

16. In Your Own Words Susan said, "When I think about solving equations using a balance puzzle, I think about doing the same thing on each side." What do you think she means?

Mixed Review

17. Probability Kayla fills a jar with 100 chips, some blue and some orange. She asks Ignacio to guess how many of each color are in the jar.

a. Ignacio reaches in without looking and removes ten chips, four blue and six orange. What reasonable guess might he make for the number of chips of each color in the jar?

b. Ignacio takes out five more chips, and all are orange. What reasonable guess might he now make for the number of chips of each color in the jar?

c. Kayla tells Ignacio there are actually three times as many orange chips as blue chips in the jar. How many of each are there?

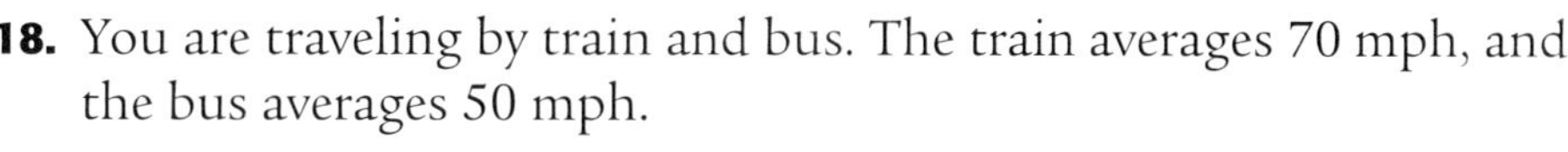

18. You are traveling by train and bus. The train averages 70 mph, and the bus averages 50 mph.

a. Let t and b represent how many hours you ride on the train and the bus, respectively. Write an expression for the total miles you travel on your trip.

b. Suppose the total distance you travel is 600 miles. Use your answer from Part a to express this as an equation.

c. If you travel 600 miles and spend five hours on the train, how many hours was the bus ride?

Find each sum or difference.

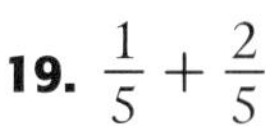

19. $\frac{1}{5} + \frac{2}{5}$ **20.** $\frac{1}{4} + \frac{3}{4}$ **21.** $\frac{7}{8} - \frac{6}{8}$

22. $\frac{3}{7} + \frac{1}{6}$ **23.** $\frac{2}{3} + \frac{5}{8}$ **24.** $\frac{3}{4} - \frac{2}{3}$

25. $\frac{1}{2} + \left(-\frac{1}{4}\right)$ **26.** $\frac{3}{7} + \left(-\frac{4}{9}\right)$ **27.** $\frac{6}{11} - \left(-\frac{2}{5}\right)$

Solve Equations

Kate solved the equation $2k + 11 = 4k + 3$ by using a balance puzzle.

Inv 1	Symbolic Solutions	461
Inv 2	Do the Same Thing to Both Sides	462
Inv 3	Solve More Equations	464
Inv 4	Inequalities	466

I can remove 2 bags from each side, and the scale will still balance. Taking 2 bags of k from each side, I have
11 = 2k + 3 left.

Kate summarized her solution in symbols.

$2k + 11$	$= 4k + 3$	
$-2k$	$-2k$	removing 2 bags from each side
11	$= 2k + 3$	
-3	-3	removing 3 blocks from each side
8	$= 2k$	
$8 \div 2$	$2k \div 2$	dividing what's left on each side into two equal groups
4	$= k$	

Explore

Think about how you would solve the following equation by using a balance puzzle.

$$14 + 3a = 7a + 2$$

For each step in your solution, use an equation to express what is left on the balance. Then summarize your solution using symbols like Kate did. If it is helpful, draw what is left on the balance after each step.

Investigation 1 Symbolic Solutions

You have solved equations by drawing and imagining balance puzzles. Now you will practice using symbols to record your solution steps.

Develop & Understand: A

Solve each equation by drawing or imagining a balance puzzle. Summarize your solution in symbols as Kate did.

1. $2x + 7 = 5x + 4$
2. $4r + 20 = 10r + 5$

For the balance at the beginning of this lesson, Kate labeled her solution to describe how the sides of the *balance* changed at each step.

Here is Kate's solution with labels describing *mathematical operations.* They show how the *equation* changed at each step.

Example

$2k + 11 = 4k + 3$	
$2k + 11 - 2k = 4k + 3 - 2k$	Subtract $2k$ from each side.
$11 = 2k + 3$	
$11 - 3 = 2k + 3 - 3$	Subtract 3 from each side.
$8 = 2k$	
$8 \div 2 = 2k \div 2$	Divide each side by 2.
$4 = k$	

Develop & Understand: B

3. Copy your symbolic solutions for Exercises 1–2. Next to each step, explain how the *equation* changed from the previous step.

Try to solve each equation below by working with just the symbols and doing the same operations on both sides. Try to make the equation simpler each time. Next to each step in your solution, explain how the equation changed from the previous step. If you have trouble, draw or imagine a balance puzzle.

4. $3y + 17 = 4y + 6$
5. $3x + 4 = x + 14$

Share & Summarize

Explain how doing the same mathematical operation to both sides of an equation is like doing the same thing to both sides of a balance puzzle.

Investigation 2 Do the Same Thing to Both Sides

Balance puzzles and equations can be solved by doing the same thing to both sides. Each step simplifies the problem until the solution is clear.

Develop & Understand: A

Try to solve each equation in Exercises 1–4 by doing the same mathematical operation to both sides without thinking about a balance puzzle. Next to each step in your solutions, explain how the equation changed from the previous step.

1. $5x = x + 4$

2. $r + 12 = 3r + 4$

3. $2P + 20 = 5P + 6$

4. $5y + 3 = 3y + 15$

5. Kenneth made up a number puzzle.

I'm thinking of a number. If I multiply the number by 5 and subtract 9, I get twice my original number.

Write an equation you can solve to find Kenneth's number. Solve your equation by doing the same thing to both sides. Next to each step in your solution, explain how the equation changed from the previous step. Be sure to check your solution.

Kenneth's number puzzle is an example of when thinking about a balance puzzle would not be helpful. This is because it is difficult to show subtracting 9 blocks from 5 bags using a balance puzzle. Try it and see.

Doing the same thing to both sides works for a great many equations. From now on, you can use this method to solve any equation, whether or not the equation fits the model of a balance puzzle.

Develop & Understand: B

Solve each equation in Exercises 6–11 by doing the same thing to both sides. If it helps, explain how the equation changed at each step. Check your solutions.

6. $2p + 7 = 4p + 10$

7. $3r - 2 = r$

8. $2a - 6 = 0$

9. $y - 3 = 2y - 6$

10. $1.5x + 3.5 = 8 + x$

11. $\frac{4}{5}x = 12$

12. Shane, Nestor, and Chantal found the correct solution to this equation.

$$5r - 16 = r$$

Shane started by subtracting $5r$ from both sides. Nestor started by adding 16 to both sides. Chantal started by dividing both sides by 5.

a. Solve the equation by starting with Shane's first step.

b. Solve the equation by starting with Nestor's first step.

c. Solve the equation by starting with Chantal's first step.

d. Whose method seems easiest? Explain why.

Share & Summarize

1. Consider the equation $2x - 1 = \frac{1}{2}x + \frac{7}{2}$.

   a. Explain why the equation is difficult to solve by thinking about a balance puzzle.

   b. Explain why the equation can be solved by doing the same thing to both sides.

2. Solve $2x - 1 = \frac{1}{2}x + \frac{7}{2}$ by doing the same thing to both sides. Explain how the equation changed at each step.

Investigation 3 Solve More Equations

You have seen that by doing the same thing to both sides, you can solve equations that are difficult to solve with balance puzzles. Now you will practice solving some more equations.

Develop & Understand: A

1. Solve the equation $5 - 2x = 3x$ by first adding the same amount to both sides. Check your solution.

2. Solve the equation $10 + 4z = 2z - 2$. Check your solution.

3. Solve the equation $\frac{1}{2}B = B - 4$ by first multiplying both sides by the same amount. Check your solution.

4. Solve the equation $\frac{k}{3} = 8 - k$.

Solve each equation by doing the same thing to both sides.

5. $4b - 2 = 6b - 4$

6. $2c + 1 = 9 + 3c$

7. $2d + 1 = 14 - d$

8. $5e + 2.5 = 2.5 - 7.5e$

9. Toya's hamster had a litter of babies. After Toya gave away three of the babies, $\frac{3}{4}$ of the litter remained. How many baby hamsters did Toya have *left?*

Real-World Link

A litter of hamsters averages 6 to 8 babies per litter. They occasionally have 15 or more. Babies are born furless, and their eyes do not open for one to two weeks.

When you solve an equation by doing the same thing to both sides, you usually try to make the equation simpler at each step. What if you made an equation *more complicated* at each step?

Example

Start by writing the solution.	$x = 7$
Multiply both sides by 3.	$3x = 21$
Add $2x$ to both sides.	$5x = 21 + 2x$
Subtract 2 from both sides.	$5x - 2 = 19 + 2x$
Divide both sides by 2.	$2.5x - 1 = 9.5 + x$

In Exercises 10–13, you will create complicated algebra equations. You will start with a simple equation and do the same thing to both sides to make it more complex. When you are satisfied that your equation is sufficiently difficult, challenge a partner to solve it.

Develop & Understand: B

10. Check to make sure the solution of the final equation in the example on the previous page is 7.

11. Explain why 7 is the solution of *each* equation in the example.

12. Create your own "complicated" equation. Start by choosing a number. Write the equation $x =$ *your number.* Then make the equation more complicated by doing the same thing to both sides three or four times. Verify that the number you first chose is the solution of your final equation. If it is not, retrace your steps to correct the error.

13. Exchange equations, and solve your partner's equation.

14. Erina's partner created the following "complicated" equation.

Start by writing the solution.	$x = 7$
Multiply both sides by 5.	$5x = 35$
Add $2x$ to both sides.	$7x = 2x + 35$
Subtract 40 from both sides.	$7x - 40 = 2x + 75$

What did Erina's partner do incorrectly in creating the equation?

Share & Summarize

Explain why the solution of an equation does not change when you alter the equation by doing the same thing to both sides.

Investigation  Inequalities

Vocabulary

inequality

solution set

Math Link

Inequalities use the following symbols.

$>$ greater than

$<$ less than

$\geq$ greater than or equal to

$\leq$ less than or equal to

As you know, an *equation* is a mathematical sentence stating that two quantities have the same value. An **inequality** is a mathematical sentence stating that two quantities have different values.

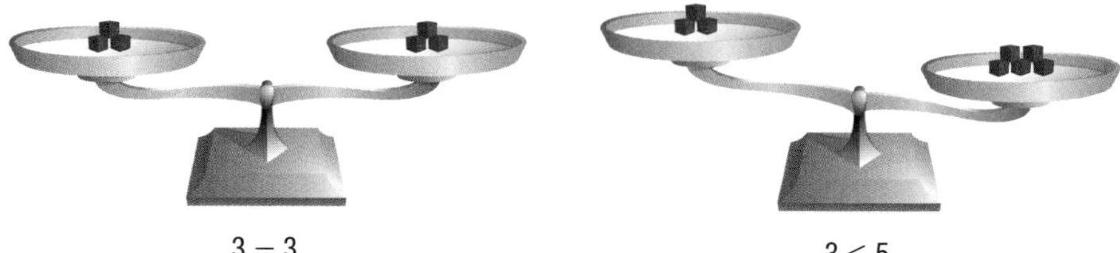

Develop & Understand: A

1. Which of the following statements are true?

$25 \geq 25$ $\qquad 3.2 < 3.14 \qquad \frac{4}{5} > \frac{5}{6}$

$\frac{5}{9} < \frac{4}{7} \qquad -5 \geq 0 \qquad -100 < -101$

In the last investigation, you solved equations by doing the same thing to both sides. If you do the same thing to both sides of an inequality, does the relationship between the two sides always stay the same?

2. Copy and complete the table below by performing the indicated operations to each side of the inequality $2 < 6$.

Original Inequality	Operation Performed	New Inequality	True/False
$2 < 6$	Add 2		
$2 < 6$	Add -2		
$2 < 6$	Subtract 2		
$2 < 6$	Subtract -2		
$2 < 6$	Multiply by 2		
$2 < 6$	Multiply by -2		
$2 < 6$	Divide by 2		
$2 < 6$	Divide by -2		

3. Use your results from Exercise 2 to write a conjecture about what happens to an inequality when you do each of the following.
   a. add the same number to both sides
   b. subtract the same number from both sides
   c. multiply both sides by the same number
   d. divide both sides by the same number

4. Test your conjectures. Write two different inequalities that are true. Test each of your rules on each of the inequalities. Do you end with similar results?

Think & Discuss

Share your conjectures from Exercise 3 and your test results from Exercise 4. Explain why you think your rules work for all inequalities.

5. How many blocks are in the bag shown to the right?

6. In the inequality $x < 10$, what does x equal?

The **solution set** to an inequality is the set of all possible values of a variable that make the inequality true. For the inequality $x \geq 10$, the solution set is 10 and all numbers greater than 10.

You can solve inequalities by *doing the same thing to both sides.* But, there is one important difference. If you multiply or divide each side of an inequality by a negative number, the direction of the inequality sign reverses.

To understand why this happens, consider the numbers -1 and 3. Multiply each number by -1. The new numbers 1 and -3 are the same distance from zero as -1 and 3. However, each new number lies on the opposite side of the number line. Figure 1 shows that -1 is less than 3. In Figure 2, the red circles represent the new numbers, 1 and -3.

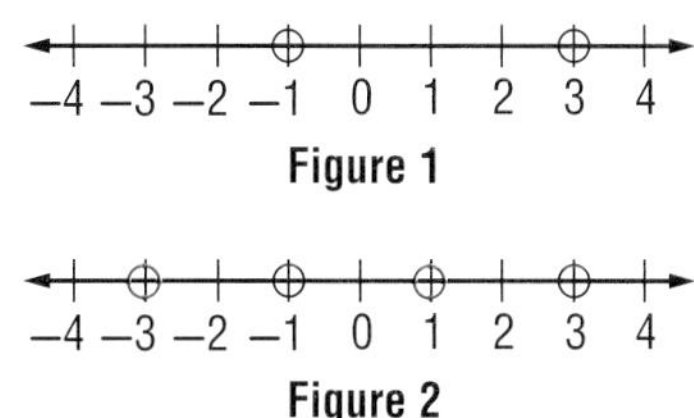

Figure 1

Figure 2

The inequality $1 > -3$ represents this new relationship. When you multiply or divide each side of an inequality by a negative number, switch the direction of the inequality sign in order to produce a true statement.

Develop & Understand: B

Solve each inequality by doing the same thing to both sides. Test each solution by choosing a number from the solution set and substituting it into the original inequality.

5. $x + 2 < 3$
6. $3z > 27$
7. $y - 3 > 2y + 5$
8. $\frac{1}{3}n + 5 \geq 4$
9. $\frac{3}{5}x + 7 > \frac{4}{5}x + 12$
10. $5q - 0.8 < 3q + 1.6$
11. Create an inequality whose solution is $x < 17$.

Example

You can graph the solution set of an equation or an inequality on a number line.

$x = 3$ –5 –4 –3 –2 –1 0 1 2 3 4 5

$x > 3$ –5 –4 –3 –2 –1 0 1 2 3 4 5

$x \geq 3$ –5 –4 –3 –2 –1 0 1 2 3 4 5

The open circle in the graph of $x > 3$ indicates that the number 3 itself is not part of the solution set. The closed circle in the graph of $x \geq 3$ indicates that 3 is included in the solution set. The arrow indicates that the solution set also includes all the numbers greater than 3.

Develop & Understand: C

12. Write the inequality whose solution is shown by the number line below.

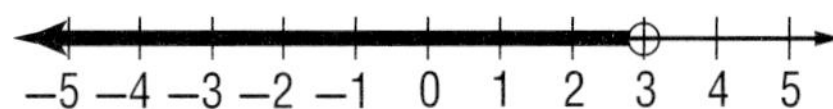

For Exercises 13–15, solve and graph each inequality.

13. $3y < -15$
14. $-4x + 5 < 8$
15. $5m + 1 \geq 7m + 10$

Share & Summarize

In what ways is an inequality similar to an equation? In what ways is it different?

How is solving an inequality similar to solving and equation? How is it different?

Practice & Apply

Solve each equation by drawing or imagining a balance puzzle. Use symbols to record your solution steps. Check your solutions.

1. $2x + 9 = 6x + 1$

2. $6z + 28 = 12z + 10$

3. $29 + 6w = 11w + 4$

4. $9 + 5v = 2v + 12$

Solve each equation by working with the symbols. Next to each step in your solution, explain how the equation changed from the previous step. Check your solutions.

5. $2a + 10 = 6a + 2$

6. $9b + 28 = 12b + 7$

Solve each equation by doing the same thing to both sides. Check your solutions.

7. $3 + 9c = 12c$

8. $18 + d = 7d + 6$

9. $2k - 6 = 4k + 5$

10. $\frac{1}{4}m - 3 = 0$

11. $9.25n + 3.3 = 1.3 - 0.75n$

12. $0.1p + 6 = 7$

13. $17r + 4 = 34r + 2$

14. $2s + 7 = 10s - 7$

15. Curtis said, "If you multiply the number of dogs that I have by 8 and then subtract 10, you get the same number as when you triple the number of dogs that I have and add 5."

a. Write an equation to find how many dogs Curtis has.

b. Solve your equation by doing the same thing to both sides. Check your solution.

16. Karl said, "If I divide my dog's age by 4 and subtract 2, I get the same result as when I divide my dog's age by 8."

a. Write an equation you can solve to find the dog's age.

b. Solve your equation by doing the same thing to both sides. Check your solution.

17. Consider the equation $27q - 100 = 2q$.

a. Solve the equation by first adding 100 to both sides.

b. Solve the equation by first subtracting $2q$ from both sides.

c. Solve the equation by first subtracting $27q$ from both sides.

d. Which of your solution methods was easiest?

Solve each equation by doing the same thing to both sides.

18. $\frac{1}{2}q + 3 = 10.5 + q$

19. $\frac{1}{3}c = 3c - 9$

20. $\frac{d}{7} = 12 + d$

21. $\frac{2}{7}d - 2 = 3d - 4\frac{5}{7}$

22. Keri is making lunch for her three best friends, but the only juice she has is in the little bottles her sister likes. She has 16 small bottles of apple juice. She fills a 10-ounce glass for each friend and one for herself, and she still has 6 bottles of juice left. Write an equation to find how many ounces of juice a bottle holds. Solve your equation.

23. Write three equations with a solution of -5.

24. For which of the following inequalities is 7 a member of the solution set?

$x \geq 7$	$x < 0$	$x \leq 10$
$x > 5$	$x < 6.9$	$x \leq 7\frac{1}{3}$

For Exercises 25–26, write the inequality whose solution set is shown by each graph.

25.

26.

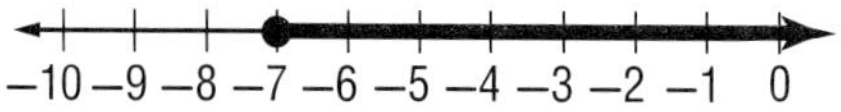

For Exercises 27–30, solve and graph each inequality.

27. $2x - 2 > 20$

28. $3y + 22 < 4y + 10$

29. $\frac{1}{6}n + 5 \leq 0$

30. $-2z + 0.6 \geq 2.4$

31. Write three different inequalities each with a solution of $x > 3$.

Connect & Extend

32. Yago tried to solve this equation using a balance puzzle.

$$5r + 7 = 13 + 2r + 3r$$

What solution do you think Yago found?

33. Find the Error Odell tried to solve the equation $6(x + 7) = 4x - 3$. He recorded the following steps.

$6(x + 7) = 4x - 3$

$6x = 4x - 10$	after subtracting 7 from both sides
$2x = -10$	after subtracting $4x$ from both sides
$x = -5$	after dividing both sides by 2

When Odell checked his solution, it did not work. What was his mistake?

Challenge Solve each equation by doing the same thing to both sides.

34. $\sqrt{x} = 3$ **35.** $\sqrt{v} - 2 = 1$ **36.** $\sqrt{5x - 9} = 4$

37. You can solve many equations by drawing two graphs and finding the point where they intersect. Consider the equation $2x + 3 = 1 + 3x$.

a. On the same set of axes, graph $y = 2x + 3$ and $y = 1 + 3x$.

b. Give the coordinates of the point where the graphs intersect.

c. Now solve $2x + 3 = 1 + 3x$ by doing the same thing to both sides.

d. How does the x-coordinate of the intersection point of the equations $y = 2x + 3$ and $y = 1 + 3x$ compare to the solution of $2x + 3 = 1 + 3x$? Explain why this makes sense.

e. Solve $3x - 4 = -2x + 6$ by drawing two graphs.

38. Kate has four packages of balloons and eight single balloons. Her brother gave her 20 more balloons and three more packages. She now has twice the number of balloons than she initially had.

a. Write and solve an equation to find how many balloons are in a package.

b. With how many balloons did Kate start?

39. **Geometry** The volume of a cylinder is given by the formula $V = \pi r^2 h$, where r is the radius and h is the height.

a. Cylinders A and B have the same volume. Write an equation to find the height of cylinder B. Explain how you wrote your equation.

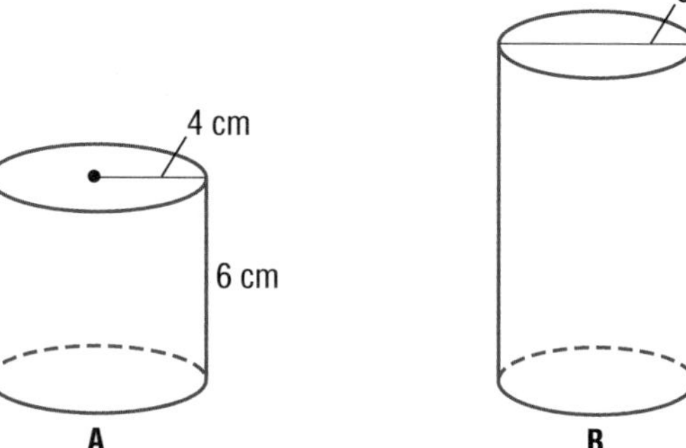

b. Solve your equation to find the height of cylinder B to the nearest tenth of a centimeter.

40. Tasha filled several bags each with the same number of marbles. She said, "There are more marbles in four bags than there are in two bags plus four extra marbles."

a. Write and solve an inequality to find the number of marbles in each bag.

b. Write a sentence describing the number of marbles in each bag.

41. Derrick works in a bookstore. On Monday, he received a shipment of books. When he put all of the books on shelves, he filled 5 shelves and had 13 books left over. On Tuesday, a bigger shipment arrived. This shipment filled 9 shelves, and almost filled a tenth shelf. There was room for 7 more books on the tenth shelf.

a. Write and solve an inequality to find the number of books on a shelf.

b. Write a sentence describing the number of books a shelf can hold. Each shelf holds more than 4 books.

c. Do all the numbers in the solution set make sense as answers in the context of this situation?

42. It is sometimes convenient to combine two inequalities into one. Consider the statement $5 < x < 35$.

a. Write a sentence that describes the meaning of the combined inequality.

b. Graph the combined inequality.

c. Describe a real-world situation that $5 < x < 35$ would represent.

43. In Your Own Words Describe two or more types of equations that are difficult to model with balance puzzles. Do you think those equations can be solved easily by doing the same thing to both sides? Explain.

Mixed Review

44. Geometry Recall that a two-dimensional figure that can be folded into a closed three-dimensional figure is called a net. For example, this is a *net* for a cube.

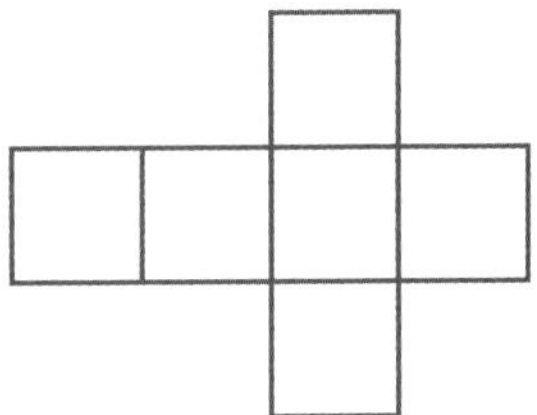

Draw another net for a cube.

45. Geometry A block prism has length 8 in., width 1 in., and height 6 in.

a. What is the prism's volume? What is its surface area?

b. Another block prism is three times as tall as the first prism. What is the volume of this prism? What is its surface area?

c. A cylinder has the same volume as the original prism and a height of 12 in. What is the area of the cylinder's base?

Find each product or quotient.

46. $\frac{3}{5} \cdot \frac{2}{7}$

47. $\frac{1}{9} \cdot \frac{6}{11}$

48. $\frac{5}{8} \div \frac{10}{3}$

49. $-\frac{1}{3} \cdot \frac{4}{9}$

50. $-\frac{3}{10} \cdot \left(-\frac{7}{8}\right)$

51. $\frac{12}{5} \div \left(-\frac{13}{6}\right)$

Solve Equations with Parentheses

Many situations that people encounter every day can be tackled by using mathematical techniques such as writing and solving algebraic equations. Sometimes, an equation that is written to solve a problem looks quite complicated. However, equations can often be simplified to make them easier to solve.

Inv 1 Build and Solve Equations 474

Inv 2 Subtract with Parentheses 477

Inv 3 More Practice with Parentheses 481

Think & Discuss

Rewrite this equation to make it easier to solve. Provide an explanation for each step.

$$2n + 3(n + 1) = 43$$

Solve your simplified equation. Check that your solution is a solution of the original equation.

Investigation 1 Build and Solve Equations

You have solved many exercises by writing and solving equations. Sometimes, after you write an equation for a situation, you can rewrite it before you solve. You already know some techniques for rewriting expressions to make them simpler.

Example

Jonna and Don collect autographs of baseball players. Jonna has five more autographs than Don. If they have 19 autographs altogether, how many does each friend have?

Here is Maya's solution.

Let n be the number of autographs Don has. Then Jonna has $n + 5$.

$n + (n + 5) = 19$	Combine Don's and Jonna's autographs.
$n + n + 5 = 19$	Remove the parentheses.
$2n + 5 = 19$	Add n and n.
$2n = 14$	Subtract 5 from both sides.
$n = 1$	Divide both sides by 2.

So, Don has 7 autographs and Jonna has 12. This adds to 19, so it checks.

Develop & Understand: A

1. Zoe thought about the situation in the example differently.

   "I'll let J stand for the number of autographs Jenna has. Don has 5 fewer than Jenna, so he must have $J - 5$ autographs."

   a. Write the equation Zoe had to solve.

   b. Do you think Zoe will get the same numbers of autographs for Jenna and Don as Maya did? Why or why not?

   c. Solve Zoe's equation. Check your solution.

2. Jorge and Talia have 48 model planes altogether. Jorge has three times as many model planes as Talia. Write and solve an equation to find how many model planes each student has.

3. Melissa has one more than twice the number of beanbag animals Rachel has. Latanya has a third as many beanbag animals as Melissa has. Together, the three friends have 38 beanbag animals. Write and solve an equation to find how many beanbag animals each friend has.

In Exercises 4–14, you will practice rewriting expressions to make them simpler and easier to handle.

Develop & Understand: B

Veronica is sorting groups of baseball cards by brand. They are Players, Best, and Atlantic. In Exercises 4–7, write and solve an equation to find the number of baseball cards in each pile. Check that your solutions make sense.

4. Veronica begins with 103 baseball cards, all Players and Best. There are 11 more Best cards than Players cards.

5. The next group Veronica sorts has a total of 98 Players and Atlantic cards. It has 36 fewer Atlantic cards than Players cards.

6. The third group has 112 cards, which Veronica divides into two piles. The Atlantic pile has three times as many cards as the Best pile.

7. The last pile, a mix of all three brands, contains 136 cards. Once Veronica sorts it into three piles, the Players pile has seven more than twice the number in the Best pile. The Atlantic pile has three times as many as the Players pile.

Real-World Link

The first baseball cards were printed in the late 1800s. In 1933, they began to be packaged with a stick of bubble gum and one card for a penny.

A bookstore clerk is sorting boxes of books to be shelved. From each box, he makes three stacks of books. They are fiction, nonfiction, and children's. In Exercises 8–10, write and solve an equation to find the number of books in each stack.

8. The first box has 31 books. The nonfiction stack has 5 fewer books than the fiction, and the children's has 15 more than the fiction.

9. From the next box, the fiction stack has three times as many as the children's, and the nonfiction has two fewer than the children's. There are 78 books altogether.

10. From a third box, the nonfiction stack has one more than double the number in the fiction stack, and the children's stack has double the number in the nonfiction stack. The difference between the numbers of books in the children's stack and the fiction stack is 47.

Develop & Understand: C

Solve each equation.

11. $3n + 2 = n$

12. $5(s - 2) - 22 = -32$

13. $7(t - 3) + 1 = 3(t + 4)$

14. $3(2x + 1) + 2(x - 3) = \frac{1}{2}(x - 10)$

Share & Summarize

1. Simplify and solve the equation $3(n + 4) + 2(n - 12) = 2n$. Explain each step in your solution.

2. Solve the equation $3x + x + 5(x + 4) = 56$. Check your solution.

Investigation  Subtract with Parentheses

Vocabulary

conjecture

At the beginning of Investigation 1, Maya wrote an equation to show how two collections of baseball players' autographs were added. She used parentheses around the sum $n + 5$, because that sum represents Jenna's number of autographs.

$$n + (n + 5) = 19$$

To add such an expression, you can simply remove the parentheses. For example, the equation above can be rewritten.

$$n + n + 5 = 19$$

In this investigation, you will explore what happens when you *subtract* expressions in parentheses.

Real-World Link

Some of the rarest and most valuable baseball cards can actually be worth less if they are signed by the player than if they are not.

Think & Discuss

How would you find the value of the expression $63 - (10 + 3)$?

Simon and Jin Lee evaluated $63 - (10 + 3)$ in different ways.

- Simon added 10 and 3 and then subtracted the sum from 63.
- Jin Lee subtracted one number at a time from 63. First, she subtracted 10, and then she subtracted 3.

Which method do you prefer? Are both approaches correct? Explain.

Simon wondered whether *any* sum could be subtracted by subtracting the numbers one at a time. After testing several cases, Simon said he thinks $a - (b + c)$ and $a - b - c$ are *equivalent expressions*. Remember that two expressions are *equivalent* if they have the same value for any values of the variables. Is he correct? Explain.

Simon's statement is a **conjecture**, a statement which he believes to be true based on some cases he has considered. If Simon's conjecture is true, the two expressions, $a - (b + c)$ and $a - b - c$, will give identical results no matter what numbers are substituted for the variables. If you can find even *one* case for which the two expressions give different results, you can conclude that they are *not* equivalent and that Simon's conjecture is false.

Develop & Understand: A

1. Test Simon's conjecture by trying at least four sets of values for a, b, and c. Record your results in a table like the one shown, which displays the results for $a = 50$, $b = 3$, and $c = 12$. Test a variety of values for a, b, and c. Consider cases in which the following occurs.
   - a is greater than $(b + c)$
   - a is less than $(b + c)$
   - c is a negative number
   - b is a negative number

a	b	c	b + c	a − (b + c)	a − b − c
50	3	12	15	35	35

Based on your research, do you think Simon's conjecture is correct?

You may already be convinced that Simon's conjecture is correct. To *prove* it is correct, you must show that it makes no difference what the values of the variables are. You cannot test every possible combination, but you can use algebra and what you know about operations with negative numbers to show that the conjecture is true for all sets of values.

You know that subtracting a number is the same as adding its opposite and that adding a number is the same as subtracting its opposite.

$$5 - (-2) = 5 + 2 \qquad 15 + (-8) = 15 - 8$$

You also know that the opposite of a number equals -1 times the number.

$$-3 = -1 \cdot 3 \qquad -(-11) = -1 \cdot (-11)$$

Think & Discuss

The series of steps below *proves* that $a - (b + c) = a - b - c$. Provide a reason for each step.

Step 1. $a - (b + c) = a + -(b + c)$

Step 2. $= a + -1(b + c)$

Step 3. $= a + (-1) \cdot b + (-1) \cdot c$

Step 4. $= \mathrm{a} - b - c$

So, $\mathrm{a} - (b + c) = a - b - c$.

Develop & Understand: B

Evaluate each expression.

2. $40 - (10 + 6)$
3. $35 - (5 + 8)$
4. $17.5 - (6.4 + 0.7)$

Rewrite each expression without parentheses. Then rewrite the expression again, if possible, to make it as simple as you can.

5. $7 - (7 + s)$
6. $12 - (r + 8)$
7. $(t - 7) + 3$

Shaunda has a strategy for quickly solving difficult subtraction expressions in her head.

Example

To compute $63 - 19$, Shaunda estimates the answer by first subtracting 20 from 63 to get 43. Then she corrects her estimate. She knows that she subtracted one too many since 20 is one greater than 19, so she adds back one to get 44 as the result.

Here is Shaunda's strategy in symbols.

$$63 - 19 = 63 - 20 + 1 = 43 + 1 = 44$$

Develop & Understand: C

8. Evaluate each expression by using Shaunda's method. First, approximate by subtracting 20 and then add back the amount needed to correct the result. Write the steps in symbols as done in the example.
   a. $63 - 18$
   b. $63 - 17$
   c. $63 - 15$
9. In these expressions, the number being subtracted is close to 200. Modify Shaunda's method and apply it to find each difference. Write the steps in symbols.
   a. $351 - 197$
   b. $351 - 195$
   c. $351 - 190$

Zach said that after looking at his answers for Exercises 8 and 9, he thinks $a - (b - c)$ is equivalent to $a - b + c$.

10. Look at the symbolic version of Shaunda's example, which is $63 - 19 = 63 - 20 + 1 = 43 + 1 = 44$. What do you think Zach would say are the values for a, b, and c?

11. Use the table to test the two expressions in Zach's conjecture with at least four sets of values for a, b, and c. One example is shown. Try cases in which the following occurs.

- a is greater than $(b - c)$
- a is less than $(b - c)$
- c is greater than b
- b is a negative number

a	b	c	$b - c$	$a - (b - c)$	$a - b + c$
50	12	3	9	41	41

Based on your results, do you think Zach's conjecture is correct?

12. Zach said the proof of his conjecture is very similar to the proof of Simon's conjecture in Think & Discuss on page 477. Here is the first step of Zach's proof.

$$a - (b - c) = a + -(b - c)$$

a. Explain how you know this first step is a true statement.

b. Finish the proof. If you need help, look back at the proof of Simon's conjecture. Be sure to provide a reason for each step in your proof.

Develop & Understand: D

Evaluate each expression.

13. $50 - (10 + 6)$ **14.** $50 - (10 - 6)$ **15.** $50 - (6 - 10)$

Rewrite each expression without parentheses. Then rewrite the expressions again, if possible, to make them as simple as you can.

16. $12 - (r - 8)$ **17.** $5 - (5 + n)$

18. $3 - (7 - t)$ **19.** $15r - (3r - s)$

Share & Summarize

Select one of the expressions from Exercises 16–19 and explain how you thought about it. Write an explanation to convince someone that your simplified version is correct.

Investigation 3 More Practice with Parentheses

As you saw in Investigation 2, the two expressions below will always be equal, no matter what the values of x and y are. But maybe you are still not sure it really makes sense that they are equal.

$$10 - (x + y) \qquad\qquad 10 - x - y$$

Think about the following. Imagine that you have \$10. You owe your parents x dollars, and you owe your best friend y dollars. How much money will you have left if you can pay both your debts?

You can add your debts together, $x + y$, and then subtract the total from \$10. The amount you have left is $10 - (x + y)$. Or, you can subtract your debts one at a time, and so the amount you have left is $10 - x - y$. In either case, you will have the same amount of money left, so these two expressions must be equal.

Develop & Understand: A

1. Leon owed his little sister x dollars. She wanted to buy a new book, so she asked Leon to pay back some of the money. He said he would give her y dollars.

   a. After paying back y dollars, how much did Leon still owe his sister?

   b. Leon had \$15 after paying his sister the y dollars that he promised. If he paid the rest of the money that he owed, how much would he have left? Find two expressions to represent this amount.

2. Ron wanted to borrow money from two of his siblings. Neither wanted to make him a loan, so he struck a deal with them. If he took more than a week to repay them, he would pay them double the amount borrowed. His brother loaned him x dollars, and his sister loaned him y dollars. Ron was not able to get any money for two weeks, when a friend paid him \$25 that she had borrowed.

   Which expressions represent the amount of money Ron had left after paying his siblings double the amount he had borrowed from them?

   a. $25 - 2(x + y)$ b. $25 - (2x + 2y)$ c. $25 - 2x - 2y$

Jared says that when he has to multiply an expression in parentheses by a quantity, he first uses the distributive property to multiply each item inside the parentheses. Then he uses the rules for adding and subtracting expressions in parentheses.

Example

Here is how Jared simplified $5 - 2(3 - 8)$.

First, he distributed the 2.	$5 - (6 - 16)$
Then he applied the rule for subtracting parentheses.	$5 - 6 + 16$
Then he subtracted and added.	15

Jared used the same approach to simplify $12k - 3(7 - 2k)$.

$$\begin{aligned} 12k - 3(7 - 2k) &= 12k - (21 - 6k) \\ &= 12k - 21 + 6k \\ &= 18k - 21 \end{aligned}$$

In order to become more skilled at simplifying expressions with parentheses, you will combine adding and subtracting with parentheses with some of the earlier skills that you developed.

Develop & Understand: B

Simplify each expression as much as possible.

3. $5r - 2(r - 8)$
4. $32L - 3(L - 7)$
5. $12n - 2(n - 4)$
6. $4(3x - 2) - 3x$
7. $-5(g - 3)$
8. $8t - \frac{1}{3}(12t - 30)$

For each expression, write an equivalent expression without parentheses.

9. $a + d(b + c)$
10. $a - d(b + c)$
11. $a + d(b - c)$
12. $a - d(b - c)$
13. In the example, you saw how Jared simplified $5 - 2(3 - 8)$. Here is another way to simplify $5 - 2(3 - 8)$. Provide a reason for each step.

    a. $5 - 2(3 - 8) = 5 + -2(3 - 8)$

    b. $= 5 + (-6 - (-16))$

    c. $= 5 + (-6 + 16)$

    d. $= 5 + -6 + 16$

    e. $= 15$

Choose any method to solve each equation. Check your answers.

14. $2(2q + 4) - 2q = 24$

15. $12 - \frac{1}{2}(2n - 6) = 7$

16. $2x - (4 - x) = 5$

17. $4r - 3(r + 1) = 0$

Real-World Link

Until the 1980s, tennis rackets were made from wood. Wood tennis rackets, first made in the U.S. in the 1870s, are now collectors' items.

18. Sunny Sports Shop sells beginner's tennis rackets for \$40 each. Most customers who buy one of these rackets also buy several cans of tennis balls, which cost \$2.50 each. The customer's bill has three parts.
    - the price of the racket
    - the total price of the balls
    - the sales tax, which is 7% of the combined price of the racket and the balls

    a. A customer buys a racket and x cans of balls. Write expressions for each of the three parts of the customer's bill.

    b. Write an expression for the total bill, which includes the racket, x cans of balls, and the sales tax.

    c. Aurelia bought a racket and some cans of balls. Her total bill came to \$53.50. Write and solve an equation to find how many cans of balls she bought.

Share & Summarize

1. Simplify and solve the following equation. Write an explanation of your work that would convince someone that your simplified equation is correct.

$$5n - \frac{1}{3}(3n - 6) = \frac{1}{4}(8n + 12)$$

2. Check your answer by substituting it into the original equation.

On Your Own Exercises
Lesson 9.4

1. Together, Lee and Joanna have 19 computer games. Joanna has 3 more games than Lee. Write and solve an equation to find the number of games each friend has. Check your answer.

2. Martin and five of his friends donated change to a school fundraiser to buy food for children in need. Together, they donated $4.40. Three of the friends gave 20¢ less than Martin, and two gave 10¢ more. Write and solve an equation to find how much Martin donated. Check your answer.

3. Peter had three piles of pickled peppers. He has a total of 52 pickled peppers. His second pile has seven more peppers than the first pile. His third pile has three fewer than the first pile. Write and solve an equation to find how many pickled peppers are in each pile. Check your answer.

4. Mr. Álverez baked 60 fruit tarts for the school bake sale. He put them on four trays. The second tray had twice as many tarts as the first. The third tray had 1 more than the second. The fourth tray had five more than the first. Write and solve an equation to find how many tarts were on each tray. Check your answer.

5. Gloria sorted her collection of Winged Liberty Head dimes into three stacks based on their quality. The second stack has twice as many coins as the first, the third stack has 36 more than the second, and the first stack has 43 fewer than the third. How many dimes are in each stack? Show how you found your answer.

Real-World Link

The Winged Liberty Head dime was minted from 1916 to 1945. The coin's designer put wings on Liberty's cap to represent freedom of thought. But the image is often mistaken for the Roman god Mercury, and thus the coin is commonly known as the Mercury dime.

Solve each equation.

6. $d + (2d - 3) = 13(d - 1)$
7. $5(2x - 1) + 3x = 2(x + 5)$

Evaluate each expression.

8. $72 - (12 + 9)$
9. $72 - (12 - 9)$
10. $12.5 - (3.3 - 9.5)$
11. $38 - (7 + 28)$
12. $291 - (84 - 45)$
13. $5 - (2.8 + 5.9)$

Rewrite each expression without parentheses. Then rewrite the expressions again, if possible, to make them as simple as you can.

14. $8 - (x - 8)$
15. $15 - (8 + y)$
16. $x - (37 + x)$
17. $6 - (12 - n)$
18. $5s - (6r + s)$
19. $8x - (-2x - 3y)$

For each expression, write an equivalent expression without parentheses.

20. $A + (B - C)$ **21.** $A - (B - C)$ **22.** $A - (B + C)$

Rewrite each expression without parentheses. Then rewrite the expressions again, if possible, to make them as simple as you can.

23. $2(a - 8) + 30$ **24.** $30 + 4(t - 7)$ **25.** $5 + 3(n + 5)$

26. $15r + 2(r + s)$ **27.** $3t + 3(10 - t)$ **28.** $7x + \frac{1}{2}(2x - 4y)$

29. $m + 4(m + 5)$ **30.** $4p + 7(2p + 2)$ **31.** $4(x + 2) + 5(3x + 2)$

Solve each equation.

32. $8 - (4 - n) = 18$

33. $8 - (n - 4) = 18$

34. $6x - 2(x + 1) = 10$

35. $2x - 3(x - 1) = 17$

36. $3y + 4(y + 2) = 78$

37. $2(x + 3) + x = 63$

38. $5(x + 2) = 6x$

39. $6 - 2(8 - 2r) = 20$

40. $3 - 2.5(m + 2) = \frac{m}{2} + 4$

41. $r + 8 = 3(1 - r) + 2$

42. GameWorld is selling SuperSpeed game systems for $400. Game cartridges are $50 each. Sales tax is 6% of the total price for all items.

a. Write an expression for the total, with tax, of one SuperSpeed system with g game cartridges.

b. One customer bought several games and had a final bill of $742. Write and solve an equation to find how many games the customer bought.

43. Create an equation about two friends who collect postage stamps. Make sure your equation can be solved so that each friend has a different whole number of stamps. Include an explanation of how you created your equation and how it can be solved.

Connect & Extend

44. In a town election, the three candidates receiving the most votes were Malone, Lawton, and Spiros. Malone received 60 more than twice as many votes as Lawton. Spiros received 362 votes more than Lawton. In all, 2,430 votes were cast. How many votes did each candidate receive? Who won the election? Show how you found your answer.

45. This conversation occurred the day after the class election.

Desiree: I heard that Evita won our class election.

Reynoldo: Yes, but it was really close! She received only one more vote than Mai.

Troy: It sure wasn't very close for third place. Mai received twice as many votes as Lu-Chan.

Laura: Everybody voted except James, who's home with chicken pox.

Of the 27 students in the class, only Evita, Mai, and Lu-Chan received votes. How many votes did each of the three students receive? Show how you found your answer.

46. Nuna is y years old. In 30 years, she will be three times as old as she is now. Write and solve an equation for this situation.

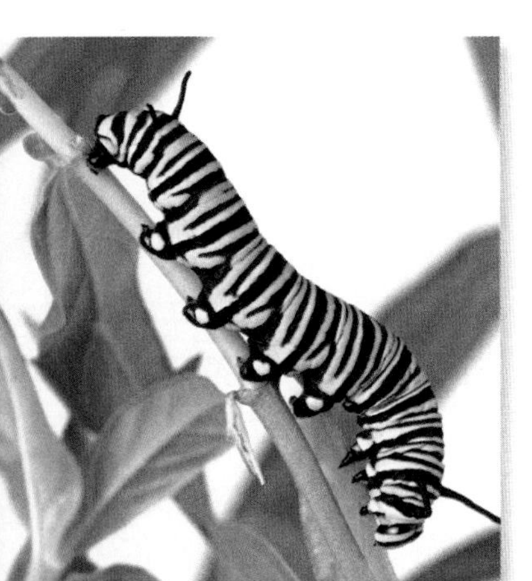

Real-World Link

Monarch caterpillars eat milkweed leaves. These leaves contain a poison that makes the caterpillars taste bad to birds and other predators.

47. Two caterpillars are crawling in and out of a hole in a fence. The caterpillars take turns being in the lead. The length of the longer caterpillar is L, and the length of the shorter caterpillar is S. The hole is 9 cm above the ground.

Look at the caterpillars on the far left. The expression $9 + L - S$ represents the height reached by the head of the lead caterpillar.

Match each of the 12 expressions with the drawings. Some positions match more than one expression, and some expressions match more than one position.

a. $9 + (S - L)$ **b.** $9 + (S + L)$ **c.** $9 - (S + L)$

d. $9 - S - L$ **e.** $9 - L - S$ **f.** $9 - L + S$

g. $9 + (L + S)$ **h.** $9 - (L + S)$ **i.** $9 - S + L$

j. $9 + S - L$ **k.** $9 - (S - L)$ **l.** $9 - (L - S)$

Investigation 1 Ratios and Rate Comparisons

You already know about ratios and rates from work in earlier grades and from your own life. You will now use that knowledge to reason about ratio and rate comparisons.

Develop & Understand: A

1. The triangular tiles on The Art Institute's wall are laid in this pattern.

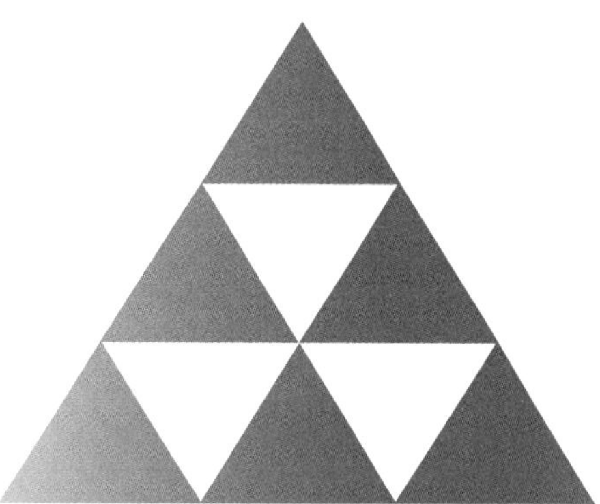

   a. What is the ratio of white tiles to blue tiles in this pattern?

   b. The entire wall contains 900 blue tiles. How many white tiles does it have?

   c. What is the ratio of white tiles to all tiles in this pattern?

   d. If an area with this pattern has 2,880 tiles in all, how many white tiles does it have?

Math Link

Ratios may be written in several ways. Here are three ways to express the ratio "four to one."

4 to 1   4 : 1 $\frac{4}{1}$

2. Tehani made this flower necklace.

   a. What is the ratio of white flowers to pink flowers?

   b. Tehani wants to make a longer necklace with flowers in the same pattern. If she plans to use 20 pink flowers, how many white flowers will she need?

   c. What is the ratio of pink flowers to all flowers in this necklace?

   d. Tehani wants to make a long necklace for her older sister in the same pattern. If she uses 27 flowers in all, how many white flowers will she need?

3. Carlita took a canoe trip with her family. The graph shows the progress of her trip.

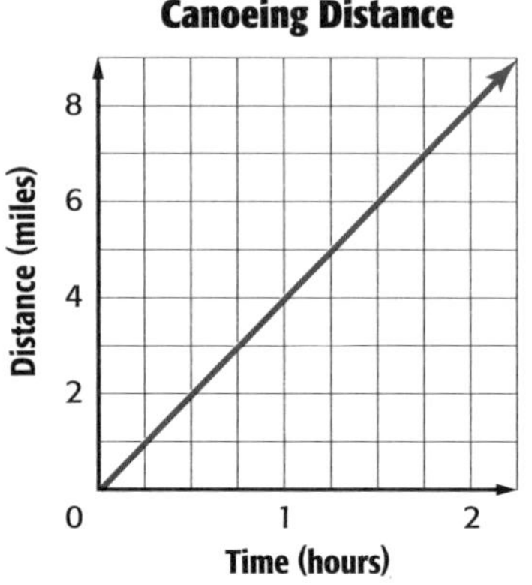

a. At what rate is Carlita traveling?

b. How many miles would you expect Carlita to canoe in 3.5 hours?

4. The two numbers on each card in this set are in the same ratio.

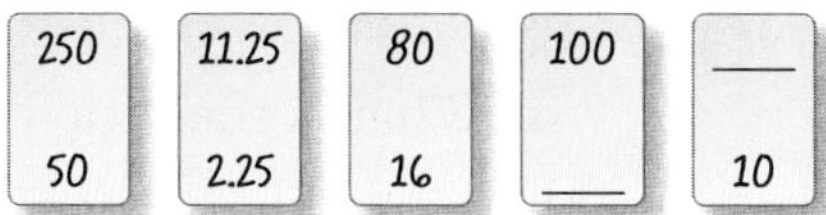

a. Find the missing numbers.

b. Draw three more cards that belong in this set.

5. The two numbers on each card in this set are in the same ratio.

a. Find the missing numbers. Explain how you found your answers.

b. What ratio expresses the relationship between the top number and the bottom number on each card?

c. Draw three more cards that belong in this set.

Share & Summarize

1. Consider this keyboard.

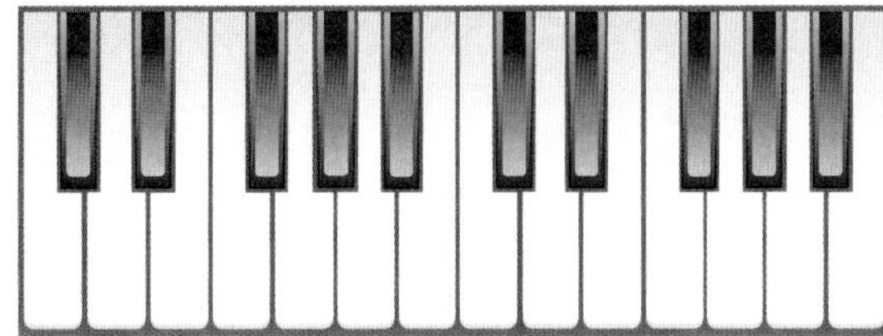

   a. What is the ratio of white keys to black keys on the keyboard?

   b. This pattern of keys is repeated on larger keyboards. How many black keys would you expect to find on a keyboard that has 42 white keys?

   c. What is the ratio of black keys to *all* keys on this keyboard?

   d. How many black keys would you expect to find on a keyboard with 72 keys in all?

Investigation 2 Compare and Scale Ratios

You have practiced writing ratios to express comparisons between quantities. Reasoning about ratios will help you solve the exercises in this investigation.

Develop & Understand: A

Researchers are developing a shade of orange paint. They are experimenting by mixing containers of red and yellow paints.

1. Below are two mixtures the researchers tested. Each red bottle represents a container of red paint. Each yellow bottle represents a container of yellow paint. All hold the same amount of paint. Which mixture is darker orange? Explain how you decided.

Mixture A

Mixture B

2. Tell which mixture is darker orange.

Mixture C

Mixture D

The researchers have found some shades they like. Now they want to make larger batches of paint.

3. The researchers call this mixture Awesome Orange. Draw a picture to show how Awesome Orange can be created using 12 containers in all.

4. This mixture is called Sunset. Draw a set of containers that could be used to create a larger batch of Sunset.

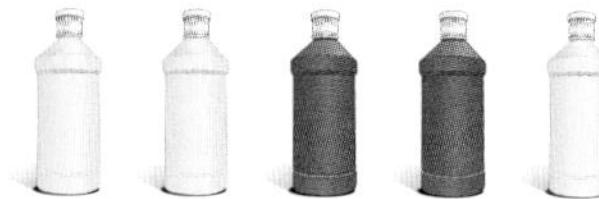

5. Compare the two mixtures above. Use ratios in your comparison.

Think & Discuss

The mixture at the right is Tropical Paradise. Maya, Simon, and Zach tried to make a batch of Tropical Paradise using 9 containers.

Simon extended the pattern.

"I'll draw copies of the 3 containers until I have 9 in all. First, I draw the 3 containers that I know. Then I draw them again to get 6 containers. I draw them once more to get 9 containers.

Maya thought about ratios.

"In the original mixture, $\frac{2}{3}$ of the containers are red. If I use 9 containers, $\frac{2}{3}$ of them must be red. Two-thirds of 9 is 6, so 6 containers must be red and 3 must be yellow."

Zach used a ratio table to record several equivalent ratios.

Red Containers	2	4	6
Total Containers	3	6	9

Did the students reason correctly? Explain any mistakes they made.

Did any of the students use reasoning similar to your own?

Develop & Understand: B

6. Complete this ratio table to show the number of red containers and the total number of containers for various batches of this shade.

Red		4	6	8	10		200
Total	3	6				90	

7. The ratio table in Exercise 6 compares the number of red containers to the total number of containers. Complete the next ratio table to compare the number of red containers to the number of yellow containers in this mixture.

Red		7	14			35
Yellow	1.5	3		9	12	

8. Soup and Salad, Inc. is catering a winter luncheon. According to the company, 8 cans of soup can feed 20 people.

a. Complete this ratio table based on this information.

People		10	15	20		
Cans of Soup	2			8		

b. How many people will two cans of soup feed?

c. If Soup and Salad, Inc. is planning to feed 25 people, how many cans of soup will they need?

Share & Summarize

1. Which mixture below is darker orange? Describe your strategy.

2. Make a ratio table to show the number of red containers and the total number of containers you would need to make five different-sized batches of Mixture X.

On Your Own Exercises
Lesson 10.1

Practice & Apply

Real-World Link

Great Smoky Mountains National Park is the most visited national park in the United States. Almost 95% of the park is forested.

source: National Park Service: www.nps.gov

1. During September 2007, Great Smoky Mountains National Park had 39,888 overnight guests. There were 3,565 overnight backcountry campers in the park. An additional 32,158 visitors stayed overnight in the campgrounds. The four comparisons below are based on this information.

i. About nine times as many visitors used the campgrounds compared to those who were backcountry campers.

ii. The ratio of those who camped in the backcountry to those who stayed in the campgrounds is 3,565 to 32,158.

iii. The ratio of overnight visitors to the number of campground visitors is about 5 to 4.

iv. Almost 10% of the overnight visitors were backcountry campers.

a. Explain why each statement above is true.

b. Suppose that Great Guides arranges backcountry camping trips. Which statement do you think they should use in an advertising brochure? Explain.

2. Mitchell and Elias went hiking. The graph shows their progress.

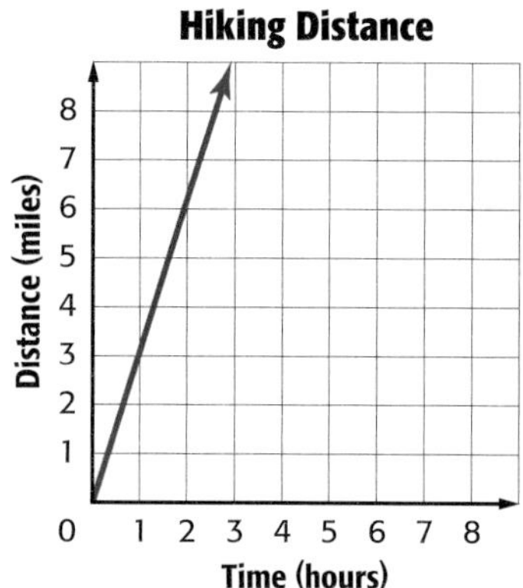

a. At what rate did the boys hike?

b. How many miles would you expect the boys to hike in 4 hours?

3. Roberta is laying out the pages of a magazine. She divides one of the pages into 20 sections and decides which sections will be devoted to titles, to text, to photographs, and to advertisements.

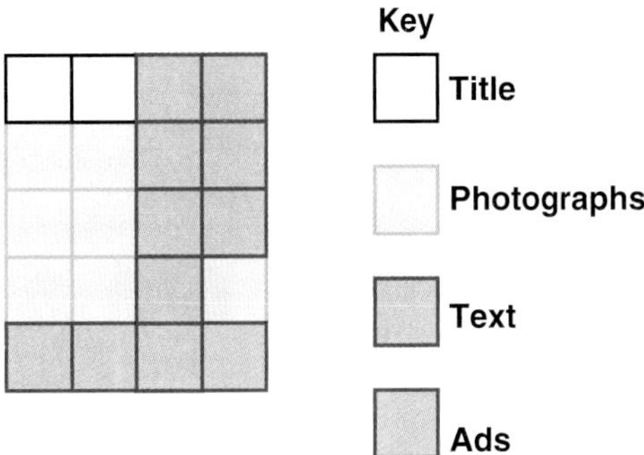

Describe something in the layout that has each given ratio.

a. 2 to 3

b. 1:4

c. $\frac{7}{8}$

Math Link

Remember that ratios can be expressed three ways.

1 to 3

1:3

$\frac{1}{3}$

4. Decide whether Mixture A or Mixture B is darker. Provide an explanation for your choice.

5. Brigade Blue paint is shown below. How many containers of blue and white paints are needed for a batch of paint that has a total of 75 containers?

For Exercises 6–8, tell which mixture is lighter green. Then draw or describe a larger batch that will make the lighter green mixture.

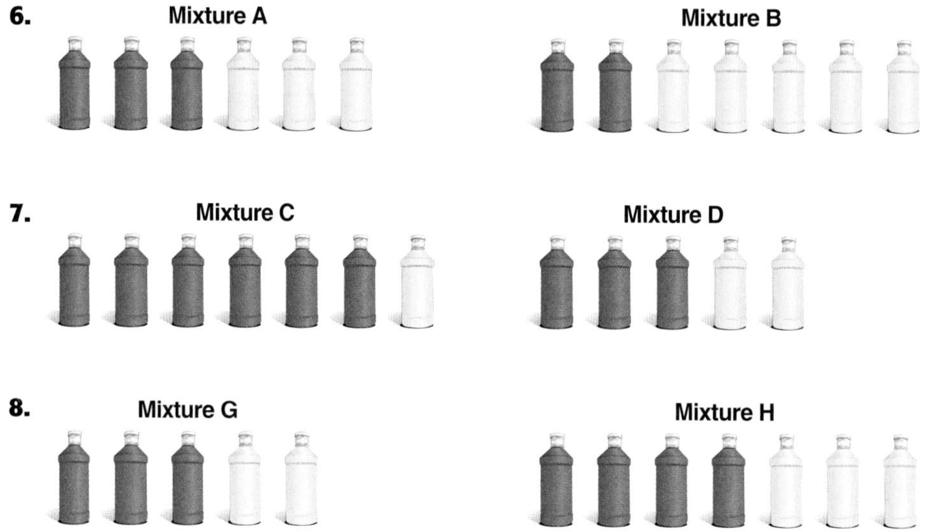

9. Complete this ratio table to show the number of blue and yellow containers for various batches of this shade.

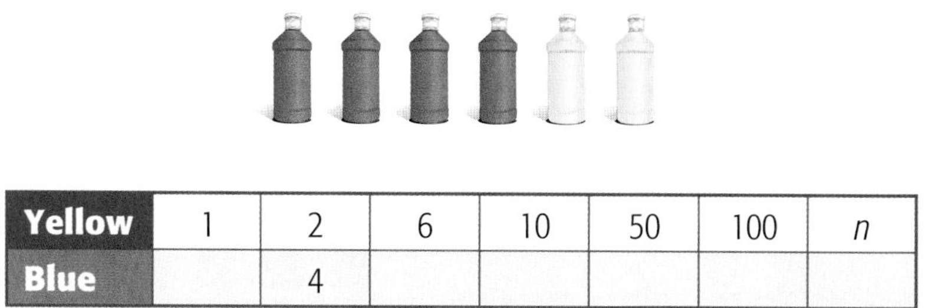

Yellow	1	2	6	10	50	100	n
Blue		4					

In Exercises 10–12, is the given table a ratio table? If so, tell the ratio. If not, explain why not.

10.

2	4	6	8	10
5	10	15	20	25

11.

1	3	5	8	10
2	5	8	11	14

12.

4	40	400	4,000	40,000
7	70	700	7,000	70,000

Connect & Extend

13. **Preview** Explore the ratios of x- and y-coordinates for some lines.
   a. Graph the line that goes through the points (4, 2), (6, 3), and (−2, −1).
   b. What is the ratio of x to y?
   c. Choose some points with an x:y ratio of $\frac{1}{5}$. Graph the points on the same axes as in Part a. Draw a line through your points.
   d. Now choose some points with an x:y ratio of $\frac{2}{3}$ and graph them. Draw a line through your points.
   e. What do you notice about the three lines you graphed?
   f. Calculate the slope of each line. How are the slopes related to the ratios?
14. Consider this mixture.

   a. Draw a mixture that is darker than this one but uses fewer containers. Do not make *all* the containers blue.
   b. Explain why you think your mixture is darker.

Real-World Link
The Golden Ratio is thought to be a central theme in the design of the mosque of Uqba's prayer space and minaret.

15. **Architecture** Architects have used a particular rectangular shape to design buildings. For rectangles of this shape, the ratio of the long side to the short side is roughly 1.6 : 1. This ratio is very close to what is known as the *Golden Ratio*. Rectangles of this shape look pleasing to the eye.
   a. Try drawing three rectangles of different sizes that look "ideal" to you. For each rectangle, measure the side lengths and find the ratio of the long side to the short side.
   b. Which of your rectangles has the ratio closest to the Golden Ratio?
16. **Measurement** Marsha always forgets how to convert miles to kilometers and back again. However, she remembers that her car's speedometer shows both miles and kilometers. She knows that traveling 25 miles per hour is the same as traveling 40 kilometers per hour. So, in one hour, she would travel 25 miles, or 40 kilometers.
   a. From this information, find the ratio of miles to kilometers in simplest terms.
   b. To cover 100 km in an hour, how fast would Marsha have to go in miles per hour? Explain how you found your answer.

17. Crafts Unlimited sells ribbon by the foot. Serena needs ribbon for an art project. The cost of the ribbon is 5 feet for $3.00.

Feet of Ribbon	1	5	10	15	
Cost (in dollars)		$3.00			

a. Complete this ratio chart based on the information provided.

b. How many feet of ribbon can Serena buy for $12.00?

c. If Serena needs 17 feet of ribbon, how much will she pay? How do you know?

18. This table is not a ratio table.

a. Graph the values in the table. Draw a line or a smooth curve through the points. Describe the shape of the graph.

x	10	20	30	40
y	1.5	2	2.5	3

b. Make a table of ratios equivalent to 10 : 1.5. Include at least four entries. Call the values corresponding to the first quantity in the ratio x and the other values y.

c. Graph the values from Part b. Draw a line or a smooth curve through the points.

d. How are the two graphs alike? How are they different?

e. Make another ratio table of x and y values using a ratio close to 10 : 1.5, such as 8 : 1.5 or 12 : 1.5. Graph the values in your table. Draw a line or a smooth curve through the points.

f. Challenge What characteristic do the graphs in Parts c and e share but the graph in Part a does not? Explain why this is so.

Mixed Review

Evaluate each expression.

19. $23 \cdot 2 - 5(3.1)$

20. $23 \cdot [2 - 5(3.1)]$

21. $23 \cdot (2 - 5)(3.1)$

22. $3^3 \cdot \frac{4}{3} - 2$

Solve each equation by doing the same thing to both sides.

23. $5x + 7 = 3x + 3$

24. $21a + 5 = 6a + 35$

25. $3t + 4 = t - 8$

26. $6.25x - 30 = 0.25x + 6$

27. Elaine told her banker, "If I put my money in Account A, which will double it and add 24 dollars, I will have the same amount as if I put my money in Account B, which will triple it and add 4 dollars."

a. Write an equation to find how much money Elaine has.

b. Solve your equation by doing the same thing to both sides.

Proportions and Similarity

Inv 1 Proportional Relationships 505

Inv 2 Equal Ratios 508

Inv 3 Solve Proportions 510

Inv 4 Map Scales 514

Inv 5 Similarity 517

Inv 6 Estimate Heights of Tall Objects 519

Theresa is making trail mix to sell for a seventh-grade fundraiser. The amount of peanuts in the recipe is proportional to the amount of raisins.

As with any proportional relationship, the graph of the relationship between peanuts and raisins is a line through the origin.

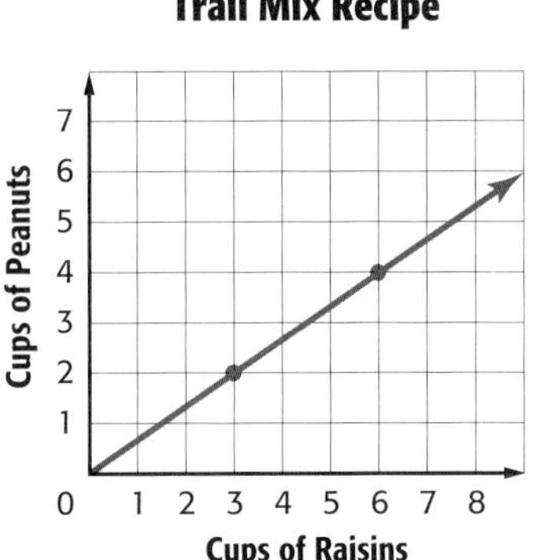

Think & Discuss

The point (6, 4) is on the graph. What does this tell you about the trail mix?

Identify two more points on the graph. Find the ratio of peanuts to raisins for all three points. How do the ratios compare?

Find the slope of the line. Remember that the slope of a line is rise divided by run. What does it tell you about the trail mix? How does it relate to the ratios you found?

Give the coordinates of two other points that are on the line but that are beyond the boundaries of the graph. Explain how you found the points.

Math Link

Proportional means that as one variable doubles the other doubles, as one variable triples the other triples, and so on.

Investigation 1 Proportional Relationships

Vocabulary

proportional relationship

A **proportional relationship** is a relationship in which all pairs of corresponding values have the same ratio. In Think & Discuss, you found that the ratio of peanuts to raisins is the same for any point on the graph. For example, (3, 2) and (6, 4) both lie on the same graph.

$$\text{So, } \frac{3}{2} = \frac{6}{4}.$$

You can test for proportionality by using cross-products or by writing fractions in lowest terms.

Cross-Products

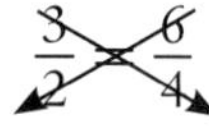

$3 \cdot 4 = 12$

$6 \cdot 2 = 12$

Lowest Terms

$$\frac{6}{4} = \frac{3 \cdot 2}{2 \cdot 2} = \frac{3}{2} \cdot \frac{2}{2} = \frac{3}{2} \cdot 1 = \frac{3}{2}$$

Develop & Understand: A

1. On the cards below, the top numbers are proportional to the bottom numbers. Find another card that belongs in this set. Explain how you know your card belongs.

2. Consider these triangles.

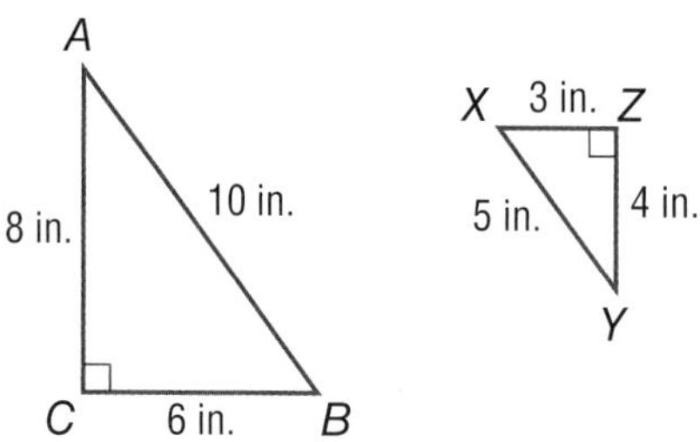

   a. Are the side lengths of triangle *ABC* proportional to those of triangle *XYZ*? Explain how you know.

   b. Remember that similar figures have the same shape but possibly different sizes. Are the triangles similar? Explain.

3. Bryan says, "I created Mixture B by adding one red container and one white container to Mixture A. Then I made Mixture C by adding one red container and one white container to Mixture B. Since I added the same amount of red and white each time, the number of red containers is proportional to the number of white containers."

   Is Bryan correct? Explain.

Mixture A **Mixture B** **Mixture C**

4. Kenyon created this tile pattern.

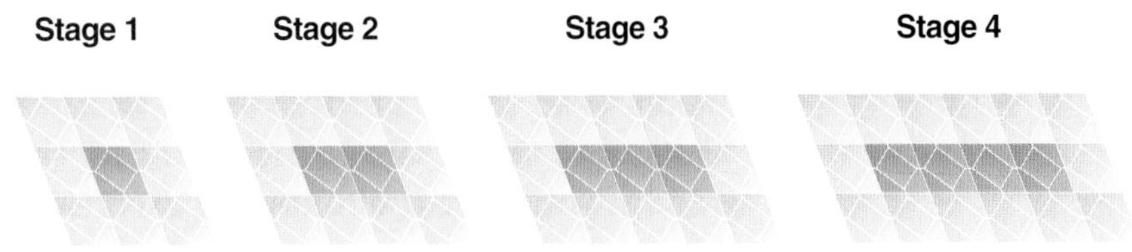

Math Link

To be a pattern, the arrangement of tiles must change in a *predictable* way from stage to stage.

a. Describe how the pattern of green and yellow tiles changes from one stage to the next.

b. For the stages shown, is the number of green tiles proportional to the number of yellow tiles? Explain.

c. Starting with stage 1 above, draw the next two stages for a tile pattern in which the number of green tiles *is* proportional to the number of yellow tiles.

5. Stores are able to make a profit by buying in bulk. They are charged less per item because they agree to buy a large number of items.

For Parts a–c, compare the two situations. Determine if the two situations are proportional or not proportional.

a. A store may purchase 100 t-shirts for \$28 or 3,000 t-shirts for \$12,180.

b. A store may purchase 5 digital cameras for \$150 or 45 digital cameras for \$1,350.

c. A store may purchase 36 packs of gum for \$7.20 or 800 packs of gum for \$96.

Share & Summarize

1. Describe at least two ways to determine whether a relationship is proportional.
2. Describe two quantities in your daily life that are proportional to each other.
3. Describe two quantities in your daily life that are not proportional to each other.
4. Use the word *proportional* to explain how you can tell whether two triangles are similar.

Investigation 2 Equal Ratios

You know that in a proportional relationship, all pairs of corresponding values are in the same ratio. You may find it helpful to think about this idea as you solve the following exercises.

Develop & Understand: A

1. The rectangles below are similar. Find the value of x.

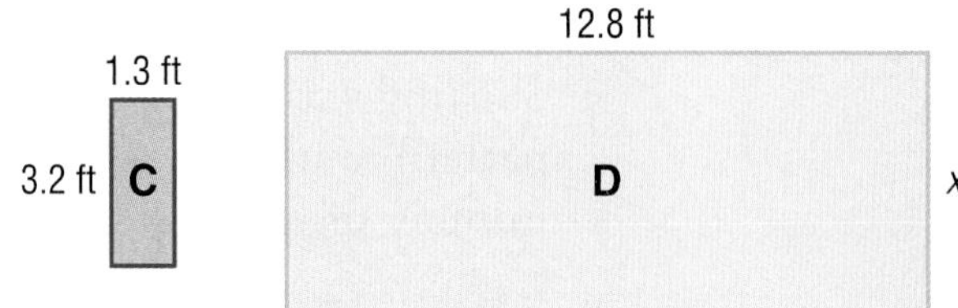

2. Ms. Palmer manages a craft shop. She uses this graph to quickly find the cost of different amounts of cotton fabric.

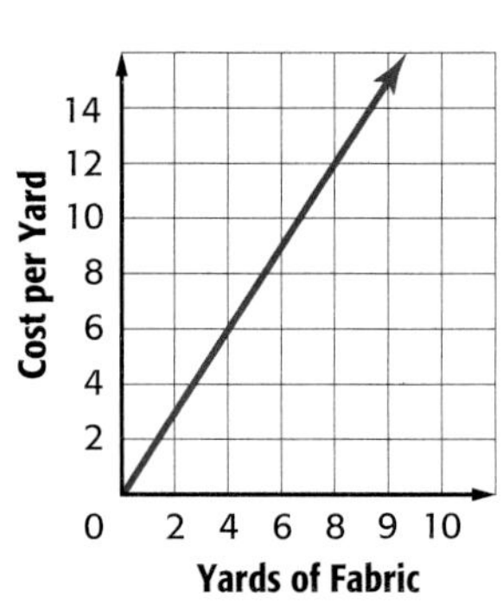

   a. Give the coordinates of two points that lie on the graph. Explain what the ordered pairs mean with respect to this real-world context.

   b. Show that your two points have the same ratio.

   c. A customer wants to buy 30 yards of fabric, but the graph does not extend that far. How much will the fabric cost?

   d. Give the coordinates of two other points that lie on the line but are beyond the boundaries of the graph. Show that the points have the same ratio.

Think & Discuss

When the seventh grade trail mix sale began, Mr. Malone, a Math teacher, bought the first six packages. From his purchase alone, the class raised \$27.00.

By the end of the first day of the sale, the class had raised \$121.50. How many packages had been sold by the end of the first day?

There is more than one way to solve an exercise involving proportional relationships. Ramiro, Mali, and Jeremy each thought about the trail mix problem in a different way, but all three found the same answer.

Ramiro used unit rates. He said, "I found the cost of one package of trail mix by solving for x. Then I divided \$121.50 by that amount."

$$\frac{\$27}{6 \text{ packages}} = x$$

$$\frac{\$121.50}{x} = \text{sold packages}$$

Mali wrote and scaled a ratio. She said, "I multiplied both parts of the ratio by the same number to find an equal ratio in the form of *\$121.50 to something.*"

$$\frac{\$27}{6 \text{ packages}} \cdot \frac{\$4.50}{\$4.50}$$

Jeremy wrote and solved an equation. He said, "I wrote an equation to find how many packages they must sell to raise \$121.50. Then I solved the equation to find the answer."

$$\frac{\$27}{6} = \frac{\$121.50}{n}$$

Develop & Understand: B

3. Complete each student's method to find the number of trail mix packages sold the first day. You should find the same answer with each method.

4. The last time Ian rented a canoe at Budget Livery, he paid \$28.95 for 3 hours. Today, he used the canoe for 4 hours 45 minutes. If the rental cost is proportional to time, what will be his rental charge?

5. Justin and his parents went to Ireland during his winter break. Before they left, Justin exchanged some money for euros. In exchange for $50 U.S., he received 74 euros. He returned with 10 euros.

   If the exchange rate is still the same, how many dollars will he receive for his euros?

6. First-Rate Paint wants to make a large batch of the shade below. The final batch must have 136 containers of blue and yellow paints. How many containers of blue paint will be needed?

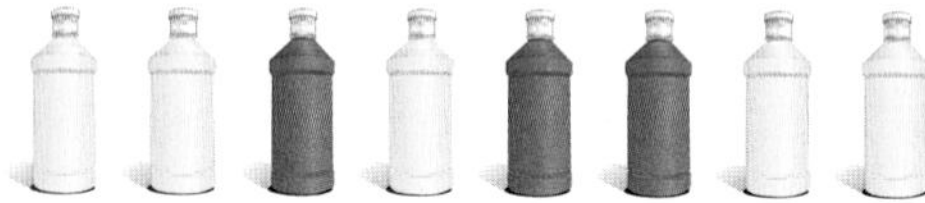

Share & Summarize

Choose one exercise from Exercises 3–6. Solve it using a different method than the one you previously used. Explain each step clearly enough that someone from another class could understand what you did.

Investigation 3 Solve Proportions

Vocabulary

proportion

You have solved many exercises involving equal ratios. An equation that states that two ratios are equal is called a **proportion**.

$$2{:}3 = 10{:}15 \qquad \text{or} \qquad \frac{n}{25} = \frac{3}{15}$$

One of the solution methods you used in Exercises 3–6 on pages 509 and 510 involved writing and solving a proportion. This method is useful for solving many different situations.

Example

Malik saw a photograph of Vincent van Gogh's famous painting *Starry Night* in a book about art history. The book stated that the actual painting is 73.7 centimeters high, but it did not give the width.

Malik decided to measure the photograph and calculate the painting's length. The photograph measured 6.9 centimeters high and 8.6 centimeters wide.

Since the photograph is a scaled version of the painting, the dimensions of the painting are multiplied by the same number to get the dimensions of the photograph.

That means that the photograph's $\frac{\text{height}}{\text{length}}$ ratio is a scaled height-to-length ratio for the painting. Since the two ratios are equivalent, Malik set up this proportion.

Actual dimensions ⟶ $\frac{73.7}{x} = \frac{6.9}{8.6}$ ⟵ **Photo dimensions**

To find x, he solved his equation.

$$\frac{73.7}{x} = \frac{6.9}{8.6}$$

$\frac{73.7}{x} = 0.8$ Divide 6.9 by 8.6.

$73.7 = 0.8x$ Multiply both sides by x.

$92.1 \approx x$ Divide both sides by 0.8.

Malik concluded that the real painting's length must be about 92.1 centimeters.

Real-World Link

The Dutch artist Vincent van Gogh (1853–1890) was a painter for only the last ten years of his life. Many of his famous works show motion through the use of swirling patterns of color.

Develop & Understand: A

1. Maya suggested to Malik that he could have used a different proportion.

$$\frac{x}{73.7} = \frac{8.6}{6.9}$$

   a. Explain why Maya's proportion is also correct.

   b. Solve Maya's proportion. Do you think it is easier or more difficult to solve than Malik's proportion?

2. Consider this situation.

   In January 2008, Aaron went to the bank and exchanged 50 Australian dollars for 56.61 U.S. dollars. Later that day, Nailah went to the same bank to exchange 72 Australian dollars for U.S. dollars. How many U.S. dollars did she receive?

   a. Set up two proportions you could use to solve this situation.

   b. Solve one of your proportions. How much money in U.S. dollars did Nailah receive?

3. Augustin earns $5 per hour mowing lawns. He wants to buy two DVDs that cost a total of $32.50. Set up and solve a proportion to determine how many hours he needs to work to earn enough money.

By now, you probably have several strategies for solving exercises involving proportional relationships and equal ratios. Now you will have a chance to practice these strategies.

Develop & Understand: B

Solve these exercises using any method you like.

4. This ratio table shows the number of cans of paint needed to cover various lengths of fence.

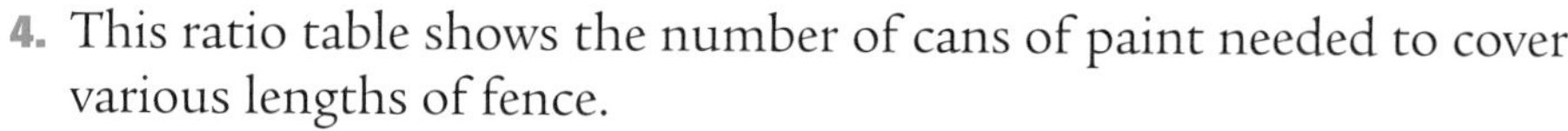

Cans of Paint	2	4	6	8	10	12
Feet of Fence	30	60	90	120	150	180

   a. Find the number of cans needed to paint 75 feet of fence.

   b. Find the number of cans needed to paint 140 feet of fence.

5. The Holmes family has two adults and three children. The family uses an average of 1,400 gallons of water per week. Last month, two houseguests stayed for a week. Assuming that the water usage is proportional to the number of people in the house, about how much water did the household use during the guests' visit?

6. The week before Thanksgiving, turkeys are on sale for $1.29 per pound at the EatMore grocery store. How much will a 14-pound turkey cost?

7. Alvin is using a trail map to plan a hiking trip. The scale indicates that $\frac{1}{2}$ inch on the map represents $1\frac{1}{2}$ miles. Alvin chooses a trail that is about $5\frac{1}{2}$ inches long on the map. How long is the actual trail?

8. Viviana loves mixed nuts. The local farmer's market sells a mix for $2.79 per pound. How many pounds can she buy for $10.85?

9. Joe is making his popular waffles for Sunday brunch. He knows that five waffles usually feed two people. If he wants to serve 11 people, how many waffles should he make?

10. At Camp Maple Leaf, there are two counselors for every 15 campers. The camp directors expect 75 campers next summer. How many counselors will they need?

Share & Summarize

Did you use the same strategy for solving all of the exercises in this investigation? If so, explain it. If not, choose one of the strategies you used and explain it.

Investigation 4 Map Scales

Vocabulary

map scale

A global positioning system, or GPS, is a device that can easily fit in your hand but provides a great deal of information about exactly where you are. You could also use a map to determine where you are in the world or in your own neighborhood.

Think & Discuss

Maps are drawn accurately by taking the actual distances of features on Earth and shrinking them.

Thinking back to stretching and shrinking machines, what are two different ways you can shrink a number?

Suppose you wanted to make a map of your room, and it needed to fit on a regular sheet of paper. How much would you need to shrink your room's dimensions?

In this investigation, you will use several strategies for finding the distance between two locations. These strategies include using a **map scale** and equivalent ratios to estimate the distance between two cities.

Explore

With a partner, measure the distance between three pairs of cities. Use the map scale to determine an estimate of the actual distance. Look up the actual distance between each pair of cities using the Internet. How close were your estimates?

What factors do you think might affect your estimate?

Suppose a map scale shows that 1 inch is equal to 150 miles in real life.

Map Key

0 mi 150 mi

You measure and find that the distance between two cities is about $3\frac{1}{4}$ inches. One way to determine actual miles is to use a ratio table.

Map Distance (in.)	$\frac{1}{4}$		1	2	3	$3\frac{1}{4}$
Actual Distance (mi)	37.5		150	300	450	487.5

So, the estimated distance between the two cities in real life is about 488 miles.

In addition to ratio tables, proportions can also be used to estimate distances. Consider the following example.

Example

The Chang family is planning the route for its annual vacation. The Changs are using a map with a scale of 1 inch equal to 50 miles. On the map, the roundtrip route measures 6.5 inches. Mrs. Chang sets up and solves a proportion to estimate the actual distance.

$$\frac{1 \text{ in.}}{6.5 \text{ in.}} = \frac{50 \text{ mi}}{x \text{ mi}}$$ Set up a proportion.

$$\frac{1}{6.5} = \frac{50}{x}$$ Divide out common units.

$$\left(\frac{6.5x}{1}\right)\left(\frac{1}{6.5}\right) = \left(\frac{6.5x}{1}\right)\left(\frac{50}{x}\right)$$ Multiply each side by the common denominator, $6.5x$.

$$x = 325$$ Simplify.

The Chang family trip will be about 325 miles.

Develop & Understand: A

Use your knowledge of ratio and proportion to find the following distances.

1. Douglas is helping his dad plan a trip from their house across town to his Uncle Warren's house. The city map scale shows that 0.5 inch is equal to 3 miles. After measuring streets and turns carefully, Douglas estimated the distance to be about 4.5 inches. How far does Douglas live from his Uncle Warren?

Use your knowledge of ratio and proportion to find the following distances.

2. Jessica wants to know about how far she lives from her grandmother. She is using a map that has a scale of 2 cm = 70 km. Jessica found the distance from her house to her grandmother's house to be about 5 centimeters.

   a. How far does Jessica live from her grandmother?

   b. Jessica will be traveling by bus to her grandmother's house. Suppose the bus travels at an average speed of 30 kilometers per hour. How long will the trip take, not including scheduled stops for gasoline and food?

3. Marquell drew a map that shows the bus route from her house to her school. She used a scale of 2 centimeters equal to 1 mile. Marquell measured the bus route distance as 9 centimeters. What is the actual distance from Marquell's house to the school?

4. The Johnsons planned their trip to the annual family reunion using a map with a scale of 1 inch equal to 25 miles. Measuring carefully, Mr. Johnson estimated the distance to be 12.5 inches. How far will the Johnsons travel on their trip?

5. To compete in a debate, the school debate team traveled 50 miles. Using a map, Joaquin measured the trip distance as 2.5 inches. On the map, how many miles did 1 inch represent?

Real-World Link

Travel on the ocean is measured in *nautical miles.* Travel on land is measured in *statute miles.* A statute mile is equal to 5,280 feet. A nautical mile is equal to 6,080 feet, or one minute of arc of a great circle of Earth.

Share & Summarize

Reading and interpreting maps are two important skills. Explain what the map scale represents to someone who has not used a map.

One map shows a distance of 1.5 inches between two cities. Another map shows a distance of 2 inches between the same two cities. Can both be correct? Explain.

Investigation 5 Similarity

Did you ever wonder how people measure the width of the Grand Canyon or the height of a giant Sequoia tree? Clearly, measurements like these cannot be made with a ruler or yardstick. Measurements of very large or very small objects or distances are often made indirectly, by using proportions.

Explore

Zach and Jin Lee are on the decorating committee for their town's Independence Day parade. They plan to hang balloons on the streetlights along the parade route. First, they want to determine how tall the streetlights are so they can find ladders of the right size.

It is late afternoon, and Zach notices that everything is casting long shadows. This gives him an idea.

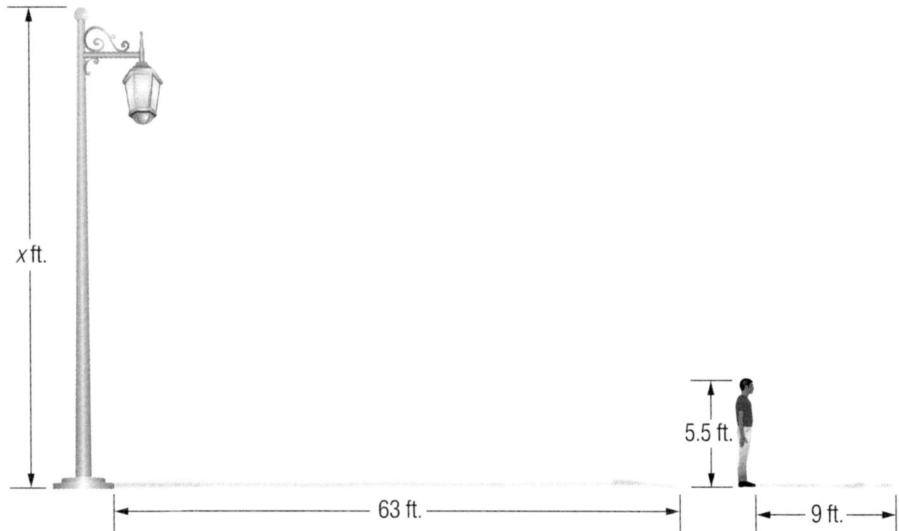

Zach asks Jin Lee to measure his shadow. She finds that it is about 9 feet long. Using a tape measure, Zach finds that the streetlight's shadow is about 63 feet long.

Zach says, "I know that my height is 5.5 feet. Since the sunlight is coming at me and the pole from the same angle, I have two similar triangles. I can figure out how tall the streetlight is!"

- What are Zach's similar triangles? Make a sketch if it helps you think about this.
- How do you know Zach's triangles are similar?
- How can you use the triangles to find the height of the streetlight?
- How tall is the streetlight?

Develop & Understand: A

Jin Lee likes Zach's method of using shadows to find heights. She shows her grandfather how to find the height of the tree in his yard.

She measures the length of the tree's shadow and finds it is 24.5 feet long. Then she holds a 12-inch ruler perpendicular to the ground and finds that it casts a 4.75-inch shadow.

1. Draw and label a simple sketch of the situation.

2. Describe two similar triangles Jin Lee can use to find the tree's height.
3. Set up a proportion Jin Lee could solve to find the tree's height.
4. How tall is the tree?
5. When Jin Lee found the tree's height, she thought all the measurements should be in the same unit. So, she converted everything to inches.

   Her grandfather asked her, "Could you have left inches as inches and feet as feet and still found the correct answer?" Explain why the answer to his question is yes.
6. Later, Jin Lee and Zach set up this proportion.

$$\frac{1 \text{ ft}}{n \text{ ft}} = \frac{4.75 \text{ in.}}{24.5 \text{ ft}}$$

   They found that the tree was only about 5 feet tall. What did they do incorrectly?

Share & Summarize

1. When setting up a proportion involving two triangles, why is it important for the triangles to be similar?
2. The rope on the flagpole at school has frayed and snapped, and a dog ran off with part of it. The principal wants to buy a new rope, but she does not know how tall the pole is.

   Write a paragraph explaining to the principal the method of indirect measurement used in this investigation. Make your explanation as complete as you can so she could use it to find the height of the pole.

Inquiry

Investigation 6 Estimate Heights of Tall Objects

Vocabulary

angle of elevation

Materials

- protractor
- string or thread
- weight (paper clip or cube)
- tape (optional)
- tape measure, yardstick, or meterstick

In this investigation, you will estimate the height of a tall object without actually measuring it. First, you will need to measure the angle of elevation.

The **angle of elevation** is the angle made by the imaginary horizontal segment extending from your eye to an object and the imaginary segment from your eye to the top of the object.

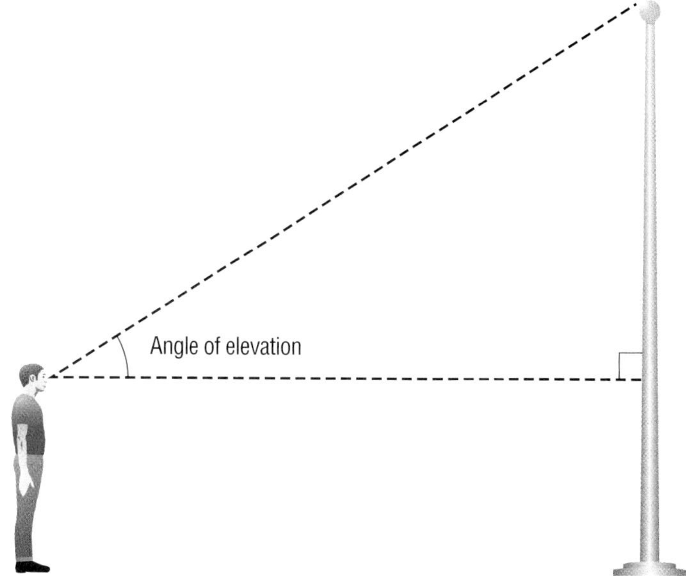

First, assemble your measurement instrument. Tie a piece of string to a protractor at the middle of the straight side. Attach a weight to the string. Now you are ready to measure the angle of elevation.

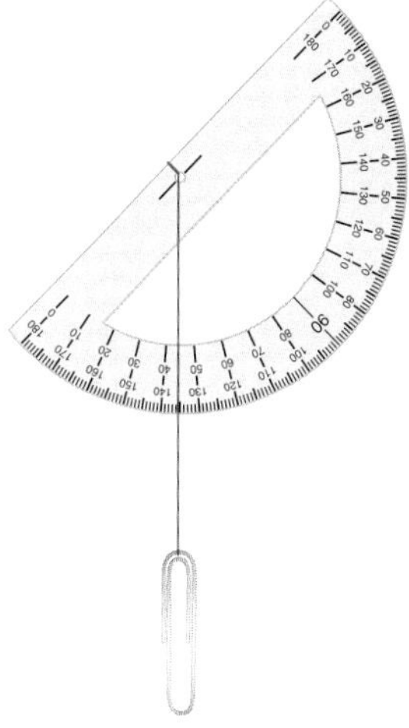

Go on

Select a position at some distance from the object your class will be measuring. Line up the straight side of your protractor with an imaginary line from your eye to the top of the object. On the drawing below, the top of the object from the person's point of view is labeled point *E*. Hold the protractor close to your eye.

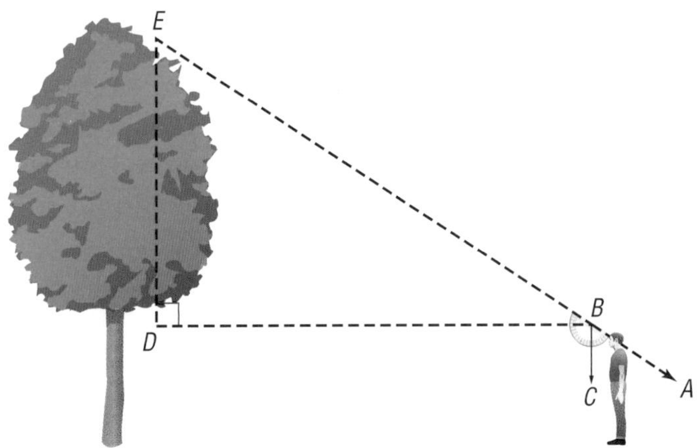

Collect Your Data

1. Look at angle *ABC* in the drawing. It is formed by the imaginary line through the top of the object and your eye and the line made by the string. Read the measure of this angle from your protractor. Record it.
2. Now measure the distance from the object to your position. This is side *DB* in the drawing above. You might need to mark where you are standing. Or, have a partner measure the distance while you remain standing where you are.

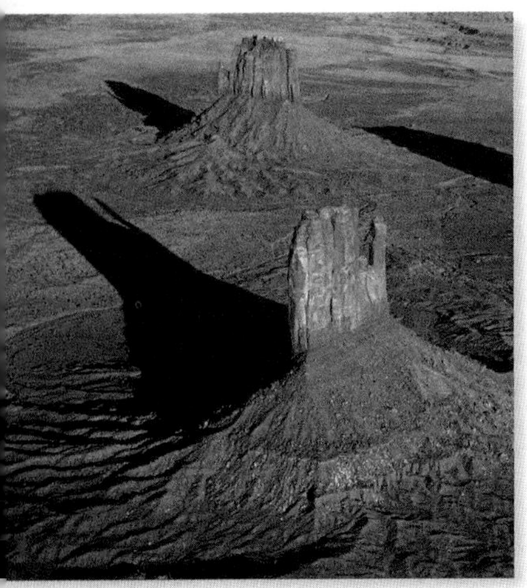

Math Link

In *similar triangles,* corresponding angles have the same measures.

Sketch the Situation

3. Make a sketch similar to the one on the previous page that represents your situation. Include the same labels for the important points. Label any lengths or angle measures you know.
4. You measured angle *ABC* with your device. Use that measurement to calculate the measure of angle *DBE,* which is the angle of elevation.
5. Use your protractor to draw a triangle similar to triangle *BDE.* Measure the sides of your triangle. Label your drawing with the side lengths.

Calculate the Height

You can use the triangle you drew for Question 5 to find the object's height.

6. Set up and solve a proportion to calculate the length of side *DE*.

7. Notice in your diagram that the length of side *DE* is not the height of the object. What is the last thing you need to do to find the height of the object? What is the object's height?

8. Compare the height you found with the heights your classmates determined. How close are they? What might account for any differences?

What Did You Learn?

9. Describe in your own words how you can use the angle of elevation to find the height of a tall object. Be sure to explain how similar triangles help.

On Your Own Exercises
Lesson 10.2

Practice & Apply

1. The Huntsville Craft Shop sells two types of pre-packaged beads. Here are the contents.

Itty Beady Beads	Bead-U-T-ful Beads
8 oz red	6 oz red
7 oz white	5 oz white
5 oz blue	4 oz blue

a. Is the amount of white beads in each package proportional to the total ounces of beads? Why or why not?

b. Is the amount of red beads in each package proportional to the total ounces of beads? Why or why not?

2. Is the number of red paint containers in these two mixtures proportional to the total number of containers? Explain how you know.

3. Ecology A hybrid car will travel many miles on a gallon of gas by using a combination of electricity and gasoline for fuel. The table shows estimates of how far the car will travel on various amounts of gas.

Gallons of Gas	0.5	1	1.5	2	2.5	3
Miles	20	40	60	80	100	120

Are the miles traveled proportional to the gallons of gas? How do you know? If so, describe how they are related and write the ratio.

4. Set up a proportion for the following situation. Solve it using any method you choose.

Grant Middle School held a recycling drive to raise funds for the class trip to Washington, D.C. The recycling center will pay \$1.20 per pound of aluminum. The school was paid \$7.20 for the aluminum cans that were collected.

How many pounds of aluminum did the school recycle?

5. Many schools have a recommended student : teacher ratio. At Quality Care Preschool, the ratio is 9 : 1. Next year, Quality Care expects enrollment to increase by 135 students. How many new teachers will need to be hired to maintain the student : teacher ratio?

6. Esperanza is following this recipe for cranberry oatmeal breakfast bars.

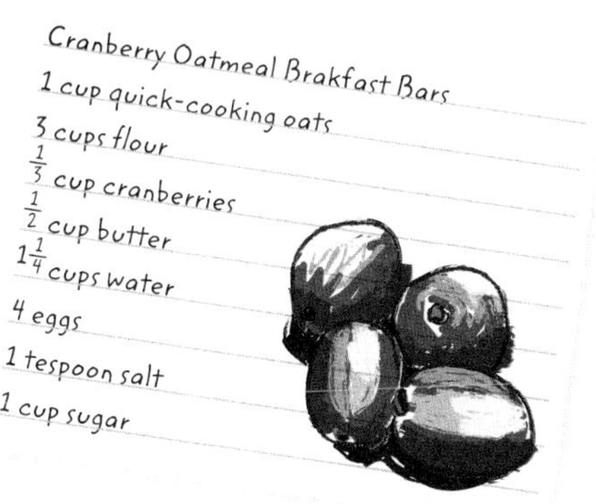

Cranberry Oatmeal Brakfast Bars
1 cup quick-cooking oats
3 cups flour
$\frac{1}{3}$ cup cranberries
$\frac{1}{2}$ cup butter
$1\frac{1}{4}$ cups water
4 eggs
1 tespoon salt
1 cup sugar

Just as Esperanza is preparing to mix the ingredients, she realizes she has only $\frac{1}{4}$ cup of cranberries, so she decides to make a smaller batch of breakfast bars. How much of each ingredient should she use?

7. One seventh-grade homeroom class has 20 students, 12 of which are girls. Another homeroom has 25 students, 14 of which are girls.

a. Do these two classes have the same proportion of girls? Explain how you know.

b. How many girls would have to be in a class of 25 students for it to have the same ratio of girls to students as the smaller class?

8. Using a map with a scale of 2 centimeters equal to 1 mile, Marcos measures the distance to his grandfather's house as 11 centimeters. Copy and complete the ratio table below to find the distance to Marcos' grandfather's house.

Map Distance (cm)	2	3	4	5	6	7	8	9	10	11
Actual Distance (mi)										

9. Confirm your answer to Exercise 8 by setting up and solving a proportion.

10. Haylee has a map that measures $8\frac{1}{2}$ inches by 11 inches. Briartown is situated in the center of the map. Suppose the map scale is $\frac{3}{4}$ inches to 40 miles.

The following cities are the given distances from Briartown. Each city is situated due north, due south, due east, or due west.

?
?
?
?
Briartown

a. Which cities could be on the same map as Briartown?

Riverside: 82 miles	Greenwood: 177 miles
Morsetown: 287 miles	Crestville: 431 miles
Everton: 197 miles	Franklin: 299 miles

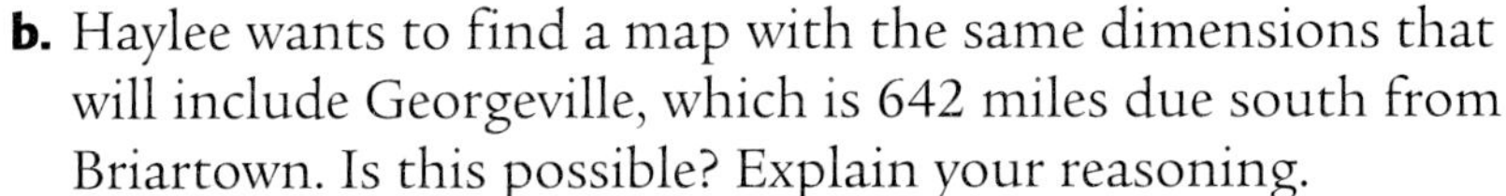

b. Haylee wants to find a map with the same dimensions that will include Georgeville, which is 642 miles due south from Briartown. Is this possible? Explain your reasoning.

11. A farmer wants to cut down three pine trees to use for a fence he is building. To decide which trees to cut, he wants to estimate their heights. He holds a 9-inch stick perpendicular to and touching the ground. He measures its shadow to be about 6 inches. If he wants to cut down trees that are about 40 feet tall, how long are the tree shadows?

12. Booker is working on a layout for the school newspaper. He is using a unit of measure commonly used in publishing called a pica. The allotted space for a photograph is 10 picas by 15 picas. The photograph he wants to use is 8 inches by 10 inches. Are the allotted space and the photograph similar? Can Booker use the photograph for the layout?

13. At 212 feet high, the tallest Ferris wheel in the United States is at Fair Park in Dallas, Texas. Suppose a man standing near the Ferris wheel casts a shadow of 3.5 feet. At the same time, the Ferris wheel's shadow is 110 feet long.

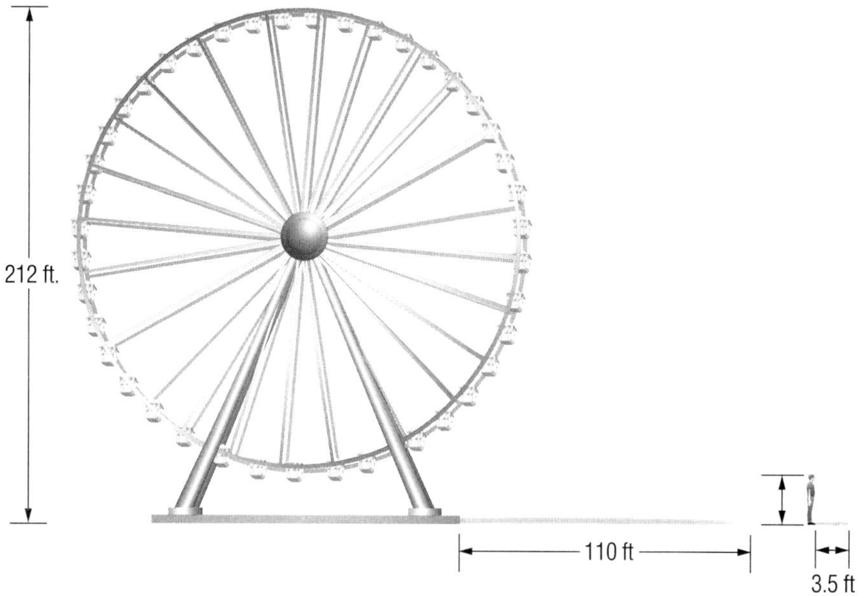

a. Set up a proportion to find the approximate height of the man.

b. Approximately how tall is the man?

Connect & Extend

14. This recipe makes twelve servings. Write a new recipe that will serve eighteen people and has correctly proportioned ingredients.

Fresh and Fancy Fruit Salad
$\frac{1}{2}$ cup white grapes
1 cup strawberries
$\frac{1}{2}$ cups peaches
$\frac{3}{4}$ cup apples
$\frac{1}{4}$ cup pineapple juice
2 oranges
1 tsp. ginger
$\frac{1}{3}$ tsp. salt

15. Mr. Mills turned 30 years old on the day of his wedding. Today, he and his wife will celebrate their 30th wedding anniversary and his 60th birthday. Mr. Mills has celebrated twice as many birthdays as wedding anniversaries. Does this mean that these numbers are proportional?

For each tile pattern, decide whether the numbers of tiles of any two colors are proportional to each other. That is, consider three ratios, white:blue, blue:red, and white:red, for each stage of the tile patterns. Explain your answers.

16.

17.

18. Renee is trying to set up a proportion to determine how much gas she will need for a 1,000-mile car trip. She averages 30 miles per gallon of gas. Here is what she wrote.

$$\frac{1{,}000 \text{ miles}}{G \text{ gallons}} = \frac{1 \text{ gallon}}{30 \text{ miles}}$$

a. Explain why Renee's "proportion" is not correct.

b. Set up a correct proportion. Solve it using any method you choose.

c. Find another proportion that Renee could use to solve this situation. Solve your proportion to check your answer to Part b.

19. Casey found a photograph of himself that was taken when he was three years old. In the photo, he is standing next to a bookcase that his parents still have. Explain how he could figure out how tall he was when the photo was taken.

20. **Challenge** Consider the map described in Exercise 10. Suppose cities can also be situated diagonally from Briartown. What is the maximum distance a city can be located from Briartown and still be on the map?

21. Economics When traveling to another country, you often have to exchange U.S. dollars for the country's currency. Exchange rates for currency fluctuate every day, depending on world economies. For this exercise, use these exchange rates.

One U.S. dollar ($) equals:

- 0.5 English pound (£)
- 0.99 Canadian dollars ($)
- 0.67 Euro (€)
- 106.7 Japanese yen (¥).

a. How many Canadian dollars would you receive for $250 U.S.?

b. How many U.S. dollars would you receive for 375 yen?

c. Suppose you are returning to the United States from a trip to England, and you have £200 left. How many U.S. dollars will you receive for your pounds?

d. Suppose you are traveling from France to England and have 1,523 euros left. How many pounds will you receive for your remaining euros?

The dollar, equal to 100 cents, is part of a decimal currency system used in several countries, including Australia, New Zealand, Liberia, and the Bahamas.

22. Challenge Alma's family took trips to England during two summer breaks. The exchange rates for both of the trips are shown below.

2006: 1 U.S. dollar equals 0.53 English pound
2007: 1 U.S. dollar equals 0.49 English pound

Alma's mother converted $1,200 into pounds for both trips. How many pounds did the family have to spend for each trip? What is the difference in the amount of money the family had for these trips in U.S. dollars, using 2007's exchange rate?

23. In Your Own Words An architect has agreed to visit a seventh grade class to talk about how architects use scale drawings in their work. She is not sure all the students will know what it means when she talks about shapes being "in proportion." Write some advice telling her how to explain what this means.

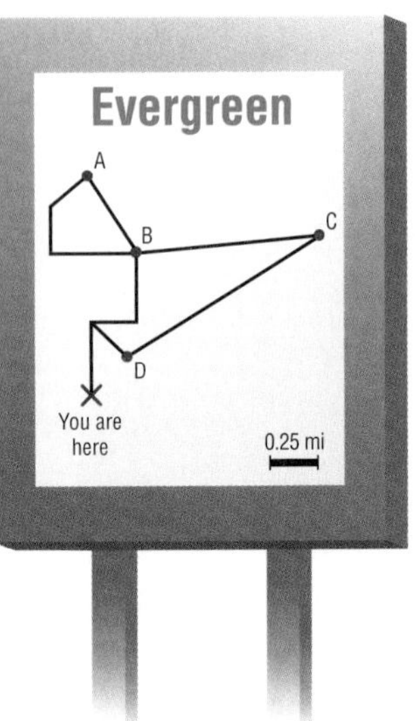

24. Della's scout troop hikes in the Evergreen Recreational Area. The troop is planning a trip, and Della wants to provide recommendations for three different hiking distances. The options are for hikes of short, medium, and long distances. She also wants to provide different sightseeing choices.

Help Della decide recommended routes by answering the following questions.

a. If Outlook Point is at point A, what is the shortest distance someone can hike to see it?

b. What is the shortest possible route that allows a hiker to see all points of interest, points A–D and return to the original starting point?

c. Assuming that hikers start and return to the "You Are Here" spot, describe a route that hikers could take that would be between five and six miles in length.

25. Nick and Maina are members of Della's troop. They have volunteered to make a map that is a smaller version of the wooden sign shown above.

a. If the map needs to fit on an 8.5″ by 11″ sheet of paper, what scale can they use and still make the map as large as possible?

b. Suppose Nick and Maina try to put the map on a 12-by-12-inch sheet of paper. What difficulty would this present? How could they overcome the difficulty?

26. Prove It! One way to check whether rectangles are similar is to place them so that one of the corners overlaps, as shown below. Then look at the diagonals of each rectangle.

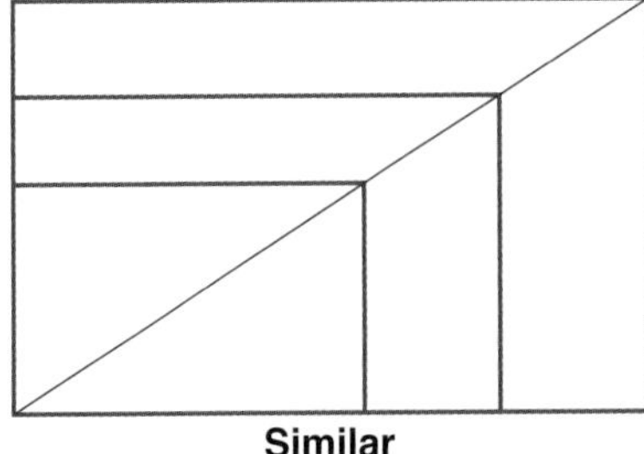
Similar

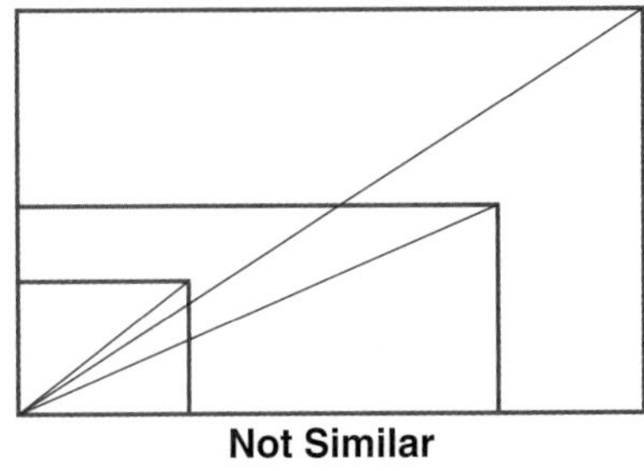
Not Similar

If the diagonals all align, as in the diagram on the left, the rectangles are similar. Prove that this is true. (Hint: Consider the triangles that are formed by the diagonals.)

27. **Challenge** These two right triangles are similar.

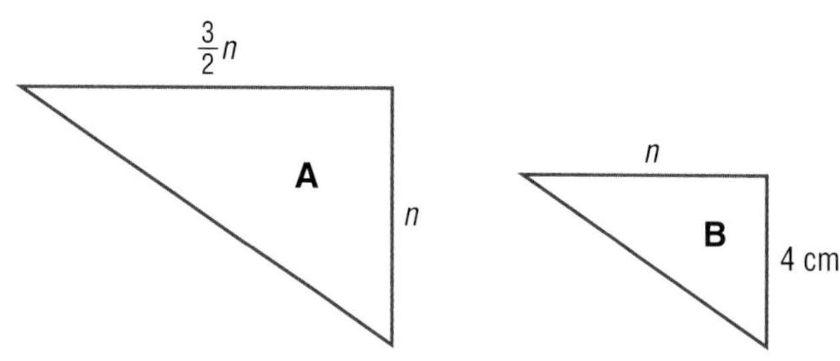

Find the lengths of the legs of both triangles. Show your work.

Mixed Review

Expand each expression.

28. $5(m + 2)$

29. $6(2y + 9)$

30. $\frac{1}{2}(6k - 30)$

31. $1.2(5p - 1.2)$

32. $3a(2a + 5b)$

33. $-7s(6r - 11s)$

Factor each expression.

34. $7t + 14$

35. $33n - 44$

36. $5c + 5d$

37. $10a^2 - 20ab$

38. Eleven times a number increased by thirty is equal to 45 less than the same number.

 a. Write an equation to solve for the missing number n.

 b. Solve the equation for n.

39. Write two equations with a solution of 6.

For Exercises 40–45, identify to which sets each of the following numbers belong. Choose from natural numbers, whole numbers, integers, and rational numbers.

40. $\frac{17}{2}$

41. 0

42. -45

43. -6.373

44. $\sqrt{16}$

45. $\sqrt{7}$

Percents and Proportions

Meredith Middle School is putting on a production of *The Music Man.* The eighth grade had 300 tickets to sell, and the seventh grade had 250 tickets to sell. One hour before the show, the eighth grade had sold 225 tickets, and the seventh grade had sold 200 tickets.

Inv 1 Find Percents 530

Inv 2 Proportions Using Percents 534

Think & Discuss

Which grade was closer to the goal of selling all its tickets? Explain your answer.

To figure out which grade was closer to its goal, you may have tried to compare the ratios 225 : 300 and 200 : 250. One way to compare ratios easily is to change them to percentages.

Investigation 1 Find Percents

Materials

- graph paper (optional)

This diagram is called a *percent diagram.* It represents the seventh and eighth grade ticket sales. The heights of the left and right bars represent the goal for each grade. The bar in the middle is called the *percent scale.* It is a common scale for the two different ratios.

Notice that the three bars are the same height even though they represent different numbers. On the percent scale, the height always represents 100%.

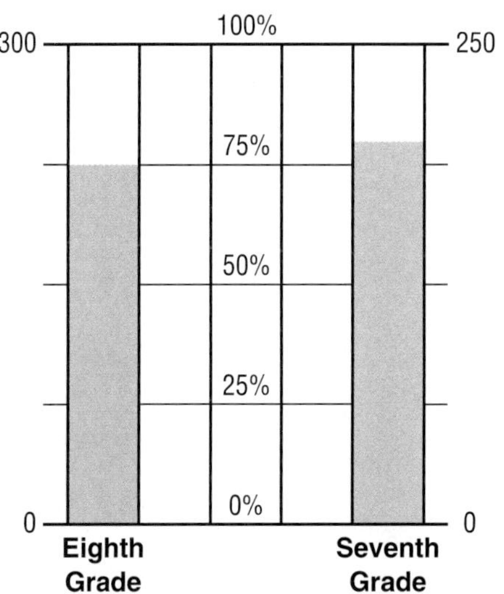

Percents make it possible to represent both grades' ticket sales as some number out of 100. For the eighth grade, 100% represents 300 tickets. For the seventh grade, 100% represents 250 tickets.

Think & Discuss

The eighth grade sold 225 tickets, and the seventh grade sold 200 tickets. Why is the bar for the seventh grade taller, when they sold fewer tickets?

Develop & Understand: A

Jefferson Middle School has 600 students, and Memorial Middle School has 450 students. For Exercises 1–6, use this percent diagram to estimate an answer.

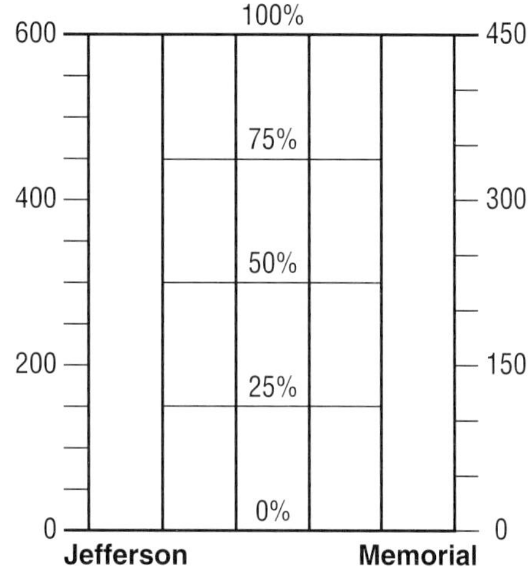

1. A survey of the two schools finds that 300 Jefferson students watch more than one hour of television every night, and 270 Memorial students watch more than one hour per night.
   a. What percentage of Jefferson students watch more than one hour of TV each night?
   b. What percentage of Memorial students watch more than one hour of TV each night?
   c. Comparing percentages, do more Jefferson students or more Memorial students watch more than one hour of TV each night?
2. Jefferson has 275 girls and Memorial has 250 girls. Which school has a greater percentage of girls? Explain.
3. At each school, 75% of the students play a musical instrument.
   a. About how many students at Jefferson play an instrument?
   b. About how many students at Memorial play an instrument?

4. The Math teachers at each school selected 50 students to represent the school at a Math competition.

   a. About what percentage of Jefferson students attended the Math competition?

   b. About what percentage of Memorial students attended the Math competition?

5. Suppose you looked for *P*% of 450 and *P*% of 600 on a percent diagram. Which would be the greater number? Explain your answer in terms of the diagram.

6. Which represents a greater percentage, *t* out of 450 or *t* out of 600? Explain your answer.

Develop & Understand: B

Look again at the ticket sales by the seventh and eighth grades at Meredith Middle School. The eighth grade had 300 tickets to sell, and the seventh grade had 250 tickets to sell.

7. At the end of the first week, each grade had sold 40% of its goal.

   a. Draw a percent diagram to represent this situation.

   b. Explain how you can use your diagram to estimate the number of tickets the seventh grade sold. Give your estimate.

   c. Use your diagram to estimate the number of tickets the eighth grade sold.

8. One hour before the show started, the seventh grade had sold 200 tickets and the eighth grade had sold 225. Melanie used her calculator to express the ratios of sold tickets to sales goal for both grades as percentages. How could you use your calculator to find $\frac{225}{300}$ as a percentage?

9. Suppose each grade was only 25 tickets short of meeting its goal.

   a. How many tickets would the eighth grade have sold?

   b. How many tickets would the seventh grade have sold?

   c. Calculate the percentages sold for each grade. Comparing percentages, which grade came closer to its goal?

10. By the time the play began, each grade had sold exactly 270 tickets.

a. What percentage of its goal did the eighth grade sell?

b. What percentage of its goal did the seventh grade sell?

c. Draw a percent diagram to represent this situation. Since the seventh grade sold *more* than its goal, you will need to extend the bar for the seventh grade.

11. Determine which ratio in each pair is greater by finding the percentage each represents.

a. $\frac{56}{69}$ or $\frac{76}{89}$

b. $\frac{106}{210}$ or $\frac{206}{310}$

c. $\frac{50}{45}$ or $\frac{210}{205}$

12. Find four ratios equivalent to 75%.

13. Find four ratios equivalent to 115%.

Share & Summarize

Here are two statements about Jefferson and Memorial Middle Schools.

a. Jefferson has 275 girls and Memorial has 250 girls, so there are more girls at Jefferson.

b. Jefferson is almost 46% girls and Memorial is almost 56% girls, so there are more girls at Memorial.

1. Which statement uses a common scale to make a comparison?

2. Which statement do you think is a better answer to the question, "Which school has more girls?" Explain your thinking.

Investigation 2 Proportions Using Percents

Materials

- graph paper (optional)

In Lesson 10.2, you used ratios to write proportions and represent real-world situations. In the last investigation, you saw that you can use percentages as a common scale to compare ratios. Percents can also be used to solve equations.

Develop & Understand: A

Work with your group to solve these exercises. Be ready to explain your reasoning.

1. Two department stores, Gracie's and Thimbles, are having holiday sales. The stores sell the same brand of men's slacks. Gracie's usually sells them for \$50 but has marked them down to 30% of that price for the sale. At Thimbles, the slacks are marked down to 40% of the usual price of \$30. Which store has the better sale price? Explain.

2. Steve and Anna went shopping during a sale at the KC Wheels bike store. Anna looked at her receipt and saw that the total before the discount was \$72, but she paid only \$40. Steve's total before the discount was \$38, but he paid only \$18. Who saved a greater percentage off the original price? Explain.

3. You are making punch for the class party. The punch will contain juice and soda with at least 70% juice. You have four gallons of juice. What is the greatest amount of punch you could make?

One way to solve a percent situation is to set up and solve a proportion. In most percent situations, you have values for two things and you want to know the value of the third. Your proportion might look like this.

$$\frac{b}{a} = \frac{n}{100}$$

For example, in Exercise 3 above, you knew the percent of juice that the punch should contain, 70%. You know how much juice you actually had, four gallons. You could have set up this proportion.

$$\frac{4}{a} = \frac{70}{100}$$

Think & Discuss

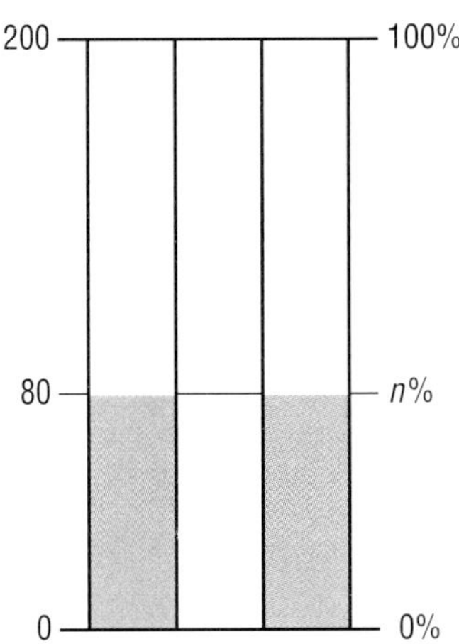

This type of percent diagram is used to help compare two numbers. State in words the situation represented by this percent diagram.

Write a proportion that is illustrated by the diagram and solve it. How did you think about setting up the proportion?

Develop & Understand: B

4. Consider this question. *What percent of 150 is 6?*
   a. Draw a percent diagram to represent the question.
   b. Express the question as a proportion and solve it.

5. Now consider this question. *What is 185% of 20?*
   a. Draw a percent diagram to represent the question.
   b. Express the question as a proportion and solve it.

6. A gardener planted 13% of his tulip bulbs in the border of his garden and the rest in the garden bed. In spring, every one of the bulbs grew into a tulip. He counted 45 tulips in the border. How many bulbs had he planted altogether? Show how you found your answer.

7. Of 75 flights leaving from Hartsfield Airport in Atlanta, 33 went to the west coast. What percentage of the 75 flights went to the west coast? Show how you found your answer.

8. The largest land animal, the African bush elephant, may weigh as much as 8 tons. However, that is only about 3.9% of the weight of the largest animal of all, the blue whale. How heavy can a blue whale be? Show how you found your answer.

Real-World Link

The enormous blue whale's primary source of food is plankton, tiny animals and plants that float near the surface of the water.

9. According to the Bureau of Labor Statistics, in 2006, the average American spent about $2,376 on entertainment, including admission fees, pets, hobbies, and entertainment equipment. They spent 363% of that amount on housing. How much did the average American spend on housing in 2006?

10. According to the U.S. Census Bureau, in 1950, the estimated world population was 2.6 billion. In 2007, it was 6.6 billion.

    a. What percentage of the 2007 population is the 1950 population?

    b. What percentage of the 1950 population is the 2007 population?

    c. In 2007, about 303 million people lived in the United States. What percentage of the world's population lived in the United States in 2007?

Share & Summarize

1. Write a proportion that expresses the statement 32% of 25 is b. Solve your proportion for b.

2. Write a proportion that expresses the statement n% of a is b. Explain why your proportion is useful for solving percent situations.

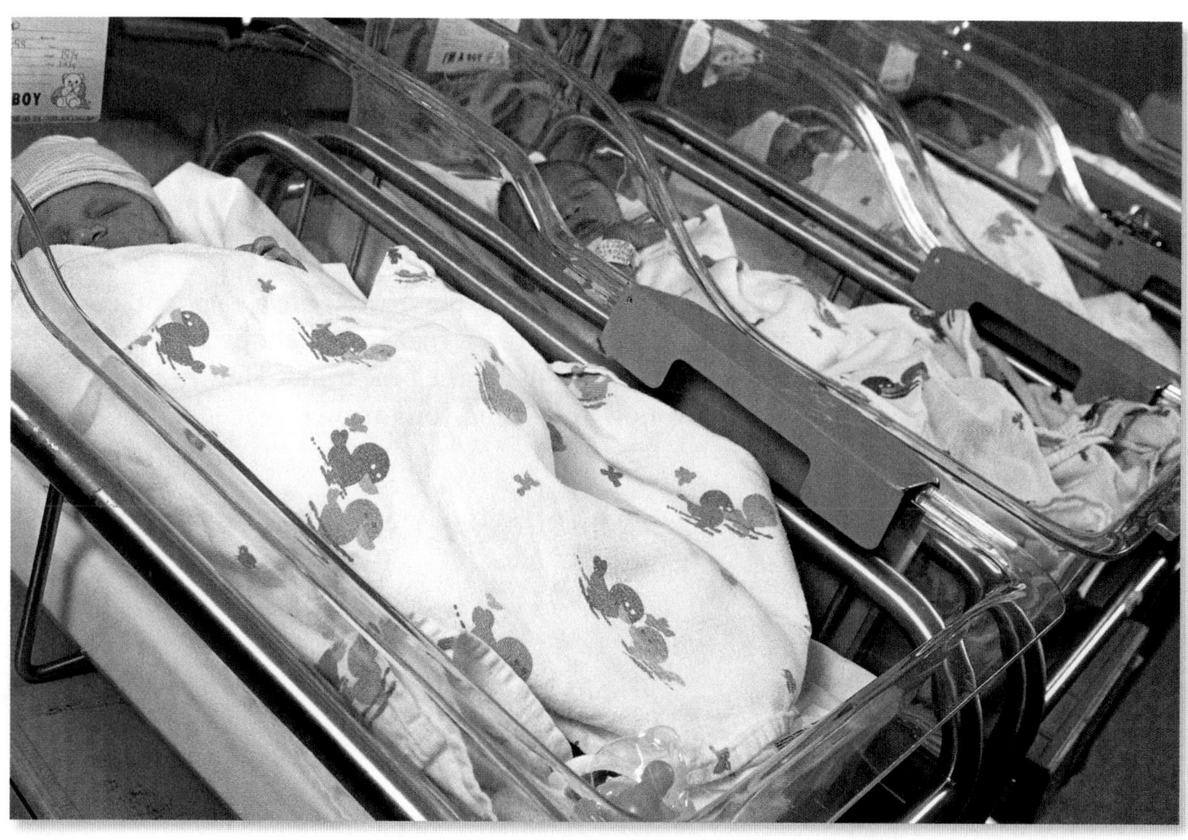

1. Travel surveys show that 68% of people who travel for pleasure do so by car. Of 338 students from Brown Middle School who traveled for pleasure last year, 237 did so by car. The same year, 134 students from Mayville Middle School traveled for pleasure, and 88 did so by car.
   a. Draw a percent diagram to illustrate this situation.
   b. From which school did a higher percentage of those who traveled for pleasure do so by car?
   c. How do the two schools compare to the survey results?

For each pair of ratios, determine which is greater by finding the percentages they represent.

2. $\frac{13}{12}$ or $\frac{65}{60}$ 3. $\frac{23}{28}$ or $\frac{44}{51}$ 4. $\frac{328}{467}$ or $\frac{16}{23}$

5. Find four ratios equivalent to 30%.
6. Find four ratios equivalent to 25%.

In Exercises 7–9, decide which is a greater number.

7. P% of 13 or P% of 23
8. P% of 368 or P% of 712
9. P% of 3 or P% of 4.7

In Exercises 10–12, decide which represents a greater percentage.

10. t out of 100 or t out of 200
11. t out of 89 or t out of 91
12. t out of 0.7 or t out of 0.5

13. Consider the question *5% of what number is 15?*
    a. Draw a percent diagram to represent the question.
    b. Express the statement as a proportion and solve it.

14. 98% of what number is 50?
15. 117% of what number is 3.5?
16. 7.5 is what percent of 9?
17. 20 is what percent of 15?

18. Seth works 12 hours a week. He recently received a 5% raise. His new weekly pay is $115.92.

a. What was Seth's weekly pay before the raise?

b. What was his hourly pay before the raise?

c. What is his new hourly rate?

19. In Woodrow Wilson Middle School, three-fourths of the students study a foreign language.

a. Suppose 540 students study a foreign language. How many students attend Woodrow Wilson Middle School?

b. Of the students studying a foreign language, 60% study Spanish. How many students study Spanish?

c. Of the students studying a foreign language, 15% study Chinese. How many students study Chinese?

20. Nora and her mother went shopping for jeans. Two stores, Jeans Are Us and Denim World, carried Nora's favorite pair of jeans. At Jeans Are Us, the jeans were discounted 20% off the regular price of $32. At Denim World, the jeans typically sold for $35. However, the jeans had been discounted an additional 10% off a 15% sale price.

From which store should Nora have purchased the jeans? Explain your reasoning.

21. Social Studies In 2000, about 20,530,000 children between the ages of 10 and 14 lived in the United States. That was about 7.3% of the entire U.S. population. Find the 2000 U.S. population.

Connect & Extend

22. Create a real-world word problem that requires finding 65% of 120. Show how to solve your situation.

23. The students at Graves Middle School held a fundraiser to help purchase new gym equipment. The school newspaper reported that the eighth grade raised the most, $234. The sixth and seventh grades each raised 30% of the total amount collected. However, the paper did not report the total amount raised by the three grades.

Use the information given to find the total amount raised and the amounts raised by the sixth and seventh grades.

24. Challenge Suppose $S > 0$ and $S = 2T$.

a. Which is greater, 80% of T or 40% of S? Explain your answer.

b. Does your answer to Part a hold if $S < 0$? Explain.

25. The table below shows data collected by a school district regarding its high school students.

	Athletes, per 100 Students	Honor Roll Students, per 100 Students
Freshmen	23	35
Sophomores	48	47
Juniors	41	54
Seniors	52	36

a. Suppose the district has about 1,350 high school students. About how many students at each grade level are athletes? About how many students are on the honor roll?

b. The entries in both columns are "per 100 students." What would be the entry for freshmen athletes if the scale was per 10 students? What if the scale was per 1,000 students?

26. In Your Own Words Describe several ways that percents are used to compare different quantities.

Mixed Review

Determine whether the following statements are *always, sometimes,* or *never* true.

27. A rational number is also an integer.

28. A natural number is also a whole number.

29. A smaller whole number can be given than the original whole number given.

30. A number can be both rational and irrational.

For Exercises 31 and 32, identify which number is greater. Explain how you found your answer.

31. $\frac{5}{8}$ or $\frac{10}{22}$

32. -9.275 or $-9\frac{6}{20}$

In Exercises 33–36, ΔABC is a right triangle, and c represents the hypotenuse. Use the Pythagorean Theorem to find each missing side. Round answers to the nearest hundredth, if necessary.

33. $a = 6$ inches, $b = 8$ inches, $c = ?$

34. $a = 5$ centimeters, $b = 7$ centimeters, $c = ?$

35. $a = 5$ feet, $b = ?$, $c = 13$ feet

36. $a = ?$, $b = 4$ units, $c = 9$ units

Rates

Rates are special types of ratios. You might not realize that you read, write, and speak about rates in everyday situations. A rate, like a *ratio*, involves division. What is unique about rates is that they represent the comparison of two related, but different quantities.

Inv 1 Unit Prices 540

Inv 2 Converting Currencies 544

When someone has a job, he or she thinks about how money and time are related. The more hours a person works, the more money he or she earns. You can say that time and money are *related*. You can write the rate as money compared to time or dollars per hour.

Think & Discuss

Can you think of more related quantities?

What about distance compared to time?

In this lesson, you will learn to read, write, and solve real-world problems involving rates.

Investigation 1 Unit Prices

Vocabulary

unit rate

Materials

- newspaper circulars with grocery store sales prices

Going to the market or grocery store is a real-life situation loaded with mathematics. To be a wise consumer, you want to learn how to get the best buy for your money. There are different factors that determine how most people spend their money, but one of the most common factors is the lowest unit rate.

In this investigation, you will explore real-world situations related to shopping and the use of unit rates.

Think & Discuss

When have you heard prices stated "per unit"?

List three different situations that involve unit rates.

Think of a cell phone plan. Name some parts of the plan that are given in a unit rate. How much is each minute you use? How much is each text message?

Example

These *are* unit rates.	These are *not* unit rates.
2 lb oranges/$1.00	3/$15.00 to purchase t-shirts
$3.55/gallon for gas	100/$5.00 for text messages
0.25 miles/$1 for a taxi	$25/8 hours for parking
$10/hour to rent a bicycle	$200/5 days of work

In Course 1, you learned that a unit rate is a term used when a quantity is compared to one unit of another quantity. In the example, each unit rate has a denominator equal to one.

Math Link

When you talk about a *unit* price, the unit must always be included so it is clear what quantities and relationships are being compared. If you say oranges are $2.25, others would not know if that is for one orange, one bag of oranges, or one box of oranges. The units make a difference.

Explore

With a partner, prepare for a trip to the market or grocery store. You only have $25 to spend. You want to make sure to spend the money wisely, so you use newspaper advertisements to comparison shop.

Your task is to create a list of six similar items that are listed in at least two of the flyers. Then determine the unit rate of each item in order to determine which items are the best buys.

Example

The cost of six cans of fruit juice is $3.20. How would you determine the cost of one can of the fruit juice?

To determine the cost of one can of fruit juice, set up the ratio $\frac{\$3.20}{6 \text{ cans}}$. By dividing the quantities, you find the cost of one can is about $0.53. So, the unit price can be written as the ratio $0.53/1 can.

You can also find unit rates by setting up and solving proportions. If six cans cost $3.20, how much does one can cost?

$$\frac{\$3.20}{6 \text{ cans}} = \frac{x}{1 \text{ can}}$$

$$1 \text{ can} \cdot \left(\frac{\$3.20}{6 \text{ cans}}\right) = 1 \text{ can} \cdot \left(\frac{x}{1 \text{ can}}\right)$$

To solve, multiply each side by 1 can, the denominator of the right side.

$$\frac{\$3.20}{6} = x$$

Divide out the common units on both sides.

$$\$0.53 = x$$

Divide $3.20 by 6. Round to the nearest cent.

One can costs $0.53.

Develop & Understand: A

1. Edwin is shopping with his mom. He sees that oranges are on sale. A bag of 15 oranges cost \$3.75. Calculate the unit rate of each orange.

2. Mandy and her twin brother Miles are looking at a sales flyer for a local discount store. They see that one of their favorite snacks is on sale at three snacks for \$4.50. If the regular cost of one bag of the snacks is \$1.95, how much will Mandy and Miles save on one bag of snacks at the sale price?

3. Mr. Hernandez is comparison-shopping for DVDs at two stores. Hi-Mart is selling three DVDs for \$41.94. Super-Saver is selling four DVDs for \$54.84. If Mr. Hernandez wants to purchase one dozen DVDs, which store is offering the better buy?

4. Parker is going to a professional baseball game with his scouting group. The regular cost of a ticket is \$23.50 per person. His scout leader bought the tickets at a special group rate. The total cost was \$300 for all 16 people to attend the game. What was the cost per ticket at the group rate? How much did the scouts save?

Unit rates can also be expressed using the opposite relationship. In Exercise 3, you found the unit rate of one DVD at Super-Saver to be \$13.71. You can also express the relationship as how many DVDs you can buy for one dollar.

Think & Discuss

How would you rewrite the unit rate for the DVDs in terms of a per dollar rate?

Why do you think that you do not typically see unit rates expressed this way?

Are there other situations where this might be useful?

Develop & Understand: B

5. Destiny wants to go roller-skating with her friends. The skating rink offers a "Terrific Tuesday" rate of \$15 for four hours of skating. How much are Destiny and her friends being charged per hour?

6. If Nina drives from Spokane, Washington, to Portland, Oregon, in 6.5 hours, what is her speed in miles per hours?

Spokane
365 mi
Portland
Springfield
Medford

7. Juan and Amelia are driving from school to a concert. The concert location is 18 miles from school. Suppose it takes Juan 33 minutes to drive from the school to the concert. It takes Amelia 25 minutes to drive the same route. How much greater is Amelia's speed?

Develop & Understand: C

Knowledge of unit rates can help budget expenses for a project.

8. Mallory and her sister want to paint the four walls of their rectangular-shaped bedroom. The bedroom's length is 10 feet, the width is 8 feet and the height is 9 feet. One gallon of paint costs $25 and covers 300 square feet. Suppose Mallory and her sister want to apply two coats of paint. Will they be able to paint the bedroom for a cost of $50? Explain your reasoning.

9. The Fuentes family is comparison shopping for a fish aquarium. A local pet store sells two sizes of aquariums. The aquariums' dimensions and prices are listed below.

	Width (in.)	Depth (in.)	Height (in.)	Price
Aquarium 1	24	12	16	$156.74
Aquarium 2	36	15	16	$239.99

Which aquarium is the better buy? Explain your thinking.

Share & Summarize

Compare these two advertisements.

Which is the better deal? Use rate, proportion, and unit rate in your explanation.

Investigation 2 Converting Currencies

Materials

- newspaper circulars or retail catalogs
- calculator
- current currency rate information from the Internet

As you discovered earlier, you will want to know the current exchange rate for the U.S. dollar if you travel to another country. *Currency* is the legal tender of a country. In the United States, it is the dollar. The exchange rate for the U.S. dollar with foreign currencies fluctuates, or changes, daily.

In this investigation, you will use your calculator to convert currencies of other countries to U.S. dollars. Comparing currencies involves ratios.

Use your newspaper circular or catalog to find the retail prices of five different items. Then, select a country that you would like to visit. Convert the price of your five items from U.S. dollars to your selected country's currency.

Real-World Link

In 2005, 49.4 million foreign visitors spent a total of 104.8 billion dollars in the United States.

Example

Mr. Marcel is in Japan. He wants to buy a watch for his daughter. The watch costs 7,696.58 Japanese yen. Mr. Marcel sees the current exchange rate posted in the store.

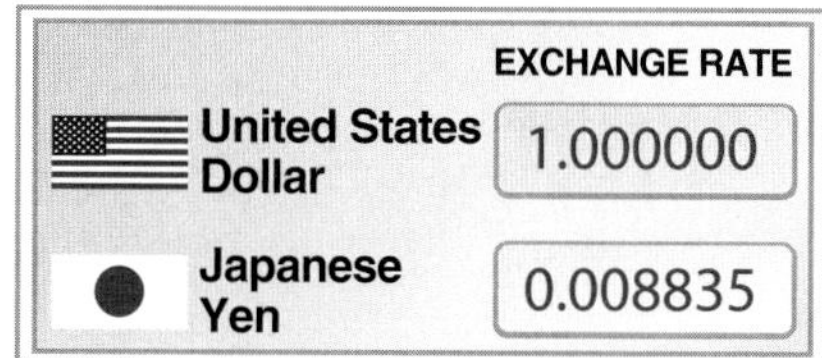

EXCHANGE RATE
United States Dollar
1.000000
Japanese Yen
0.008835

What is the cost of the watch in U.S. dollars? To find out, you can set up and solve a proportion.

- Set up your proportion.

$$\frac{1 \text{ Japanese yen}}{0.008835 \text{ U.S. dollars}} = \frac{7{,}696.58 \text{ Japanese yen}}{x \text{ U.S. dollars}}$$

- Multiply both sides by x U.S. dollars.

$$\frac{1 \text{ Japanese yen} \cdot (x \text{ U.S. dollars})}{0.008835\ 7 \text{ U.S. dollars}} = \frac{7{,}696.58 \text{ Japanese yen} \cdot (x \text{ U.S. dollars})}{x \text{ U.S. dollars}}$$

- Simplify each side.

$$\frac{1 \text{ Japanese yen} \cdot (x)}{0.008835} = 7{,}696.58 \text{ Japanese yen}$$

- Divide both sides by 1 Japanese yen.

$$\frac{1 \text{ Japanese yen} \cdot (x)}{(1 \text{ Japanese yen}) \cdot 0.008835} = \frac{7{,}696.58 \text{ Japanese yen}}{1 \text{ Japanese yen}}$$

- Simplify each side.

$$\frac{x}{0.008835} = 7{,}696.58$$

- Multiply both sides by 0.008835.

$$x \approx 68$$

So, $x \approx 68$. The watch costs about $68.

Develop & Understand: A

Real-World Link

In 2007, global tourism reached a record level with 898 million people visiting a foreign land.

1. Kara went shopping with her sister Sarah in Ireland. The currency in Ireland is the euro. The exchange rate was 1 U.S. dollar equals 0.69638 euro. The girls paid 179.50 euros for a digital camera. What was the approximate cost of the digital camera in U.S. dollars?

2. Zhen lives in Mexico. For a summer vacation, she went to China to visit family members. While visiting, she bought a bracelet that cost 34.50 pesos. The bracelet would cost about 23.48796 Chinese yuan renminbi. What was the approximate exchange rate of pesos to yuan?

3. Vincent traveled abroad with several of the school's foreign language club members. The students visited Switzerland, the United Kingdom, and Romania. Each student had to raise \$800 for travel money. How much would the \$800 be worth in each country?

   1 U.S. dollar = 0.69638 euro and 0.69638 euro (EUR) = 1.12326 Switzerland francs (CHF)

   1 U.S. dollar = 0.69638 euro and 0.69638 euro (EUR) = 0.51962 United Kingdom pounds (GBP)

   1 U.S. dollar = 0.69638 euro and 0.69638 euro (EUR) = 2.53557 Romania new leu (RON)

In addition to setting up a proportion to find an exchange rate, you can use your knowledge of equations and graphs to convert currencies from different countries.

Think & Discuss

Suppose you know that the current exchange rate between U.S. dollars and Mexican pesos is \$1.00 to 10.75. How can you set up an equation that shows this relationship?

Graph your equation. Discuss what you can determine from the graph.

Develop & Understand: B

In Exercises 4 and 5, a relationship is given between currencies. Describe each relationship with an equation. Graph the equation.

4. 1 euro equals 1.45 U.S. dollars.

5. 1 U.S. dollar equals 113.185 Japanese yen.

6. You have previously converted Fahrenheit and Celsius temperatures. What similarity is there between currency conversions and temperature conversions?

Share & Summarize

Discuss with a partner why it is important to know the correct daily currency exchange rate when you are traveling abroad.

How could you use the daily rates if you are in traveling in another country?

Practice & Apply

1. Taryn is shopping with her mother. Bananas are on sale at two different stores. The Best Cost Market advertised price is 4 pounds for \$1.25, while the Super Saver Market is advertising their bananas 3 pounds for \$0.75.

   Which store has the better price? If the other store agrees to match the sale price of the less expensive store, what will be that store's loss on its original price?

2. The Chang family is remodeling its kitchen and wants to tile the floor. A box of 18 slate tiles costs \$54.86. A dozen granite tiles cost \$42.36. Which is the better buy based on unit cost?

3. TaKeisha and Jeffery are tiling a mural. The cost of 142 blue tiles was \$156.00. The cost of 81 white tiles was \$74.00. What was the unit cost of each of the tiles?

4. Belinda wants to buy several strings of miniature lights. One package contains 150 lights and costs \$4.65. Another package contains 100 lights and costs \$3.87. Belinda needs 600 lights. Which package of lights should she purchase? Explain your reasoning.

5. Mrs. Ramirez is shopping for marigolds to plant in her garden. She needs 48 total plants. A six-pack flat of marigolds costs \$1.95. A 48-pack flat of marigolds costs \$10.80.

   a. Which flat is the better buy? Explain your reasoning.

   b. Describe a situation in which Mrs. Ramirez may be willing to pay a higher unit cost for the marigolds of her choice.

6. Weston is planning a spring break trip to Venezuela. Venezuelan bolivar is worth 2,148.80 to 1.00 U.S. dollar. Weston plans on taking \$850 with him on the trip. If Weston converts $\frac{3}{4}$ of his money to bolivars, how many bolivars will he receive in return?

7. Yogi is returning from India where he visited his grandmother this summer. He wants to convert 450 Indian rupees he has back into euros, but there are two ways to do this. Yogi could convert rupees to pounds and then pounds to euros.

   1 Indian rupee is equal to 0.01281 British pounds.

   1 British pound is equal to 1.37259 euros.

   Or, Yogi could convert rupees directly into euros?

   1 Indian rupee is equal to 0.01758 euros.

   Which way do you recommend? Why?

8. Twins Andrew and Megan are in college. They are going to Russia as exchange students for the fall semester. They found the following information on the Internet about the value of Russian currency.

$$1 \text{ US dollar} = 24.6432 \text{ Russia rubles (RUB)}$$

Andrew has saved $1,350.00 to take on the trip, and Megan has saved $\frac{7}{8}$ that amount for the trip. If the twins convert all of their U.S. dollars into the Russian ruble, how much money will each have to spend while in Russia?

Connect & Extend

9. The graph below shows the approximate relationship between the costs of four popular brands of dog food. Write a sentence comparing the unit costs for each brand of dog food.

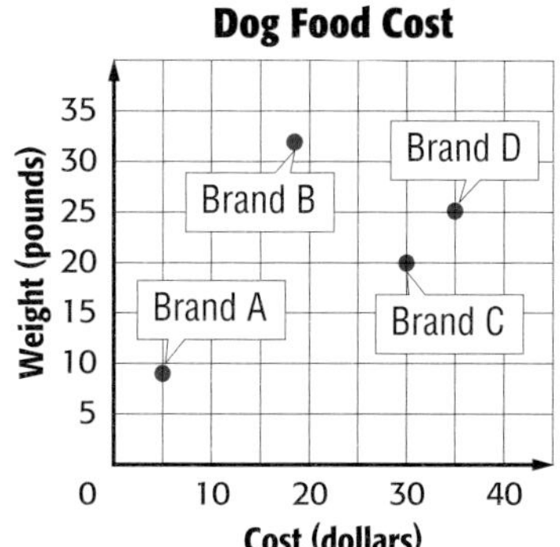

10. At Super Cost Market, ground beef has three different prices. Beef that is $\frac{85}{15}$, containing 15% fat, costs $13.85 for 15 pounds. The ground beef that is $\frac{90}{10}$ costs $14.85 for a 12.5 pound package. The ground beef that is $\frac{95}{5}$ costs $17.36 for a 14-pound package.

    a. Identify the unit cost of each package of ground beef.

    b. Find the difference in cost between the most expensive unit cost and the least expensive unit cost.

    c. Find the difference between the most expensive unit cost and the middle unit cost.

    d. Make a conjecture about why the $\frac{95}{5}$ ground beef cost more than the $\frac{90}{10}$ and the $\frac{85}{15}$ ground beef.

11. Peyton is pricing video games for his new game system. Game A has six game variations and costs $27.99. Game B has five game variations and costs $25.39. Peyton is also considering game C, which costs $31.49 and has a total of eight game variations.

    If Peyton likes each of the games equally, which would be the better buy? Explain your reasoning.

12. Shannon is purchasing a new tank for her fish. She wants to buy the largest tank for the money. The tank shown below on the left costs $100. The tank shown below on the right costs $114. Which tank would be the better buy for Shannon? Explain your reasoning.

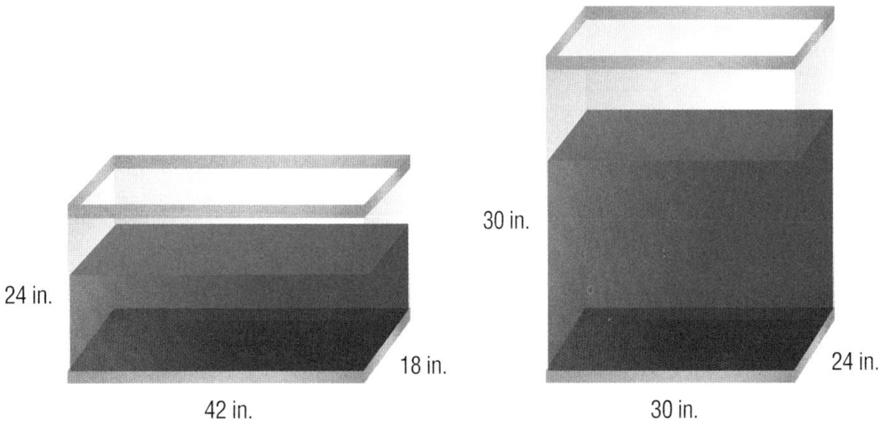

13. Ruby is getting ready for her Sweet Sixteen party. She paid $11.60 before tax for eight 2-liter bottles of beverages. Later that day, she saw an advertisement for 12-packs of beverages in 12-ounce cans. The twelve packs are on sale for $2.75 per pack.

a. If Ruby wants approximately the same amount of beverages, would the 12-packs have been a better buy? Explain your reasoning process.

b. Is there any reason Ruby might have decided that her original purchase was a better idea? Explain your reasoning.

14. Jamal wants to purchase perfume for his mother's birthday. A 3-ounce bottle of perfume is on sale for one-third the cost of a 5.4 ounce bottle. The 5.4 ounce bottle costs $36. Which bottle is the more economical purchase in terms of unit cost? How do you know?

15. Study the following relationships.

1 Canadian dollar = 1.01595 U.S. dollar

1 U.S. dollar (USD) = 0.98430 Canadian dollar (CAD)

Based on this information, would you make more profit selling a home that cost $260,000 in Canadian dollars or a $263,000 home in U.S. dollars? Explain your reasoning.

16. Camille's class is studying scientific notation. Her teacher wrote the following on the board.

$$1 \text{ Mexican peso} = 6.417 \times 10^{-2} \text{ euro}$$

What is the value of 235 Mexican pesos in euros? Write the solution in scientific notation. Round to the thousandths place.

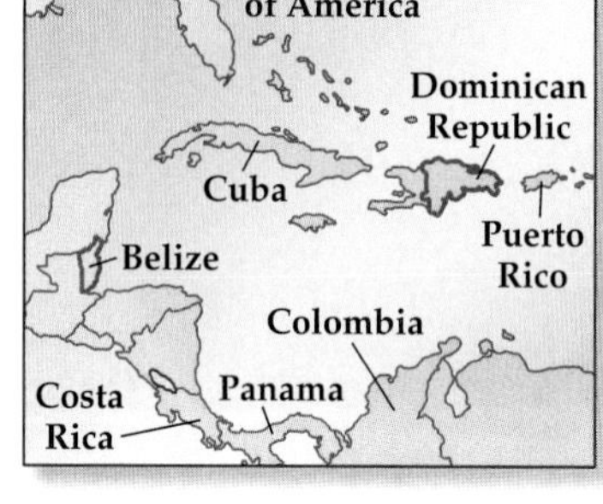

17. Mr. Sanchez was moving from Belize to the Dominican Republic. He learned that the currency exchange rate are as follows.

$$3{,}500 \text{ Belize dollar} = 60{,}576.9 \text{ Dominican R. peso}$$

$$3{,}500 \text{ Dominican R. peso (DOP)} = 202.222 \text{ Belize dollar (BZD)}$$

What is the unit rate for Belize dollars to Dominican pesos? What is the unit rate for Dominican pesos to Belize dollars based on this information? Explain your reasoning.

18. In Your Own Words Explain how you can use unit rates to choose the better buy between two food items when grocery shopping.

Mixed Review

19. Suppose Laurie invests \$200 on January 1, 2000, at 8% interest per year. Her original investment and the yearly interest are reinvested each year for ten years.

a. How much money is in Laurie's account at the end of 2000 (to be reinvested for 2001)?

b. Make a graph, on axes like those at right, showing the growth of the account's value for the ten years the money is invested.

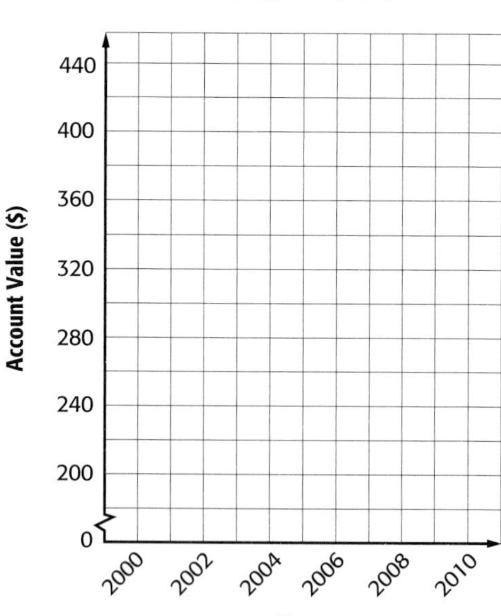

Solve each equation or inequality.

20. $5x + 3 = 2x + 10$

21. $\frac{1}{2}y + \frac{2}{5} = \frac{1}{4}y - \frac{3}{2}$

22. $12n - 9 > 4n + 1$

23. $\frac{5}{3}t + 7 \geq \frac{8}{3}t - 10$

Review & Self-Assessment

Vocabulary

angle of elevation

map scale

proportion

proportional relationship

Chapter Summary

Comparisons can take many forms, including ratios, rates, and percents. In this chapter, you compared ratios using equivalent ratios, unit rates, and ratio tables.

You learned how to use proportions to solve real-world problems involving ratios and percents. You also used proportionality when working with similar figures and map scales.

Next, you used percent diagrams to show a common scale for two different ratios. In addition, you used percents to solve equations representing real-world situations.

Rates are special types of ratios. You used rates to find unit prices and calculate currency conversions.

Strategies and Applications

The questions in this section will help you review and apply the important ideas and strategies developed in this chapter.

Comparing and scaling ratios and rates

1. The Quick Shop grocery store sells four brands of yogurt.

Brand	Maddie's	Quick Shop	Rockyfarm	Shannon
Price	2 for \$1.50	3 for \$2	\$0.75 each	\$0.80 each
Size	8 oz	6 oz	6 oz	8 oz

a. For each brand, find the ratio of price to ounces.

b. For each brand, find the unit price.

c. Use the unit rates to list the brands from least expensive to most expensive.

d. Explain how you could use the ratios in Part a instead of the unit rates to list the brands by how expensive they are.

2. Erin is making punch for her friends. To make the drink, she mixes three quarts of fruit juice with $\frac{1}{2}$ quart of ginger ale.

   a. Find the ratio of fruit juice to ginger ale in Erin's punch.

   b. Find the ratio of fruit juice to punch, that is, to the final drink.

   c. Suppose Erin has only two quarts of fruit juice. How much ginger ale should she add?

   d. For a party, Erin wants to make seven quarts of punch. How much fruit juice and ginger ale does she need?

   e. Explain how you found your answer for Part d.

Writing and solving proportions

3. Mac found a box full of paperclips. He wanted to know about how many paperclips were in the box without counting them one by one. All the paperclips were silver, so Mac had an idea.

   Mac added 20 colored paperclips, mixed all the paperclips in the box thoroughly, and then, without looking, drew out 20 paperclips. He had drawn 3 colored paperclips and 17 silver paperclips.

   a. Write a proportion that Mac could use to estimate how many paperclips are in the box.

   b. Solve your proportion.

   c. How many paperclips should Mac guess are in the box?

Using proportions to explore similarity

4. Nicolas wanted to know the height of the flagpole in front of his school. He measured his friend's height and the lengths of the shadows cast by his friend and the flagpole's shadow. Nicolas' measurements are listed below.

   Height of friend: 5 feet

   Friend's shadow: 2 feet

   Flagpole's shadow: 10 feet

   a. Write a proportion Nicolas could use to calculate the height of the flagpole.

   b. Solve your proportion.

Using percents to make comparisons

5. Social Studies Public elementary and secondary schools in the United States receive money from different levels of government. The table shows National Education Association estimates for three states for the 2005–06 school year.

School Funding Sources (Thousands of Dollars)

State	Total	Federal	State	Local and Intermediate
California	69,054,472	7,256,396	39,302,232	17,495,944
Maine	2,309,974	221,601	941,399	1,146,974
Mississippi	3,768,462	571,554	2,038,039	1,158,869

a. Which of these states received the most federal money?

b. Which of these states relies most on federal money?

c. Explain why Part b is a different question from Part a.

Demonstrating Skills

6. Examine the tile pattern.

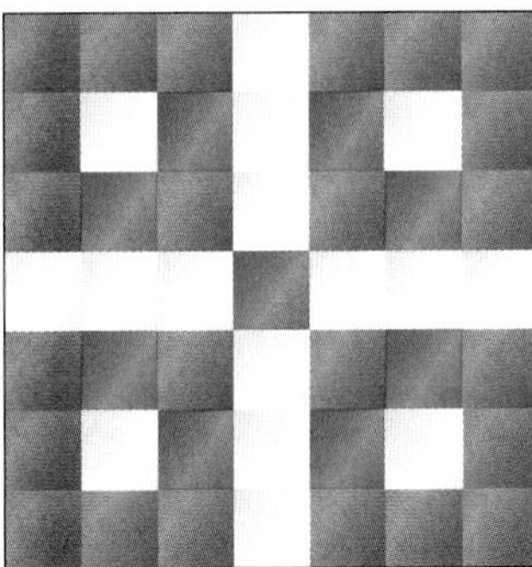

a. Write the ratio of white tiles to purple tiles.

b. Write at least five ratios for the tile pattern.

7. Paul wants to tile a large area using this pattern. Make a ratio table of possible numbers of purple and white tiles that he could use.

Find the value of the variable in each proportion.

8. $\frac{12}{5} = \frac{x}{9}$

9. $\frac{3.2}{7} = \frac{4}{7}$

10. $\frac{92}{36} = \frac{23}{w}$

11. $\frac{a}{4.7} = \frac{13}{61.1}$

Find the missing measures for these similar figures.

12.

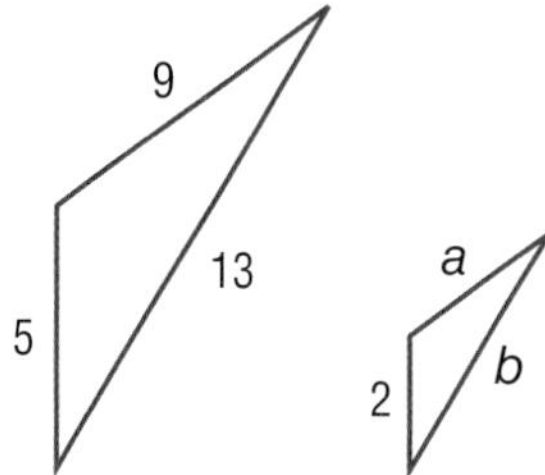

13.

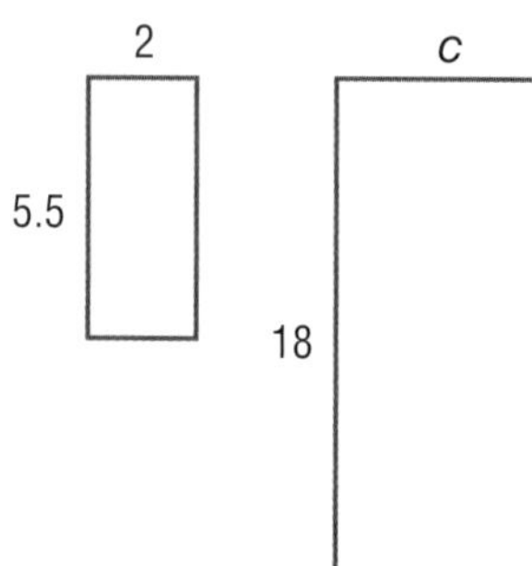

For Exercises 14–24, set up and solve a proportion to answer each question.

14. 17 is 5% of what number?

15. What percent of 450 is 25?

16. What is 32% of 85?

17. Jocelyn finds bouncy balls on sale. A box of 12 bouncy balls costs $1.44. Calculate the unit rate of each ball.

18. Jeremiah is looking for juggling balls for his school's juggling club. A Catch Above the Rest sells their juggling balls for 3 for $4.50. Just Jugglers sells their juggling balls for 5 for $10. Which store offers the better deal?

19. Jesse wants to go to the pool. The pool offers a special rate after 5 P.M. of four hours for $10. How much will Jesse pay per hour of swimming?

20. Susan and Alexis are going to a spa. The spa charges $150 for four hours. How much will Susan and Alexis pay per hour at the spa?

21. Bi-Rite grocery store has cans of chicken soup on sale for $2.40 for six cans. The cans are normally priced $0.60 a can. How much will you save per can with the sale?

22. One U.S. dollar equals 0.99688 Canadian dollars. How many U.S. dollars equal one Canadian dollar?

23. One British pound equals 208.817 Japanese yen. How many British pounds equal one Japanese yen?

24. One U.S. dollar equals 0.69058 euros. How many euros equal 1 U.S. dollar?

Test-Taking Practice

SHORT RESPONSE

1 A tall building casts a shadow of 15 meters. Stacie, who is 1.3 meters tall, casts a shadow of 15 centimeters. How tall is the building? Watch your units of measure.

Show your work.

Answer ____________________

MULTIPLE CHOICE

2 Which of the following shows the cross products of $\frac{7}{12} = \frac{x}{8}$?

A $7(12) = 8x$
B $12x = 7(8)$
C $7x = 12(8)$
D $x = 7(8)$

3 A map shows the following scale. How many kilometers would be represented by 3.2 centimeters?

1 centimeter = 8 kilometers

F 25.6
G 32
H 256
J 320

4 A candidate earned 100 out of the first 125 votes cast in an election. Suppose the candidate earned the same ratio of total votes cast. How many votes did the candidate receive if a total of 720 votes were cast?

A 695
B 576
C 100
D 80

5 The cost for 3 pounds of apples is $4.95. At the same rate, which of the following would be true?

F 1 pound costs $1.60
G 2 pounds costs $3.40
H 4 pounds costs $6.60
J 5 pounds costs $7.95

6 Samantha bought a new DVD player for $50 off its original price. The $50 was a 20% discount. Find the original cost of the DVD player.

A $40
B $100
C $250
D $300

7 In a tile floor, there are 5 blue squares for every 2 green squares. Which of the following is *not* an equivalent ratio?

F 10:4
G 10:7
H 15:6
J 100:40

Glossary/Glosario

Math Online A mathematics multilingual glossary is available at glencoe.com. The glossary includes the following languages.

Arabic	Cantonese	Hmong	Russian	Urdu
Bengali	English	Korean	Spanish	Vietnamese
Brazilian Portuguese	Haitian Creole	Portuguese	Tagalog	

Cómo usar el glosario en español:

1. Bussca el término en inglés que desees encontrar.
2. El término en español, junto con la definición, se encuentran en la columna de la derecha.

English

Español

A

absolute value **(p. 127)** The absolute value of a number is its distance from 0 on the number line, and is indicated by drawing a bar on each side of the number. For example, $|-20|$ means "the absolute value of -20." Since -20 and 20 are each 20 units from 0 on the number line, $|20| = 20$ and $|-20| = 20$.

valor absoluto **(pág. 127)** El valor absoluto de un número es su distancia desde 0 en la recta numérica, lo cual se indica trazando una barra en cada lado del número. Por ejemplo, $|-20|$ significa "el valor absoluto de -20". Como -20 y 20 se encuentran a 20 unidades de 0 en la recta numérica, $|20| = 20$ y $|-20| = 20$.

B

backtracking **(p. 19)** A method of solving algebraic equations by working backwards from the known answer to figure out the value of the variable.

vuelta atrás **(pág. 19)** Método en que se empieza con la respuesta y se trabaja de atrás hacia adelante para despejar la variable y resolver ecuaciones algebraicas.

base **(p. 93)** The number in an exponential expression that is multiplied by itself. For example, in t^3, t is the base; in 10^4, 10 is the base.

base **(pág. 93)** El número en una expresión exponencial que se multiplica por sí misma. Por ejemplo: en t^3, t es la base; en 10^4, 10 es la base.

base **(p. 216)** The parallel faces of a prism.

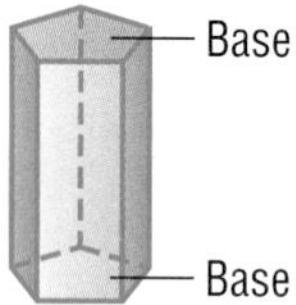

base **(pág. 216)** Las caras paralelas de un prisma.

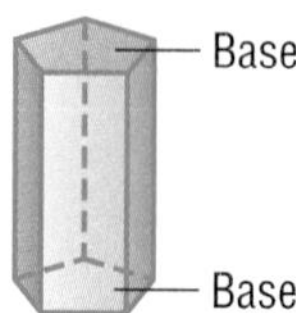

C

common factor **(p. 80)** A common factor of two or more numbers is a number that is a factor of all the numbers. For example, 5 is a common factor of 15, 25, and 40.

common multiple **(p. 83)** A common multiple of a set of numbers is a multiple of all the numbers. For example, 24 is a common multiple of 3, 8, and 12.

composite number **(p. 78)** A whole number greater than 1 with more than two factors. For example, 12 is a composite number since it is greater than 1 and has more than two factors. In fact, it has six factors: 1, 2, 3, 4, 6, and 12.

conjecture **(p. 477)** A statement that is believed to be true but has not yet been proven.

customary system **(p. 247)** The system of measurement used in the United States to measure length in inches, feet, yards, and miles; capacity in cups, pints, quarts, and gallons; weight in ounces, pounds, and tons; and temperature in degrees Fahrenheit.

cylinder **(p. 217)** A figure that is like a prism, but its two bases are circles.

Cylinder

factor común **(pág. 80)** Un factor común de dos o más números es un número que es un factor de todos los números. Por ejemplo, 5 es un factor común de 15, 25 y 40.

múltiplo común **(pág. 83)** Un múltiplo común de un conjunto de números es un múltiplo de todos los números. Por ejemplo, 24 es un múltiplo común de 3, 8 y 12.

número compuesto **(pág. 78)** Un número entero mayor que 1 con más de dos factores. Por ejemplo, 12 es un número compuesto dado que es mayor que 1 y tiene más de dos factores. De hecho, tiene seis factores: 1, 2, 3, 4, 6 y 12.

conjetura **(pág. 477)** Enunciado que se cree que es verdadero, pero el cual no se ha probado todavía.

sistema inglés **(pág. 247)** Sistema de medidas que se usa en Estados Unidos para medir la longitud en pulgadas, pies, yardas y millas; la capacidad en tazas, pintas, cuartos de galón y galones; el peso en onzas, libras y toneladas; y la temperatura en grados Fahrenheit.

cilindro **(pág. 217)** Figura que parece un prisma, pero que tiene un par de bases circulares.

Cilindro

D

distance formula **(p. 353)** The symbolic rule for calculating the distance between any two points, (x_1, y_1) and (x_2, y_2), in the coordinate plane: distance $= \sqrt{(x^2 - x^1)^2 + (y^2 - y^1)^2}$.

distributive property **(p. 49)** The distributive property of multiplication over addition states that for any numbers n, a, and b, $n(a + b) = na + nb$. The distributive property of multiplication over subtraction states that for any numbers n, a, and b, $n(a - b) = na - nb$.

fórmula de la distancia **(pág. 353)** Regla simbólica para calcular la distancia entre cualquier par de puntos, (x_1, y_1) y (x_2, y_2), en el plano the coordenadas: distancia $= \sqrt{(x^2 - x^1)^2 + (y^2 - y^1)^2}$.

propiedad distributiva **(pág. 49)** La propiedad distributiva de la multiplicación sobre la adición establece que para cualquier número n, a, y b, $n(a + b) = na + nb$. La propiedad distributiva de la multiplicación sobre la sustracción establece que para cualquier número n, a, y b, $n(a - b) = na - nb$.

double-bar graph (p. 296) A graphical display that uses paired horizontal or vertical bars to compare data.

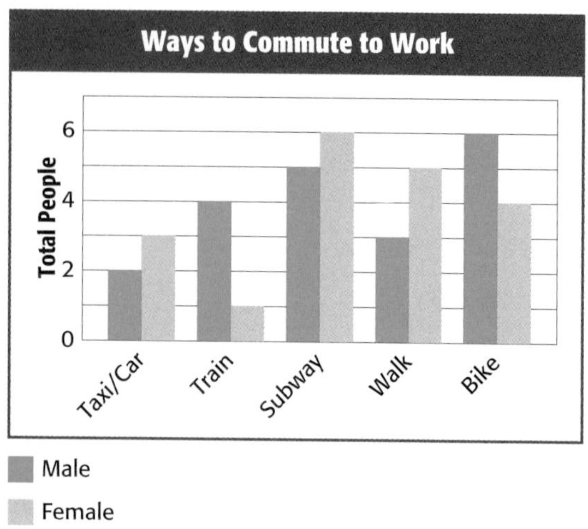

gráfico de barras dobles (pág. 296) Representación gráfica que emplea pares de barras horizontales o verticales para comparar datos.

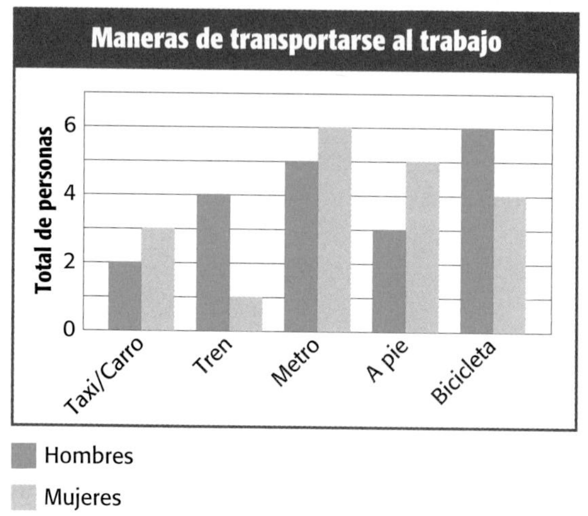

double-line graph (p. 298) A graphical display that uses paired horizontal lines to compare data.

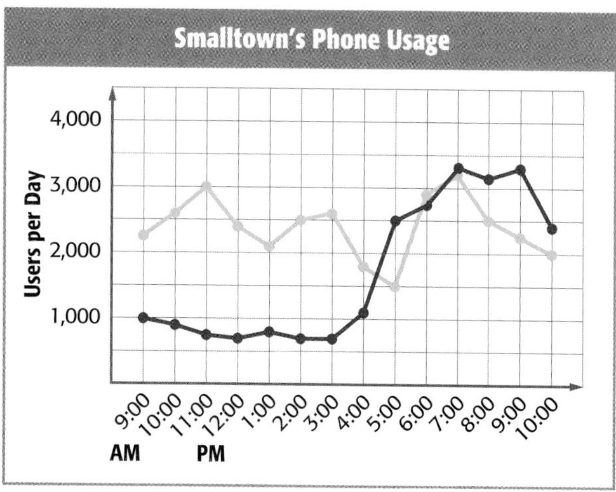

gráfico de lineal dobles (pág. 298) Representación gráfica que emplea pares de lineal horizontales para comprar datos.

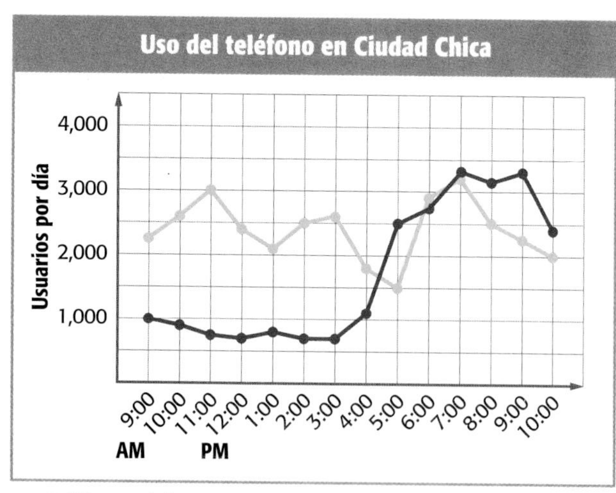

E

equivalent expressions (p. 54) Expressions that always give the same result when the same values are substituted for the variables. For example, $2k + 6$ is equivalent to $2(k + 3)$.

expand (p. 56) To use the distributive property to remove parentheses.

expresiones equivalentes (pág. 54) Expresiones que siempre dan el mismo resultado cuando los mismos valores se reemplazan con las variables. Por ejemplo: $2k + 6$ es equivalente a $2(k + 3)$.

expandir (pág. 56) Use la propiedad distributiva para eliminar los paréntesis.

exponent **(p. 12)** A symbol written above and to the right of a quantity that tells how many times the quantity is multiplied by itself. For example: $t \times t \times t$ is written t^3.

exponente **(pág. 12)** Símbolo que se escribe arriba y a la derecha de una cantidad y el cual indica cuántas veces la cantidad se multiplica por sí misma. Por ejemplo: $t \times t \times t$ se escribe t^3.

F

factor **(p. 56)** To use the distributive property to insert parentheses.

factor **(pág. 56)** El uso de la propiedad distributiva para agregar paréntesis.

factor **(p. 76)** A factor of a whole number is another whole number that divides into it without a remainder. For example, 1, 2, 3, 4, 6, 8, 12, and 24 are factors of 24.

factor **(pág. 76)** Un factor de un número entero es otro número entero que lo divide sin que quede un residuo. Por ejemplo, 1, 2, 3, 4, 6, 8, 12 y 24 son factores de 24.

factor pair **(p. 76)** A factor pair for a number is two factors whose product equals that number. For example, the factor pairs for 24 are 1 and 24, 2 and 12, 3 and 8, and 4 and 6.

par de factores **(pág. 76)** Un par de factores para un número consta de dos factores cuyo producto es igual al número. Por ejemplo, los pares de factores para 24 son 1 y 24; 2 y 12; 3 y 8; y 4 y 6.

flowchart **(p. 18)** A visual diagram that shows each step in evaluating an algebraic expression.

flujograma **(pág. 18)** Diagrama visual que muestra cada paso en la evaluación de una expresión algebraica.

formula **(p. 35)** An algebraic "recipe" that shows how to calculate a particular quantity. For example, $F = \frac{9}{5}C + 32$ is the formula for converting Celsius temperatures to Fahrenheit temperatures.

fórmula **(pág. 35)** Una "receta" algebraica que muestra cómo calcular una cantidad dada. Por ejemplo: $F = \frac{9}{5}C + 32$ es la fórmula para convertir temperaturas Celsius en temperaturas Fahrenheit.

G

gram **(p. 241)** A unit of mass in the metric system. 1000 grams = 1 kilogram

gramo **(pág. 241)** Unidad de masa del sistema métrico. 1000 gramos = 1 kilogramo

greater **(p. 178)** A relationship showing that one term or expression has a value larger than a second term or expression.

mayor que **(pág. 178)** Relación que muestra que un término o expresión tiene un valor más grande que un segundo término o expresión.

greatest common factor, GCF **(p. 81)** The greatest common factor (often abbreviated GCF) of two or more numbers is the greatest of their common factors. For example, the greatest common factor (or GCF) of 24 and 36 is 12.

máximo común divisor, MCD **(pág. 81)** El máximo común divisor (a menudo abreviado, MCD) de dos números es el mayor de sus factores comunes. Por ejemplo, el máximo común divisor (o MCD) de 24 y 36 es 12.

H

hypotenuse **(p. 343)** The side opposite the right angle in a right triangle. The hypotenuse is the longest side of a right triangle.

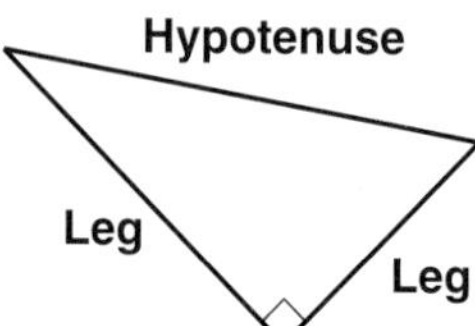

hipotenusa **(pág. 343)** El lado opuesto al ángulo recto en un triángulo rectángulo. La hipotenusa es el lado más largo de un triángulo rectángulo.

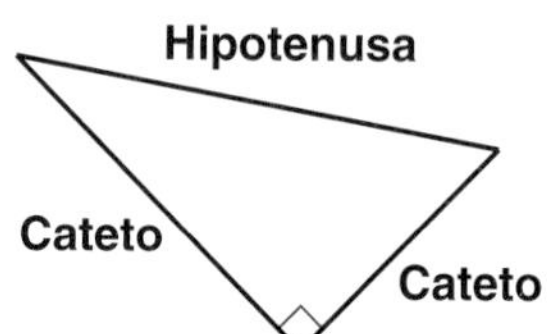

I

inequality **(p. 466)** A mathematical sentence stating that two quantities have different values. For example, $5 + 9 > 12$ is an inequality. The symbols $\neq$, $<$, and $>$ are used in writing inequalities.

desigualdad **(pág. 466)** Enunciado matemático que establece que dos cantidades tienen distintos valores. Por ejemplo, $5 + 9 > 12$ es una desigualdad. Los símbolos $\neq$, $<$, y $>$ se usan para escribir desigualdades.

irrational numbers **(p. 336)** Numbers that cannot be written as ratios of two integers. In decimal form, irrational numbers are non-terminating and non-repeating. Examples of irrational numbers include π, $\sqrt{17}$, and $3\sqrt{2}$.

números irracionales **(pág. 336)** Números que no se pueden escribir como razones de dos enteros. En forma decimal, los números irracionales son decimales no terminales y no periódicos. Ejemplos de números irracionales incluyen π, $\sqrt{17}$, y $3\sqrt{2}$.

L

least common multiple, LCM **(p. 83)** The least common multiple (often abbreviated LCM) of two or more numbers is the smallest of their common multiples. For example, the least common multiple (or LCM) of 6 and 15 is 30.

mínimo común múltiplo, MCM **(pág. 83)** El mínimo común múltiplo (a menudo abreviado MCM) de dos o más números es el menor de sus múltiplos comunes. Por ejemplo, el mínimo común múltiplo de 6 y 15 es 30.

leg **(p. 343)** One of the sides of a right triangle that is not the hypotenuse, or one of the shorter two sides of a right triangle. The figure below shows the two legs and hypotenuse of a right triangle.

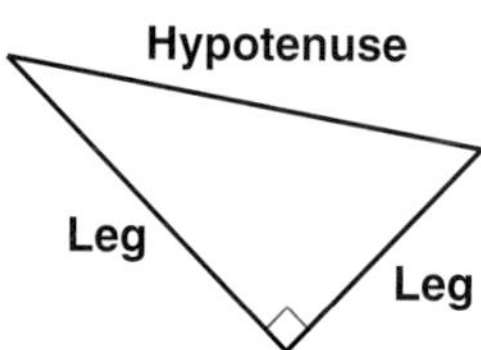

cateto **(pág. 343)** Uno de los lados de un triángulo rectángulo que no es la hipotenusa o uno de los dos lados más cortos de un triángulo rectángulo. La siguiente figura muestra los dos catetos y la hipotenusa de un triángulo rectángulo.

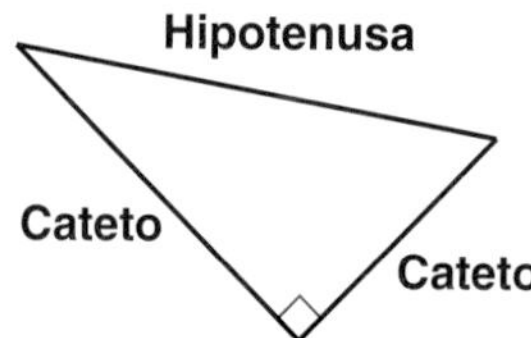

lesser **(p. 178)** A relationship showing that one term or expression has a value smaller than a second term or expression.

menor que **(pág. 178)** Relación que muestra que un término o expresión tiene un valor más pequeño que un segundo término o expresión.

like terms **(p. 62)** In an algebraic expression, terms with the same variables raised to the same powers. For example, in the expression $x + 3 - 7x + 8x^2 - 2x^2 + 1$, $8x^2$ and $-2x^2$ are like terms, x and $-7x$ are like terms, and 3 and 1 are like terms.

términos semejantes **(pág. 62)** En una expresión algebraica, los términos con las mismas variables elevadas a las mismas potencias. Por ejemplo: en la expresión $x + 3 - 7x + 8x^2 - 2x^2 + 1$, $8x^2$ y $-2x^2$ son términos semejantes, x y $-7x$ son términos semejantes y 3 y 1 son términos semejantes.

linear relationship **(p. 371)** A relationship whose graph is a straight line. Linear relationships have a constant rate of change. As one variable changes by 1 unit, the other variable changes by a set amount. For example, $m = 5t$ shows that m changes 5 units per 1-unit change in t.

relación lineal **(pág. 371)** Relación cuya gráfica es una recta. Las relaciones lineales muestran una tasa constante de cambio. A medida que una variable cambia en 1 unidad, la otra variable cambia en la misma cantidad. Por ejemplo: $m = 5t$ muestra que m cambia 5 unidades por 1 unidad de cambio en t.

M

map scale **(p. 514)** A ratio of a single unit of distance on the map to the equivalent distance on the ground.

escala de mapa **(pág. 514)** Razón de una sola unidad de distancia en el mapa a la distancia equivalente en la tierra.

mass **(p. 240)** A measurement of the amount of matter in an object.

masa **(pág. 240)** Medida de la cantidad de materia en un cuerpo.

metric system **(p. 241)** A base-ten system of measurement using the basic units: meter for length, gram for mass, and liter for capacity.

sistema métrico **(pág. 241)** Sistema de medición decimal que usa las siguientes unidades básicas: metro para longitud, gramo para masa y litro para capacidad.

model **(p. 446)** Something that has the key characteristics of something else. For example, in mathematics you could use a balance to model an equation.

modelo **(pág. 446)** Algo que tiene una característica clave de algo más. Por ejemplo: en matemáticas podrías usar una ba-lan-za para hacer un modelo de una ecuación.

monomial **(p. 61)** A number, a variable, or a product of a number and one or more variables. Examples: 3, y, $2x$, $5xy^2$

monomio **(pág. 61)** Número, variable o producto de un número y de una o más variables. Ejemplos: 3, y, $2x$, $5xy^2$

multiple **(p. 83)** A multiple of a whole number is the product of that number and another whole number. For example, 35 is a multiple of 7 since $35 = 7 \times 5$.

múltiplo **(pág. 83)** Un múltiplo de un número entero es el producto de ese número y otro número entero. Por ejemplo, 35 es un múltiplo de 7 dado que $35 = 7 \times 5$.

N

natural numbers **(p. 320)** The set {1, 2, 3, ... }.

net **(p. 228)** A flat figure that can be folded to form a closed, three-dimensional object called a solid.

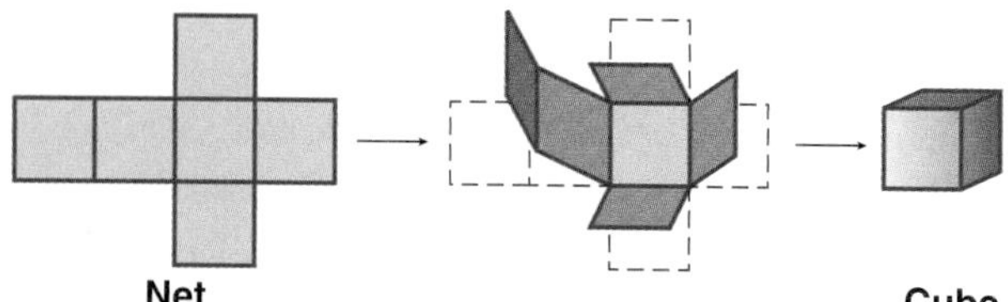

número naturales **(pág. 320)** Conjunto {1, 2, 3, ... }.

red **(pág. 228)** Figura plana que al doblarse forma un cuerpo tridimensional cerrado llamado sólido.

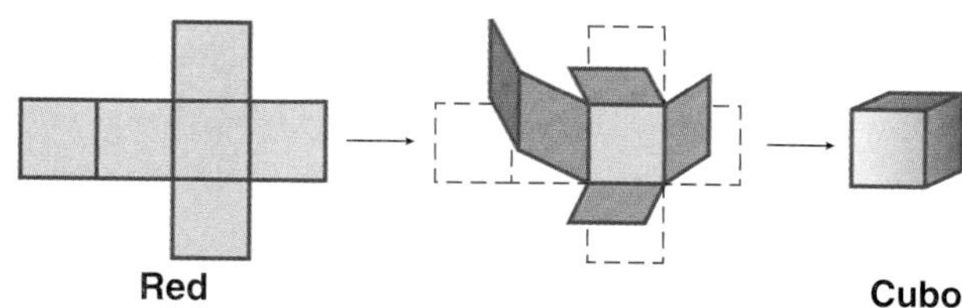

O

order of operations **(p. 99)** A convention for reading and evaluating expressions. The order of operations says that expressions should be evaluated in this order:

- Evaluate any expressions inside parentheses and above and below fraction bars.
- Evaluate all exponents, including squares.
- Do multiplications and divisions from left to right.
- Do additions and subtractions from left to right.

For example, to evaluate $5 + 3 \times 7$, you multiply first and then add: $5 + 3 \cdot 7 = 5 + 21 = 26$.
To evaluate $10^2 - 6 \div 3$, you evaluate the exponent first, then divide, then subtract: $10^2 - 6 \div 3 = 100 - 2 = 98$.

ounce **(p. 247)** A customary unit of weight. 16 ounces = 1 pound.

orden de las operaciones **(pág. 99)** Una convención para leer y evaluar expresiones. El orden de las operaciones indica que las expresiones se deben evaluar en el siguiente orden:

- Evalúa cualquier expresión entre paréntesis y sobre y debajo de barras de fracciones.
- Evalúa todos los exponentes, incluyendo los cuadrados.
- Efectúa las multiplicaciones y las divisiones de izquierda a derecha.
- Efectúa las sumas y las restas de izquierda a derecha.

Por ejemplo, para evaluar $5 + 3 \times 7$, multiplica primero y luego suma. $5 + 3 \cdot 7 = 5 + 21 = 26$.
Para evaluar $10^2 - 6 \div 3$, evalúa primero el exponente, luego divide y por último resta.
$10^2 - 6 \div 3 = 100 - 2 = 98$.

onza **(pág. 247)** Unidad tradicional de peso. 16 onzas = 1 libra.

P

polynomial **(p. 61)** A monomial or sum of monomials. Examples: $5x$, $3x^2 + 2$.

population **(p. 279)** A larger group from which a sample is taken.

pound **(p. 247)** A customary unit of weight. 1 pound = 16 ounces.

polinomio **(pág. 61)** Monomio o suma de monomios. Ejemplos: $5x$, $3x^2 + 2$.

población **(pág. 279)** Grupo grande del cual se toma una muestra.

libra **(pág. 247)** Unidad tradicional de peso. 1 libra = 16 onzas.

power **(p. 93)** A number that is written using an exponent.

potencia **(pág. 93)** Número que se escribe usando un exponente.

prime factorization **(p. 78)** The prime factorization of a composite number shows that number written as a product of prime numbers. For example, the prime factorization of 98 is 2 • 7 • 7.

factorización prima **(pág. 78)** La factorización prima de un número compuesto muestra ese número escrito como un producto de números primos. Por ejemplo, la factorización prima de 98 es 2 • 7 • 7.

prime number **(p. 78)** A whole number greater than 1 with only two factors: itself and 1. For example, 13 is a prime number since it only has two factors, 13 and 1.

número primo **(pág. 78)** Número entero mayor que 1 cuyos únicos factores son 1 y el número mismo. Por ejemplo, 13 es un número primo dado que sólo tiene dos factores: 13 y 1.

prism **(p. 212)** A figure that has two identical, parallel faces that are polygons, and other faces that are parallelograms.

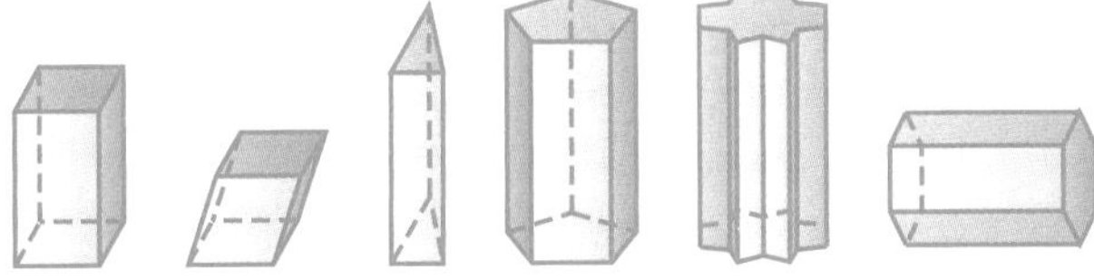

prisma **(pág. 212)** Figura con dos caras paralelas idénticas, las cuales son polígonos, y otras dos caras que son paralelogramos.

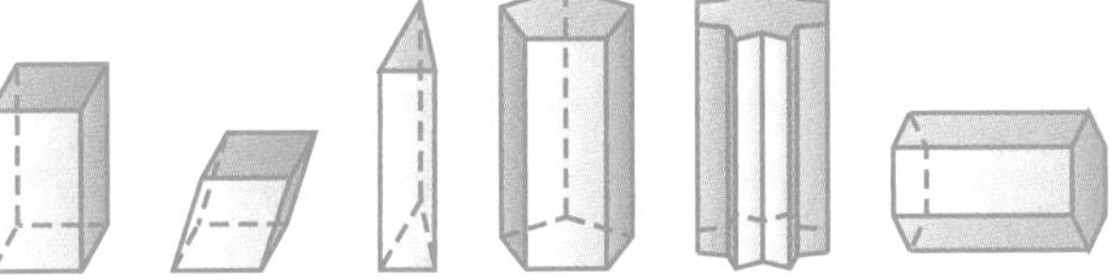

product law of exponents **(p. 98)** Law that states when the bases of two factors are the same, the power of their product is the sum of their exponents. Example: $x^a x^b = x^{a+b}$

ley del producto de exponentes **(pág. 98)** Ley que establece que cuando las bases de dos factores son iguales, la potencia de su producto es la suma de sus exponentes. Ejemplo: $x^a x^b = x^{a+b}$

proportion **(p. 510)** A statement that two ratios are equal. For example, 2:3 = 6:9.

proporción **(pág. 510)** Ecuación que establece que dos razones son iguales. Por ejemplo: 2:3 = 6:9.

proportional **(p. 377)** Used to describe the relationship between two variables in which the value of one variable is multiplied by a number, the value of the other variable is multiplied by the same number. For example, when someone is paid an hourly rate and works double the hours, that person gets double the pay. When someone works triple the hours, that person gets triple the pay. The hours worked and rate of pay per hour are proportional to each other.

proporcional **(pág. 377)** Se usa para describir la relación entre dos variables en las cuales, cuando el valor de una variable se multiplica por un número, el valor de la otra variable se multiplica por el mismo número. Por ejemplo: cuando a alguien se le paga un sueldo a cierta tasa por hora y esa persona trabaja el doble del número de horas, dicha persona obtiene el doble del pago. Cuando alguien trabaja el triple de las horas, esa persona obtiene el triple del pago. Las horas trabajadas y la tasa de pago por hora son proporcionales entre sí.

proportional relationship **(p. 506)** A relationship in which all pairs of corresponding values have the same ratio.

relación proporcional **(pág. 506)** Relación en que todos los pares de valores correspondientes tienen la misma razón.

pythagorean theorem **(p. 345)** If a and b are the measures of the legs of a right triangle and c is the measure of the hypotenuse, then $c^2 = a^2 + b^2$.

teorema de pitágoras **(pág. 345)** Si a y b son las longitudes de los catetos de un triángulo rectángulo y c es la longitud de la hipotenusa, entonces $c^2 = a^2 + b^2$.

R

rate **(p. 368)** Describes how two unlike quantities are related or how they can be compared.

tasa **(pág. 368)** Describe la relación entre dos cantidades diferentes o la manera de comparar dichas cantidades.

rational numbers **(p. 320)** Numbers that can be written as ratios of two integers. In decimal form, rational numbers are terminating or repeating. For example, 5, -0.274, and $0.\overline{3}$ are rational numbers.

números racionales **(pág. 320)** Números que se pueden escribir como razones de dos enteros. En forma decimal, los números racionales son números terminales o periódicos. Por ejemplo: 5, -0.274 y $0.\overline{3}$ son números racionales.

real numbers **(p. 336)** The set of rational numbers and irrational numbers together.

números reales **(pág. 336)** Conjunto de números racionales y números irracionales.

relative error **(p. 185)** The ratio of the half-unit difference in precision to the entire measure, expressed as a percent.

error relativo **(pág. 185)** Relación entre la diferencia de media unidad en precisión y la medida entera, expresada en forma porcentual.

relatively prime **(p. 81)** Two or more numbers are relatively prime if their only common factor is 1. For example, 7 and 9 are relatively prime.

relativamente primo **(pág. 81)** Dos o más números son relativamente primos si su único factor común es 1. Por ejemplo, 7 y 9 son relativamente primos.

representative sample **(p. 284)** A part of a population that has approximately the same proportions as the whole population with respect to the characteristic being studied.

muestra representativa **(pág. 284)** Una parte de la población que tiene aproximadamente las mismas proporciones que la población entera con respecto a las características bajo estudio.

right triangle **(p. 343)** A triangle that has one $90°$ angle.

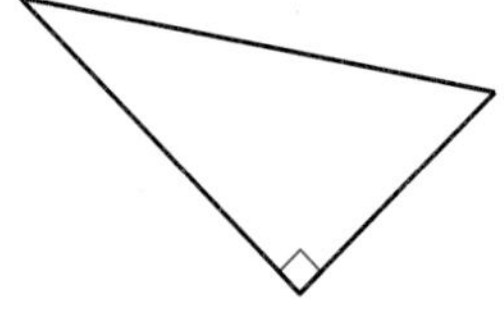

triángulo rectángulo **(pág. 343)** Triángulo que posee un ángulo de $90°$.

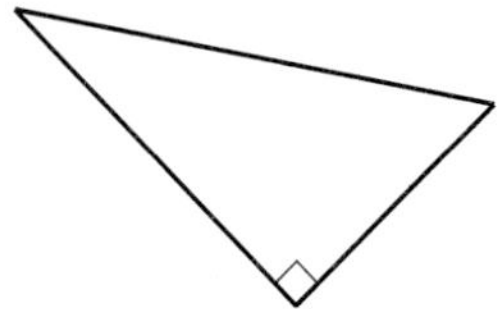

S

sample **(p. 279)** A smaller group taken from a population that is used to represent the larger group.

muestra **(pág. 279)** Un grupo más pequeño que se toma de la población y el cual se usa para representar el grupo más grande.

scientific notation **(p. 178)** A number that is expressed as the product of a number greater than or equal to 1 but less than 10, and a power of 10. For example, 5,000,000 written in scientific notation is 5×10^6.

notación científica **(pág. 178)** Número que se expresa como el producto de un número mayor que o igual a 1, pero menor que 10 y una potencia de 10. Por ejemplo: 5,000,000 escrito en notación científica es 5×10^6.

slope **(p. 392)** The steepness of a line.

pendiente **(pág. 392)** El grado de inclinación de una recta.

solid **(p. 228)** A three-dimensional figure formed by intersecting planes.

sólido geométrico **(pág. 228)** Figura tridimensional formada por planos intersecantes.

solution set **(p. 467)** The set of elements from the replacement set that make an open sentence true.

conjunto de soluciones **(pág. 467)** Conjunto de elementos del conjunto de reemplazo que hace que un enunciado abierto sea verdadero.

speed **(p. 397)** How fast an object is going (always positive).

rapidez **(pág. 397)** El grado de velocidad con que viaja un cuerpo (es siempre un número positivo).

square root **(p. 333)** The number you need to square to get a given number. For example, the square root of 36 is 6.

raíz cuadrada **(pág. 333)** El número que debes elevar al cuadrado para obtener un número dado. Por ejemplo, la raíz cuadrada de 36 es 6.

surface area **(p. 213)** The area of the exterior surface of an object, measured in square units.

área de superficie **(pág. 213)** El área de las superficies exteriores de un cuerpo, medida en unidades cuadradas.

T

ton **(p. 247)** A customary unit of weight. 1 ton = 2000 pounds.

tonelada **(pág. 247)** Unidad tradicional de peso. 1 tonelada = 2000 libras.

U

unit rate **(p. 541)** Term used when one of two quantities being compared is given in terms of one unit. Example: 65 miles per hour or $1.99 per pound.

tasa unitaria **(pág. 541)** Término que se usa cuando una de dos cantidades bajo comparación se da en términos de una unidad. Ejemplo: 65 millas por hora o $1.99 por libra.

V

variable **(p. 10)** A quantity that can change or vary, or an unknown quantity.

velocity **(p. 397)** The rate at which an object is moving from or toward a designated point (can be positive or negative).

volume **(p. 213)** The space inside a three-dimensional object, measured in cubic units.

variable **(pág. 10)** Cantidad que puede cambiar o variar, o una cantidad desconocida.

velocidad **(pág. 397)** Tasa a la cual se mueve un cuerpo desde un punto o hacia un punto designado (puede ser positiva o negativa).

volumen **(pág. 213)** El espacio dentro de un cuerpo tridimen-sio-nal, medido en unidades cúbicas.

W

weight **(p. 240)** The pull of gravity on a amount of matter.

whole numbers **(p. 320)** The set {0, 1, 2, 3, ...}.

peso **(pág. 240)** Atracción de la gravedad sobre una cantidad de materia.

números enteros **(pág. 320)** Conjunto {0, 1, 2, 3, ... }.

Y

***y*-intercept** **(p. 401)** The point at which a graph intersects the y-axis.

intersección *y* **(pág. 401)** El punto en que una gráfica interseca el eje y.

Index

A

Absolute value, 151
Acute, 357
Acute triangle, 357
Add and subtract
- integers, 128–130
- on number line, 134–139
- signed numbers, 126–153

Algebra
- application of, 150, 151, 152
- distributive property, 49–68
- expressions and formulas, 30–48
- in the strangest places, 2
- variables and expressions, 4–29
- *See also* Algebraic expressions; Equations

Algebraic expressions
- bags and blocks, 51–52
- combine like terms, 60–62
- divide with exponents, 109–110
- equivalent, 54
- evaluate, 12–17
- evaluate with negative exponents, 198–199
- multiply with same base, 96–100
- multiply with same exponent, 100–103
- same and different, 52–55
- use parentheses in, 56–59
- write, 9–12
- write to describe balance puzzles, 450–452
- write to match situations, 32–34
- *See also* Expressions

Angle measure, 300
Angle of elevation, 519
Applications
- algebra, 150, 151, 152
- astronomy, 190, 407
- ecology, 522
- economics, 22, 193, 340, 527
- electronics, 44
- food, 238
- geography, 404
- geology, 45
- geometry, 46, 47, 48, 88, 192, 277, 293, 388, 409, 445, 459, 472, 473, 488
- graphs, 487
- history, 89
- hobbies, 206
- literature, 288
- mass, 456
- measurement, 21, 22, 27, 45, 311, 382, 385, 386, 503, 529
- national parks, 500
- number games, 330
- patterns, 458
- physical science, 166, 356, 406
- physics, 45, 205
- probability, 68, 91, 119, 459
- ratio, 48
- social studies, 90, 227, 293, 538
- sports, 313, 316, 403
- statistics, 119, 151, 153
- travel, 459
- winter ice, 239
- world cultures, 67

Area
- circle, 217
- defined, 331
- parallelogram, 223
- square, 332
- surface, 212
- triangle, 223, 232

Assessment. *See* Review & Self-Assessment; Test-Taking Practice

B

Backtracking
- defined, 19
- solving equations by, 20, 69, 437–439
- using flowchart, 436–437

Balance puzzles, 447–454
Bar graph, 294
Base, 93, 216

C

Calculator
- big numbers, 182
- calculate error, 187
- evaluate expressions, 13, 94, 183, 190
- scientific notation, 182–184
- simulate coin toss, 265
- square roots, 333, 335

Calibrating, 243
Celsius and Fahrenheit, 7–8, 35
Centigrams, 241
Chapter Summary, 69, 120, 168, 207, 256, 314, 361, 431, 489, 551
Circle graphs
- and averages, 303–305
- and budgets, 312
- defined and discussed, 299–301
- make using protractor, 301
- and survey data, 299

Combinations, probability and, 262–264
Combining like terms, 60–62
Common factors, 80–82
Common multiple, 83
Compare
- rational numbers, 324–327
- ratios, 497–499

Composite number, 78

Index

Computer
- simulate coin toss with, 265
- *See also* Spreadsheets

Cone, 409

Conjecture, 477–480

Connect & Extend, 27, 46, 66, 88, 118, 151, 166, 191, 204, 224, 237, 275, 289, 339, 367, 385, 406, 427, 444, 456, 470, 485, 503, 525, 548, 252

Conversions
- Celsius and Fahrenheit, 7–8, 35
- currencies, 544–546
- pounds to kilograms, 254
- weights, 247–248

Cube, 39

Cups, 372

Curriculum Connections
- algebra, 150, 151, 152
- astronomy, 190, 407
- ecology, 522
- economics, 22, 193, 340, 527
- geography, 404
- geology, 45
- geometry, 46, 47, 48, 88, 192, 277, 293, 388, 409, 445, 459, 472, 473, 488
- history, 89
- literature, 288
- physical science, 166, 356, 406
- physics, 45, 205
- social studies, 90, 227, 293, 538
- statistics, 119, 151, 153

Currency, converting, 544–546

Customary, 247

Cylinders, 211, 216–219

D

Data
- graphs and. *See* Graphs
- histogram and, 294
- probability and. *See* Probability
- range of, 162
- signed numbers and, 162–164
- survey, 299

Data set, 162

Decimal numbers, 334

Decimals
- nonrepeating, 337
- nonterminating, 337
- terminating, 337

Decomposing figures/shapes
- nets, 228–234
- surface area, 39, 212–215

Diameter, 37

Differences, predict signs of, 146–147

Distance, time and, 394–396

Distance formula, 351–354

Distributive property, 49, 62

Divide
- expressions with exponents, 109–110
- division machines with exponents, 107–108
- with negative numbers, 160–161

Double-bar graph, 296

Double-line graph, 298

E

Equal ratios, 508–510

Equations
- do the same thing to both sides, 462–463
- inequalities and, 466–468
- parentheses with, 474–482, 481–483
- solve, 460–465
  - choose method to solve, 437–439
  - symbolic solutions, 461
  - with balance puzzle, 447–454
  - with model, 446–460
  - with spreadsheet, 440–443
- subtract with parentheses, 477–480

Equivalent expressions, 54

Equivalent operations, 140–142

Eratosthenes, 89

Error, relative, 185–187

Estimate
- heights of tall object, 519–521
- mass, 244–247

Expanding, 56

Exponents
- common factors, 80–82
- defined, 12, 79
- divide expressions with, 109–110
- division machines with, 107–108
- exponent machines, 92–119
- expressions with negative, 198–199
- factors, 75–77
- laws of, 111–113, 200–201
- model, 93–96
- model negative, 194–198
- multiples, 83–85
- multiply expressions with same base, 96–100
- multiply expressions with same exponent, 100–103
- negative, 194–206
- power law of, 111–113
- prime numbers, 77–79

Expressions
- bags and blocks, 51–52
- combine like terms, 60–62
- divide with exponents, 109–110
- equivalent, 54
- evaluate, 12–17
- evaluate with negative exponents, 198–199
- expanding, 64
- multiply with same base, 96–100
- multiply with same exponent, 100–103
- same and different, 52–55
- use parentheses in, 56–59
- write, 9–12
- write to match situations, 32–34
- *See also* Algebraic expressions

F

Factor tree, 78
Factoring, 56
Factors
- common, 80–82
- defined and discussed, 75–77
- greatest common, 81
- of whole number, 76

Factor pair, 76
Fahrenheit. *See* Celsius and Fahrenheit
Family Letter, 3, 73, 125, 173, 211, 261, 319, 367, 435, 493
Feet per second (ft/s), 389
Flowchart
- backtrack using, 436–437
- defined, 18
- express a quantity using, 51
- evaluate expressions using, 17–20
- solve equations using, 436

Fluid ounces, 247
Formulas, 35–39
- area of circle, 217
- area of parallelogram, 223
- area of triangle, 223, 232
- defined, 35
- distance, 351–354
- divide expressions with the same base, 109
- divide expressions with the same exponent, 110
- multiply expressions with same base, 98
- multiply expressions with same exponents, 102
- pi (π), 221
- Pythagorean Theorem, 345–350
- raise a power to a power, 113
- and spreadsheets, 40–42
- surface area and volume, 212
- volume of prism and cylinder, 216–219
- volume of sphere and cone, 222

G

Games
- Captain's Game, 131
- Detective, 418
- Greatest Common Factor (GCF), 82
- Hidden Prize, 270
- 1, 2, 3 Show! (fairness of games), 268–269
- What's in the Bag?, 278

GCF. *See* Greatest Common Factor
Geometry
- applications, 46, 47, 48, 88, 192, 277, 293, 388, 409, 445, 459, 472, 473, 488
- cylinders, 216–219
- mass, 241–247
- nets, 229–233
- prisms, 213–219
- solids, 230–233
- surface area, 39, 213–215
- volume, 213, 216–219
- weight, 247–250

Gigabyte, 184
Global positioning system (GPS), 514
Goldbach's conjecture, 88
Golden Ratio, 503
Googol, 181
Grams, 241
Grain, as unit of weight, 248
Graphs
- describe, 396–397
- double-bar graph, 295–296
- double-line graph, 298
- circle graph, 299–301
- misleading statistics and, 306–308
- from patterns, 414–416
- from rules to, 417–419
- stem-and-leaf plots, 302–305

Greatest common factor (GCF), 81
Guess-check-and-improve
- in solving equations, 437–439
- use spreadsheet to, 440–443

H

Hectograms, 241
Hemisphere, 409
Hexagonal prism, 409
Histogram, 294
Home Activities, 3, 73, 125, 173, 211, 261, 319, 367, 435, 493
Home Connection. *See* Family Letter
Hundred weight, 249
Hypotenuse, 343

I

In Your Own Words, 29, 47, 68, 91, 106, 119, 153, 168, 193, 205, 226, 238, 254, 291, 330, 341, 360, 385, 406, 445, 458, 473, 488, 528, 539, 550
Inequality, 466–468
- defined, 466
- and negative numbers, 143–145

Inquiry Investigation
- distance formula, 351–354
- estimate heights of tall objects, 519–521
- formulas and spreadsheets, 40–42
- hidden prize, 270–271
- number line model, 131–133
- package design, 220–222
- relative error, 185–187
- rolling along, 380–381
- spreadsheet to guess-check-and-improve, 440–443
- Tower of Hanoi, 114–116

Integers
- adding and subtracting, 128–130
- two-color chip model, 128–130

Inverse operations, 333
Irrational Numbers
- analyze, 336–338
- defined, 319, 334, 336

Index

K

Key Concepts
- addition of signed numbers, 125
- algebraic expressions and variables, 3
- exponents, 73
- nets, 211
- ordering rational numbers, 319
- scientific notation, 173
- similarity, 493
- solving equations, 435
- statistical sampling, 261
- tables or algebraic rules, 367

Kilograms, 241

Kilometers per hour (km/h or kph), 389

L

Larger than, 218

Least common multiple (LCM), 83

Legs, of right triangle, 343

Length, 176

Lightyear, 178

Like terms, 60–62

Line graph, 294

Linear relationships
- defined, 371
- distance and time, 394–396
- graphs and, 396–397
- patterns in, 411–416
- rates, 368–388
- recognize, 410–421
- from rules to graphs, 417–420
- secret rules, 420–422
- speed and slope and, 389–409
- walk and jog, 390–393
- *y*-intercept, 398–401

M

Machines
- division, 107–108
- exponent, 92–116
- repeater, 73, 93–102
- shrinking, 94–95, 107–108
- stretching, 73–79

Magnitude of numbers. *See* Scientific notation

Map scales, 514–516

Mass
- defined, 240
- estimate, 244–247
- measure, 241–244

Math Link
- absolute value, 127, 151, 185, 276
- answers in spreadsheet, 441
- area of circle, 217
- area of parallelogram, 223
- area of triangle, 223, 232
- base and exponent of expressions, 12
- big numbers, 175, 181
- centimeters, 188
- circumference of circle, 231
- compare two things, 398
- consecutive whole numbers, 90
- cubic units, 216
- diameter, 37
- digit repeats, 337
- distance, rate, and time rule, 392
- equals sign in spreadsheet, 40
- equivalent expressions, 487
- expand an expression, 64
- expressions, 17
- express ratios three ways, 501
- factors, 76
- feet-mile, 118
- final outcomes, 267
- find the average, 132
- four ways to write same expression, 50
- graphs, 419
- hundredweight, 249
- inches, 191
- inequality, 143
- inequality symbols, 466
- integers, 128, 320
- irrational numbers, 334
- kilogram, 242
- kilogram/grams, 244
- locate $\frac{1}{3}$ and $\frac{2}{3}$ with a ruler, 327
- mean, 289
- mile/feet, 402, 407
- mono and poly, 61
- negative, positive, and principle roots, 333
- next greatest perfect number, 88
- obtuse, 358
- order of operations, 99
- origin, 379
- ounce, 247
- parentheses styles, 57
- patterns, 507
- pi (π), 221
- plot points of table data, 382
- prime factorization, 108
- proportional, 505
- range of data, 162
- ratios, 495
- repeating numbers, 336
- scientific notation, 179, 201
- similar triangles, 520
- simulate coin toss, 265
- spreadsheet cell, 40
- square units, 216
- squaring the number, 331
- subscripted variables, 353
- sum of triangle measures, 439
- unit price, 541
- units of mass abbreviations, 250
- Venn diagram, 322
- What's My Rule, 420

Math Online, 2, 72, 124, 172, 210, 260, 318, 366, 434, 492

Mean, 162

Measure

mass, 241–244

prisms, 213–215

weight, 247–249

Measurement, applications, 21, 22, 27, 45, 311, 382, 385, 386, 503, 529

Median, 162

Meters per second (m/s), 389

Miles per hour (mph), 389

Milliliters, 369

Milligrams, 241

Millimeters per second (mm/s), 389

Mixed Review, 29, 48, 68, 91, 106, 119, 153, 168, 193, 206, 227, 239, 255, 293, 313, 330, 342, 360, 388, 409, 430, 445, 473, 488, 504, 529, 539, 550

Mode, 162

Model, for solving equations, 446

Monomial, 61

Multiples, 83–85

Multiply

expressions with same base, 96–100

expressions with same exponent, 100–103

positive and negative numbers, 155–156

two negative numbers, 157–159

Muon, 205

N

Natural numbers, 320

Nautical miles, 515

Negative exponents

evaluate expressions with, 198–199

laws of and scientific notation, 200–201

model, 194–198

Negative numbers

add and subtract with, 126–153

divide with, 160–161

and inequalities, 143–145

multiply positive and, 155–156

multiply and divide with, 154–159

multiply two negatives, 157–159

predict signs of sums and differences, 146–147

Net

beverage can shape, 233–234

defined, 211, 228

use a, 229–230

use to investigate solids, 230–233

Nonillion, 175

Novemdecillion, 181

Number lines

add and subtract on, 134–139

compare rational numbers with, 324–325, 327

equivalent operations and, 140–142

graph rational numbers on, 327

model, 131–133

Numbers

irrational, 331–342

magnitude of. *See* Scientific notation

negative. *See* Negative numbers

odd, 322

perfect, 88

prime, 77–79

rational, 320–330

real. *See* Real numbers

signed. *See* Signed numbers

whole, 320

Number sense, 29, 48, 150

Number sets, 320–323

O

Oblique prisms, 218

Obtuse, 358

Obtuse triangle, 357

Octadecillion, 181

Octahedron, 409

Octillion, 175

Odd numbers, 322

On Your Own Exercises, 21–29, 43–48, 63–68, 86–91, 104–106, 117–119, 148–153, 165–168, 188–193, 202–206, 223–228, 235–239, 250–255, 272–277, 286–293, 309–313, 328–330, 339–342, 355–360, 382–388, 402–409, 423–430, 444–445, 455–459, 469–473, 484–488, 500–504, 522–529, 537–539, 547–550

Operations

equivalent, 140–142

order of, 99

Ounces, 247

P

Parallelogram, 223

Parentheses

in expressions, 56–59

practice with, 481–483

solve equations with, 474–476

subtract with, 477–480

Patterns

explore and describe, 411–413

from rules to graphs, 417–419

graphs and rules from, 414–416

predictable, 507

secret rules and, 420–422

Percent diagram, 530

Percents

find, 530–533

proportions using, 534–536

Percent scale, 530

Perfect number, 88

Perfect square, 332

Pi (π), 221, 338

Pint, 370

Polygon, 61

Polynomial, 61

Population, 279

Pounds, 247, 372

Power, 93

raising a power to a, 113

of 10, 175–177

Power law of exponents, 111–113

Practice & Apply, 21, 43, 63, 86, 105, 117, 148, 165, 188, 202, 223, 235, 250, 272, 286, 309, 328, 339, 355, 382, 402, 423, 444, 455, 469, 484, 500, 522, 537, 547

Predictions

making, 114, 132, 185, 220

samples and, 279–285

signs of sums and differences and, 146–147

Prime factorization, 78

Prime numbers, 77–79

Principle roots, 333

Prisms

defined, 211, 212

measure, 213–215

oblique, 218

right, 218

volume of, 216–219

Probability

and chance, 68, 119, 264–267

and combinations, 262–264

and dependence, 262–271

fairness of games, 268–269

guess color of chips, 459

heads or tails?, 264–267

hidden prize investigation, 270–271

make predictions, 294–313

and percent, 91

odds and, 262–264, 270–271, 459

Problem solving

backtracking, 20, 69, 436, 437

choose a method, 437–439

guess-check-and-improve, 437, 440–443

Strategies and Applications, 69, 120, 121, 169, 207, 256, 314, 361, 431, 489, 551

Product laws of exponents, 98

Proper factors, 88

Proportion

as equation, 493

defined, 510

equal ratios, 508–510

as linear relationship, 371

map scales, 514–516

percent and, 378–379, 530–539

proportional relationships, 377–379, 505–507

rate of pay and, 374–376

rates, 377–379, 540–550

ratios, 454–504

similarity and, 517–518

solve, 510–513

Proportionality, *See* Proportion

Prove It!, 59, 110, 151, 205, 528

Puzzles

balance, 447–448

solve problems with, 453–454

Pythagoras, 344

Pythagorean Theorem

defined, 345

distance formula, 351–354

right triangles and squares, 344–347

use, 347–350

Q

Quadrillion, 175

Quart, 370

Quattuordecillion, 181

Quintillion, 175

R

Radical sign, 333

Range, 162

Rate of pay, 374

Rates

defined, 368

describe, 373–376

proportional relationships and, 377–378

understand, 369–373

unit prices and, 540–543

Rational numbers

compare and order, 324–327

defined, 320

number sets, 320–323

Ratios

applications in, 48

compare and scale, 497–499

equal, 508–510

and rate comparisons, 495–497

rates as, 540–543

Real-Life Math

Algebra in the Strangest Places, 2

Astronomical Figures, 72

Astronomical Information, 172

At Any Rate, 366

Follow the Bike Path, 318

Empire State Building as project, 492

For Your Amusement, 434

Patterns and Plans, 210

Soaring to New Heights, 124

What Do Most Americans Think?, 260

Real numbers

irrational numbers, 331–342

Pythagorean Theorem and, 343–350

rational numbers, 320–330

Real-World Link

acute, meaning of, 357

average human eye blinks per year, 177

average length of newborn, 10

baseball batting averages, 304

baseball terms, 47

blue whale, 535

calibrating scientific instruments, 243

carbonation in water, 234

car insurance and student discounts, 308
Centaurus system, 178
check readability, 282
chewing gum, 111
clothing donations and tax returns, 60
color of human eye, 292
Constitution and senators, 90
cubic units and cargo room, 215
dollar, 527
earth's rotation, 183
Egyptians and Pythagorean Theorem, 348
Emily Dickinson, 288
Eratosthenes, 89
fastest human and animal, 391
first baseball cards, 475
first movie theaters, 386
flowcharts, 436
foreign visitor spending in U.S., 544
Gallup Polls, 293
glasses and size, 226
global tourism, 545
Golden Ratio, 503
graphs and charts in newspapers, 295
greatest prime number, 79
Great Smoky Mountain National Park, 500
hamster litters, 464
Houston Oilers, average yards per carry in 1980, 164
kite fighting, 350
licorice candy, 74
marathon, 403
maximum depth for scuba diving, 156
metric system, 385
metric unit of milligrams, 248
Monarch caterpillars, 486
most-watched sports, 300
nautical miles/statute miles, 515
origins of chewing gum, 95
origins of tea drinking, 6
paper used per year in U.S., 245
pi (π), history of approximating, 338
positive and negative bank transactions, 129
probability in real world, 263
profit and income, 28
Pythagoras, 344
rare baseball cards, 477
record low temperature in U.S., 163
Robert Frost, 281
rollers and the pyramids, 380
runs batted in, 305
soccer's World Cup, 299
spider webs, 369
spreadsheets, 442
televisions, 270
tennis rackets, 483
tile making, 410
Toronto's population, 8
tortoises and turtles, 38
U.S. Library of Congress, 184
U.S. presidential elections, 186
Venus, 190
Vincent van Gogh, 511
water as most popular beverage, 298
water pressure on divers, 161
weight and mass, 240
wind chill factor formula, 166
Winged Liberty Head dime, 484
woodpecker rates of peck per second, 370
world record in 100-meter dash, 307
world-wide wheat flour consumption, 36

Rectangular prism, 212, 219
Relationships
- proportional, 377–379, 505–507
- linear. *See* Linear relationships

Relative error, 185–187
Relatively prime, 81
Repeater machines, 73–103
Representative sample, 284
Reversible prime, 89
Review & Self-Assessment, 69–71, 120–123, 169–171, 207–209, 256–259, 314–313, 361–365, 431–433, 489–491, 551–554
Right prisms, 218
Right triangles, 343–347
Rule(s)
- ambiguous, 5
- distance, rate, and time, 392
- graphs and, 417–420
- sequences and, 5–8

S

Samples
- defined, 279
- and predictions, 279–280
- representative, 283–285
- sizes, 280–282

Scales
- calibrating, 243
- map, 514–516
- toring, 342

Scale factor. *See* Map scales.
Scientific notation
- on calculator, 182–184
- defined, 178
- laws of, 200–201
- number as, 178
- powers of 10, 175–177
- relative error, 185–187
- work with, 178–181

Septillion, 175
Sequences and rules, 5–8
Sextillion, 175
Shrinking machine, 94–95
Sieve of Eratosthenes, 89

Index

Signed numbers
- add and subtract with, 126–153
- and data, 162–164
- divide with, 160–161
- and inequalities, 143–145
- multiply and divide with, 154–159
- multiply positive and, 155–156
- multiply two negatives, 157–159
- predict signs of sums and differences and, 146–147

Similar triangles, 520

Similarity, proportion and, 517–518

Slope, 392–393. *See also* Linear relationship

Solid, 211, 228

Solution set, 467–468

Speed
- change the starting point, 398–401
- describe graphs, 396–397
- distance and time, 394–395
- units of, 389
- walk and jog, 390–393

Spreadsheets
- and formulas, 40–42
- use to guess-check-and-improve, 440–443

Square pyramid, 235, 409

Squares and square Roots, 331–335

Stable particles, 205

Statistics
- misleading, 306–308
- representative samples, 283–285
- samples and predictions, 279–280

Stem-and-leaf plots, 302–305

Stretching machines, 73

Strategies and Applications, 69, 120–121, 169, 207, 256, 314, 361, 431, 489, 551

Subscriptive variables, 353

Subtract
- with parentheses, 477–480
- *See also* Add and subtract

Sums, predict signs of, 146–147

Surface area, 39, 212–215

T

Teaspoon, 369

Technology Connection
- calculator, 13, 182–184, 187, 190, 265, 333, 335
- big numbers, 182
- calculate error, 187
- evaluate expressions, 13, 94, 104, 183, 190
- scientific notation, 182–184
- simulate coin toss, 265
- square roots, 335
- Math Online, 2, 72, 124, 172, 210, 260, 318, 366, 434, 492
- spreadsheets
- formulas and, 40–42
- for guess-check-and-improve, 440–443

Test-Taking Practice, 71, 123, 171, 209, 259, 317, 365, 433, 491, 554

Tetrahedron, 238, 409

Think About It, 2, 72, 124, 172, 210, 260, 318, 366, 434, 492

Time, and distance, 394–396

Tons, 247

Toring, 243

Tower of Hanoi, 123

Tree diagram, 267

Triangular prism, 212

Triangular pyramid, 237

Two-color chip model, 125, 128–130

U

Unit rate, 540–542

Units of capacity, 247

V

Variables
- in algebraic expressions, 32–34
- defined, 3, 10

Velocity, 397

Venn diagram, 294
- rational and irrational numbers, 336
- use, 322

Vigintillion, 181

Vocabulary, 169
- chapter, 3, 73, 120, 125, 173, 211, 261, 319, 367, 435, 493
- list, 9, 12, 17, 35, 49, 53, 56, 60, 75, 77, 80, 83, 93, 96, 126, 212, 216, 228, 240, 241, 247, 295, 302, 320, 331, 336, 343, 344, 351, 368, 369, 396, 398, 446, 466, 477, 506, 510, 514, 540
- review, 69, 207, 256, 314, 361, 431, 489, 551

Volume
- defined, 213
- prisms and cylinders, 216–219

W

Weight
- defined, 240
- measure, 247–249

Whole numbers, 320

Writing in Math. *See* In Your Own Words

Y

***y*-intercept,** 401

Photo Credits

While every effort has been made to secure permission for reproduction of copyright material, there may have been cases where we are unable to trace the copyright holder. Upon written notification, the publisher will gladly correct this in future printings.

viii Gail Shumway; **ix** NASA/EPA/Corbis; **xi** Jupiter Images/Brand X/Alamy; **xii** Jose Luis Pelaez Inc; **xiii** Jack Hollingsworth; **xiv** John Kirkpatrick/Alamy; **xv** Peter Bowater/Alamy; **xvi** Lester Lefkowitz; **xvii** Hermann Erber/Getty Images; **1** Jerry Driendl; **2** Gail Shumway; **3** Redmond Durrell/Alamy; **5** PhotoSpin, Inc/Alamy; **6** (l)D. Hurst/Alamy, (r)Ralph Wetmore; **7** Dave King; **8** Peter Gridley; **9** Redfx/Alamy; **10** Bon Appetit/Alamy; **11** Alan SCHEIN/Alamy; **22** Anna Peisl/Zefa/Corbis; **23** Ilian Animal/Alamy; **24** Supapixx/Alamy; **28** Kuttig-People/Alamy; **32** (t)Tony Cordoza/Alamy, (b)D. Hurst/Alamy; **34** Willie Sator/Alamy; **35** Randy Lincks/Alamy; **37** Tim Hill/Alamy; **39** Jerry Young; **42** Mazer Creative Services; **43** Comstock/Corbis; **44** Haruyoshi Yamaguchi/Corbis; **45** Image Source Pink/Alamy; **47** Don Tremain/Getty Images; **52** Sumos/Alamy; **60** Ros Drinkwater/Alamy; **65** Tom Grill/Corbis; **72** NASA/EPA/Corbis; **74** Wildlife GMBH/Alamy; **79** Kevin Britland/Alamy; **80** Mazer Creative Services; **85** Photodisc/Alamy; **86** Thomas Northcut; **89** Reuters/Corbis; **90** Visions of America/Joe Sohm; **95** Ladi Kirn/Alamy; **100** Ashley Cooper/Corbis; **108** UpperCut Images/Alamy; **111** Nordicphotos/Alamy; **124** Hermann Erber/Getty Images; **125** Alaska Stock LLC/Alamy; **129** William Whitehurst/Corbis; **132** Stuart Westmorland/Corbis; **133** Mazer Creative Services; **134** Image Source Pink/Alamy; **139** Corbis; **154** Richard Broadwell/Alamy; **161** Ian Cartwright/Getty Images; **163** Design Pics Inc./Alamy; **166** Ingram Publishing/AGE Fotostock; **172** Stocktrek Images/Alamy; **173** Jupiter Images/Brand X/Alamy; **176** Mazer Creative Services; **180** Space Frontiers; **184** Andersen Ross; **186** Mazer Creative Services/Texas Instruments; **191** Michael Houghton/StudiOhio; **212** Jose Luis Pelaez Inc; **213** Siede Preis/Getty Images; **214** Richard Drury; **215** Owaki-Kulla/Corbis; **217, 220** Pixtal/SuperStock; **222** Mazer Creative Services/Texas Instruments; **223** Photodisc/PunchStock; **226** Bruce Laurance; **227** Ken Cedeno/Corbis; **236** Stockbyte/PictureQuest; **240** Burke/Triolo/Brand X Pictures/Jupiter Images; **242** Westend61/Alamy; **243** Stephen Wisbauer; **245** David Seed Photography; **246** Renee Lynn/Corbis; **247** Getty Images; **248** Digital Archive Japan/Alamy; **257** (t)PhotoDisc/Getty Images, (b)The McGraw-Hill Companies, Inc/Ken Karp; **260** Jack Hollingsworth; **263** Stockbyte/Alamy; **265, 267** United States Mint; **268** Mazer Creative Services; **270** Judith Collins/Alamy; **271** Mazer Creative Services; **273** LHB Photo/Alamy; **274** Zia Soleil; **278** Mazer Creative Services; **279** James Ingram/Alamy; **281** Dmitri Kessel; **283** Corbis Premium RF/Alamy; **284** Photodisc/Getty Images; **286** Mazer Creative Services; **287** D. Hurst/Alamy; **288** Mary Evans Picture Library/Alamy; **289** Image Source/PunchStock; **290** Burke/Triolo/Brand X Pictures/Jupiter Images; **292** K-PHOTOS/Alamy; **299** Brand X Pictures/PunchStock; **304** Image Source/PunchStock; **307** Michael Steele; **308** Jupiter Images/Brand X/Alamy; **314** United States Mint; **318** John Kirkpatrick/Alamy; **319, 338** D. Hurst/Alamy; **342** C Squared Studios/Getty Images; **345** Nicholas Pitt/Alamy; **349** PhotoLink/Getty Images; **350** Photo Japan/Alamy; **352** Mazer Creative Services; **354** Richard T. Nowitz/Corbis; **356** Ingram Publishing/Alamy; **362** Peter Bowater/Alamy; **363** Judith Collins/Alamy; **365** Gary Pearl; **366** Arthur Morris/Corbis; **367** Ian Shaw/Alamy; **370** Andersen Ross; **372** ColorBlind Images; **374** PhotoAlto/Alix Minde; **376** John Elk III; **377** Mazer Creative Services; **380** Paul Springett/Alamy; **382** Archive Holdings Inc.; **383** Jose Luis Pelaez, Inc./Blend Images/Corbis; **385** Guillermo Hung; **387** Martin Harvey; **388** Nick Dolding; **389** UK Stock Images Ltd/Alamy; **394** Kim Karpeles/Alamy; **397** Kate Kunath; **398** Jim Craigmyle/Corbis; **399** David Madison; **399** RubberBall Productions/Getty Images; **402** Gandee Vasan; **403** Stocktrek/Corbis; **406** Reuters/Corbis; **408** PhotoDisc/Getty Images; **414** Ralf Schultheiss/zefa/Corbis; **417** Mark Gibson/Alamy; **422** Stockbyte; **432** Lester Lefkowitz; **433** Trolley Dodger/Corbis; **436** Sheer Photo, Inc.; **438** Tom Grill/Corbis; **441** Mazer Creative Services; **443** Sean Gallup; **445** image100/Corbis; **446** Jack Hollingsworth; **451** Stockbyte; **452** Still Images; **455** (l)Creatas/PunchStock, (r)D. Hurst/Alamy; **456** Peter Gridley; **457** Dominic Burke/Alamy; **462** vario images GmbH & Co.KG/Alamy; **468** Emilio Ereza/Alamy; **473** Douglas Pulsipher/Alamy; **474** Brand X Pictures/PunchStock; **479** Tetra Images/Corbis; **481** Patrick Steel/Alamy; **482** Jupiter Images/Comstock Images/Alamy; **483** Mario Tama; **484** The McGraw-Hill Companies, Inc./Ken Cavanagh Photographer; **487** DAJ/Getty Images; **492** Jerry Driendl; **493** Corbis; **499** Photodisc; **500** Dr. Parvinder Sethi; **503** Perry Mastrovito/Corbis; **504** Stockdisc/PunchStock; **509** S. Wanke/PhotoLink/Getty Images; **510** Paul Thompson/Corbis; **511** *Starry Night,* 1889. Vincent Van Gogh. Oil on canvas, 73 x 92 cm. Museum of Modern Art, New York City, New York; **512** B&Y Photography/Alamy; **513** Richard Hamilton Smith/CORBIS; **514** ScotStock/Alamy; **515** Creatas/PunchStock; **517** Joseph Sohm; Visions of America/CORBIS; **518** George Diebold; **520** A & L Sinibaldi; **521** Mazer Creative Services; **523** Stockbyte; **527** PhotoLink/Getty Images; **531** Gabe Palmer/CORBIS; **532** C Squared Studios/Getty Images; **534** Oleksiy Maksymenko/Alamy; **536** Photodisc Collection/Getty Images; **537** Digital Vision; **538** D. Hurst/Alamy; **541** Digital Vision/Getty Images; **544** (l) Directphoto.org/Alamy; **546** (br) Brand X Pictures; **547** (tl) Burke Triolo Productions/Getty Images.

Symbols

Number and Operations

$+$	plus or positive
$-$	minus or negative
$a \cdot b$, $a \times b$, ab or $a(b)$	a times b
$\div$	divided by
$\pm$	plus or minus
$=$	is equal to
$\neq$	is not equal to
$>$	is greater than
$<$	is less than
$\geq$	is greater than or equal to
$\leq$	is less than or equal to
$\approx$	is approximately equal to
%	percent
$a:b$	the ratio of a to b, or $\frac{a}{b}$
$0.7\overline{5}$	repeating decimal 0.75555…

Algebra and Functions

$-a$	opposite or additive inverse of a
a^n	a to the nth power
a^{-n}	$\frac{1}{a^n}$
$\lvert x \rvert$	absolute value of x
$\sqrt{x}$	principal (positive) square root of x
$f(n)$	function, f of n

Geometry and Measurement

$\cong$	is congruent to
$\sim$	is similar to
$^\circ$	degree(s)
$\overleftrightarrow{AB}$	line AB
$\overrightarrow{AB}$	ray AB
$\overline{AB}$	line segment AB
AB	length of $\overline{AB}$
⊾	right angle
$\perp$	is perpendicular to
$\parallel$	is parallel to
$\angle A$	angle A
$m\angle A$	measure of angle A
$\triangle ABC$	triangle ABC
(a, b)	ordered pair with x-coordinate a and y-coordinate b
O	origin
π	pi $\left(\text{approximately } 3.14 \text{ or } \frac{22}{7}\right)$

Probability and Statistics

$P(A)$	probability of event A

Formulas

Perimeter	square	$P = 4s$
	rectangle	$P = 2\ell + 2w$ or $P = 2(\ell + w)$
Circumference	circle	$C = 2\pi r$ or $C = \pi d$
Area	square	$A = s^2$
	rectangle	$A = \ell w$
	parallelogram	$A = bh$
	triangle	$A = \frac{1}{2}bh$
	trapezoid	$A = \frac{1}{2}h(b_1 + b_2)$
	circle	$A = \pi r^2$
Surface Area	cube	$S = 6s^2$
	rectangular prism	$S = 2\ell w + 2\ell h + 2wh$
	cylinder	$S = 2\pi rh + 2\pi r^2$
Volume	cube	$V = s^3$
	prism	$V = \ell wh$ or Bh
	cylinder	$V = \pi r^2 h$ or Bh
	pyramid	$V = \frac{1}{3}Bh$
	cone	$V = \frac{1}{3}\pi r^2 h$ or $\frac{1}{3}Bh$
Pythagorean Theorem	right triangle	$a^2 + b^2 = c^2$
Temperature	Fahrenheit to Celsius	$C = \frac{5}{9}(F - 32)$
	Celsius to Fahrenheit	$F = \frac{9}{5}C + 32$

Measurement Conversions

Length	1 kilometer (km) = 1,000 meters (m) 1 meter = 100 centimeters (cm) 1 centimeter = 10 millimeters (mm)	1 foot (ft) = 12 inches (in.) 1 yard (yd) = 3 feet or 36 inches 1 mile (mi) = 1,760 yards or 5,280 feet
Volume and Capacity	1 liter (L) = 1,000 milliliters (mL) 1 kiloliter (kL) = 1,000 liters	1 cup (c) = 8 fluid ounces (fl oz) 1 pint (pt) = 2 cups 1 quart (qt) = 2 pints 1 gallon (gal) = 4 quarts
Weight and Mass	1 kilogram (kg) = 1,000 grams (g) 1 gram = 1,000 milligrams (mg) 1 metric ton = 1,000 kilograms	1 pound (lb) = 16 ounces (oz) 1 ton (T) = 2,000 pounds
Time	1 minute (min) = 60 seconds (s) 1 hour (h) = 60 minutes 1 day (d) = 24 hours	1 week (wk) = 7 days 1 year (yr) = 12 months (mo) or 52 weeks or 365 days 1 leap year = 366 days
Metric to Customary	1 meter ≈ 39.37 inches 1 kilometer ≈ 0.62 mile 1 centimeter ≈ 0.39 inch	1 kilogram ≈ 2.2 pounds 1 gram ≈ 0.035 ounce 1 liter ≈ 1.057 quarts